Handbook of Flexible and Stretchable Electronics

Handbook of Flexible and Stretchable Electronics

Edited by Nathan Shaw

CLANRYE INTERNATIONAL
www.clanryeinternational.com

Clanrye International,
750 Third Avenue, 9th Floor,
New York, NY 10017, USA

ISBN: 978-1-64726-673-8

Cataloging-in-Publication Data

Handbook of flexible and stretchable electronics / edited by Nathan Shaw.
p. cm.
Includes bibliographical references and index.
ISBN 978-1-64726-673-8
1. Flexible electronics. 2. Stretchable semiconductors. 3. Electronics--Materials.
4. Flexible printed circuits. 5. Wearable technology. I. Shaw, Nathan.
TK7872.F54 H36 2023
621.381--dc23

For information on all Clanrye International publications
visit our website at www.clanryeinternational.com

CLANRYE INTERNATIONAL

Contents

Preface

Flexible electronics is a technology that is used to produce electronic devices that are constructed from stretchable or conformable substrates, from materials such as plastic, flex glass, metal foil and paper. Stretchable electronics are devices that are made up of electronic circuits or materials that are assembled onto stretchable substrates. Flexible and stretchable electronics is emerging as a new technology and attaining a wide range of applications, because its components can be twisted, and compressed to intricate non-planar surfaces. These electronic devices have the potential to be used in a variety of fields, including biomedical engineering, soft robotics, human machine interfaces, epidermal electronic devices, electronic skins and healthcare monitoring. This book explores all the important aspects of flexible and stretchable electronics in the present day scenario. It presents researches that have transformed the research in this field and have aided its advancement. The readers would gain knowledge that would broaden their perspective about this area of electronics through this book.

The information contained in this book is the result of intensive hard work done by researchers in this field. All due efforts have been made to make this book serve as a complete guiding source for students and researchers. The topics in this book have been comprehensively explained to help readers understand the growing trends in the field.

I would like to thank the entire group of writers who made sincere efforts in this book and my family who supported me in my efforts of working on this book. I take this opportunity to thank all those who have been a guiding force throughout my life.

Editor

1

Direct Silver Micro Circuit Patterning on Transparent Polyethylene Terephthalate Film using Laser-Induced Photothermochemical Synthesis

Chen-Jui Lan, Song-Ling Tsai and Ming-Tsang Lee *

Department of Mechanical Engineering, National Chung Hsing University, Taichung 402, Taiwan; lan.andy@gmail.com (C.-J.L.); s3677265@gmail.com (S.-L.T.)
* Correspondence: mtlee@nchu.edu.tw

Academic Editors: Seung Hwan Ko, Daeho Lee and Zhigang Wu

Abstract: This study presents a new and improved approach to the rapid and green fabrication of highly conductive microscale silver structures on low-cost transparent polyethylene terephthalate (PET) flexible substrate. In this new laser direct synthesis and pattering (LDSP) process, silver microstructures are simultaneously synthesized and laid down in a predetermined pattern using a low power continuous wave (CW) laser. The silver ion processing solution, which is transparent and reactive, contains a red azo dye as the absorbing material. The silver pattern is formed by photothermochemical reduction of the silver ions induced by the focused CW laser beam. In this improved LDSP process, the non-toxic additive in the transparent ionic solution absorbs energy from a low cost CW visible laser without the need for the introduction of any hazardous chemical process. Tests were carried out to determine the durability of the conductive patterns, and numerical analyses of the thermal and fluid transport were performed to investigate the morphology of the deposited patterns. This technology is an advanced method for preparing micro-scale circuitry on an inexpensive, flexible, and transparent polymer substrate that is fast, environmentally benign, and shows potential for Roll-to-Roll manufacture.

Keywords: laser direct synthesis and patterning; flexible electronics; laser direct write; transparent conductors

1. Introduction

The market and applications of flexible, stretchable, and wearable electronics have increased remarkably in recent years. The search for rapid and cost-effective ways to combine electronics with flexible and stretchable substrates has also grown enormously. One central issue that challenges the fabrication of flexible electronics is that most flexible substrates, such as polymers, are unsuitable for the conventional photolithography used in microelectronic fabrication because they have poor chemical and thermal resistance. The thermal issue also leads to poor adhesion between the metallic materials and the polymer substrate caused by significantly large differences in thermal and mechanical properties [1]. To overcome these problems, especially the thermally related limitations, several advanced low-temperature manufacturing techniques have been developed that allow high resolution patterning [2]. These include nano-imprinting [3], transfer printing [4], precision deposition and low temperature process using nanoparticle ink [5], and laser-assisted direct sintering and patterning of nanomaterials [6–8]. These techniques take advantage of the melting point depression effect of nanomaterials, which allows the nanoparticles of metal to be easily sintered at moderate temperature [9]. The material is deposited on the substrate from a solution followed by low temperature sintering to form highly conductive nano-/microscale structures. The pattern can be produced by selective deposition or selectively sintering. The nanomaterial-based fabrication technique is fast, maskless, and gives reasonably good resolution mainly depending on the laser parameters and optics. Large area processing for Roll-to-Roll manufacturing is also possible by the laser-nanomaterial-based microfabrication technique.

However, all these nanomaterial-based processes require the synthesis of metal nanomaterials, which is usually a complex, time-consuming, and costly procedure. The coating and recycling of nanomaterial on a polymer substrate is also very challenging. The novel laser direct synthesis and pattering (LDSP) technology presented here provides a low temperature, green, nanoparticle-free process that does not require masks, or vacuum, and is ideal for the rapid fabrication of flexible electronics [10,11]. In contrast to the nanomaterial-based laser-assisted manufacturing processes such as laser direct write that utilizes a pre-synthesized metal nanomaterial solution to deposit the material followed by laser sintering, a transparent and particle-free metallic ion reaction solution is applied to the polymer substrate in the LDSP process. A focused continuous wave (CW) laser beam is scanned across the polymer surface to create a pattern. The laser energy absorbed by the tinted polymer substrate heats the reaction solution near the interface. Metallic ions are reacted and precipitate to form the predetermined pattern on the substrate surface. LDSP uses a nanoparticle-free metal ion precursor—not metal nanoparticle ink—which makes the process significantly faster. The process is simpler and green when compared to conventional laser direct sintering of metal nanoparticles. However, one limitation of the LDSP process is that the polymer substrate needs to absorb the scanned laser light. In other words, if an economic CW visible laser is to be used as a heat source, then the polymer substrate has to be tinted. However, for applications needing highly transparent and flexible electrodes, such as touch panels and flexible solar cells, a transparent polymer substrate has to be used. In such a case, the reactive ionic solution itself must absorb the incident laser light. A possible solution is to pre-deposit an absorbing layer on the substrate [12]. However, this could induce the complexity of the fabrication process and could also affect the transparency of the substrate. Another straightforward solution involves mixing the transparent ionic solution with a highly absorptive dye. The laser beam can then be directed through the transparent substrate to the top surface from below where it will be absorbed by the adjacent solution to initiate a thermochemical reaction and cause silver synthesis and patterning.

Although conceptually promising, the feasibility and the consequences of this approach remain to be proved. In addition, for realization of economic flexible electronics, low-cost transparent substrates such as polyethylene terephthalate (PET) film should be used. However, this kind of substrate is usually vulnerable to elevated temperature during the fabrication process. The thermal impact on the PET substrate subjected to the modified LDSP process also needs to be investigated. In this study, an improved LDSP process has been modified as described. Specifically, a red dye was added to the transparent nanoparticle-free reactive ion solution. The solution was applied to the transparent PET substrate, after which the LDSP process, where a focused laser beam was directed through the underside of the transparent PET substrate, was carried out. The red-colored reactive solution serves as the absorber of photon energy and converts it to thermal energy for the chemical reaction. The effects of the additive materials, concentration and the number of laser scans were all investigated. Electrical conductivities and the mechanical durability of the resulting nano-/microscale conductive patterns on the transparent PET substrate were measured and discussed. It is emphasized that, and shown in the results, using the red dye to make the reactive solution absorptive is a simple and cost-effective way to expand the application of the LDSP technology to include transparent polymer substrates. Unlike the other possible types of additives where nano-powders are used to form adsorptive colloids, the reactive solution mixed with red dye used in this study is free from precipitation and unwanted chemical reactions caused by the use of additives. Numerical simulation was also carried out on the heat and flow transport near the laser focal spot. Discussions of the impact on the resulting silver line structures from the temperature and velocity fields are also provided.

2. Materials and Methods

2.1. Processing Solution Preparation

The transparent reactive silver ion solution was prepared following a reported procedure [13] with modification. In brief, 1.0 g of silver acetate (anhydrous 99%, Alfa Aesar, Ward Hill, MA, USA) was dissolved in 2.5 mL of aqueous ammonium hydroxide (28%~30%, ACS Reagent, J.T. Baker, Phillipsburg, NJ, USA) at room temperature with vigorous stirring, followed by the drop-wise addition of 0.2 mL of formic acid (88%, ACS Reagent, J.T. Baker). The solution was filtered through a 220 nm poly(vinylidene fluoride) (PVDF) filter. The resulting particle-free clear ionic solution contains approximately 22% silver. Separately, a red dye, Allura red AC, was dissolved in ethylene glycol (EG, 99%, Alfa Aesar) under sonication to give solutions with molarities of 8 mM, 16 mM, and 32 mM. The red-EG was then mixed with the ionic solution at a volumetric ratio of 1:1 to complete the preparation of the EG–silver ion reaction solutions. Consequently, process solutions with red-dye concentration of 4 mM, 8 mM, and 16 mM were utilized in the current study. It should be pointed out that the red dye was chosen in the current study owing to its effective absorption to the green light laser energy, as also to be shown and discussed in the following section regarding the optical property of the red dye.

2.2. The LDSP Process

The diagram in Figure 1 illustrates the setup of the experimental apparatus used in the LDSP process. The 50-μm-thick PET film with an area of 5 cm × 5 cm used for this experiment was cleaned by rinsing with ethanol and DI water consecutively. The film was then treated with oxygen plasma in an Atmospheric Pressure Plasma Cleaner (Output voltage: 10,000 V to 50,000 V, output frequency: 4 MHz to 5 MHz) for 10 s to make the surface hydrophilic and improve adhesion of the silver to the polymer surface. The PET substrate was then placed carefully on a glass slide sample holder. The process solution was pipetted onto the substrate surface to form a liquid film typically 2~3 mm thick. The thickness of the process solution was controlled by controlling the amount of the solution being dispensed. It should be noted that the LDSP process may be sensitive to liquid film thickness as discussed in our previous studies [10,14]. As also to be shown from the numerical simulation results, the range of temperature and fluid flow in the process solution affected by the focused laser is confined within 1 mm surround the focal point. Therefore, 2~3-mm-thick process solution in the current study is sufficient to prevent the impact from the liquid-air boundary. The laser beam from a continuous wave (CW) diode laser (λ = 532 nm) was aligned and focused to 50 μm in diameter on the top surface of the PET substrate from underneath with the assistance of a beam expander and beam profile analyzer (BeamGage® Beam Profiler SP620U, Ophir Optronics®, Jerusalem, Israel). The focused laser beam was then directed by a programmable galvanometer scanner system (SCANLAB hurryScan® II-7, SCANLAB GmbH, Munich, Germany) to selectively scan a predetermined light pattern through the transparent PET substrate, the energy of which was absorbed by the process solution. Since the PET substrate is transparent and thin, there was no significant extinction of the laser light and the process solution absorbed the laser energy very effectively. Figure 2 shows the absorption coefficient of the process solutions with respect to the concentration of the red dye. Measurements were carried out using a UV-Vis spectrophotometer (Genesys 10S, Thermo Scientific, Waltham, MA, USA). Due to the limited measurement range of the spectrophotometer, three diluted solution concentrations were used. Assuming negligible scattering and from extrapolation of the measured absorption coefficient using the fitted curve as shown in Figure 2, the penetration distance attributed to absorption by the 16 mM process solution was approximately 65 μm [15]. This estimated result suggests that the absorption of the incident laser energy, and consequent heating of the process solution, would be confined to a vicinity of ~$O(10^2$ μm) to the PET surface. A small and well controlled thermal impact area is essential for the LDSP process to ensure good resolution as well as the recovery of unreacted process solution. After the LDSP process was completed, the PET surface was washed carefully with deionized water and ethanol and air-dried to remove any remaining process solution.

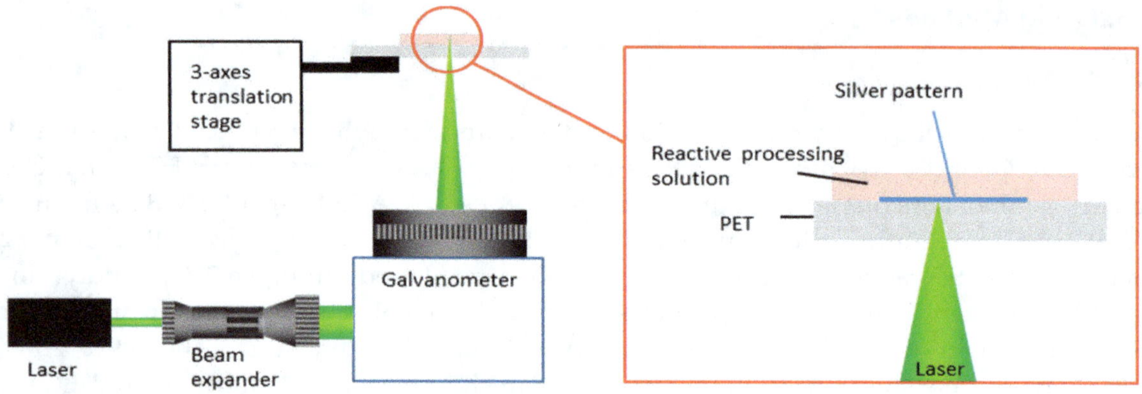

Figure 1. Experimental apparatus for the modified laser direct synthesis and pattering (LDSP) process.

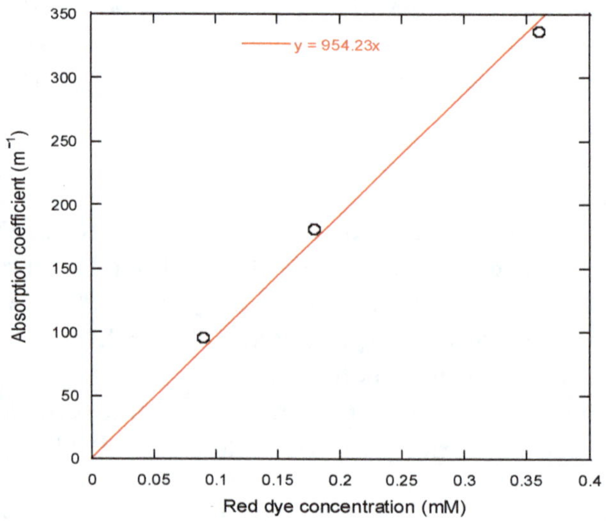

Figure 2. The absorption coefficient with respect to the red dye concentrations in the process (Ag–EG) solution.

3. Results

In this study, the laser power was fixed at 100 mW, and the laser scan speed was 10 mm/s. The number of laser scans and the red-dye concentrations were varied to conduct the parametric studies. Figure 3 shows the silver lines fabricated by the LDSP process on PET substrate with a range of different laser scans (10×, 20×, 30×) and red-dye concentrations (4 mM, 8 mM, 16 mM, based on the mixed process solution). It can be clearly seen that the 4 mM red-dye process solution did not yield a continuous silver line even when the maximum number of scans (30×) was applied. The formation of such broken lines also indicates that the thermochemical reduction of silver ions on the PET surface is not uniform and steady. The temperature and the reactant concentration are two major factors affecting the chemical reaction rate and it is possible that the concentration of reactants and the red dye was not uniform along the laser scan path. In these experiments, the red dye in the reactive solution absorbs laser energy and participates in radiative heat transfer. The reaction temperature and thermally induced convection is affected by the absorption of the red dye in the solution and although the process solution appears to be homogenous, there could still be microscale fluctuations in concentration. Thermally induced fluid motion such as buoyancy-driven flow, the Soret effect, and Brownian motion will also affect the concentration of species in the solution [16]. These fluctuations are highly random, especially for solutions of low red-dye concentration, and increasing the number of laser scans can build the line incrementally. Nevertheless, for a real application, the number of scans should be minimal and the red-dye concentration was increased to improve uniformity and the absorption of laser light. As can be seen in Figure 3, the silver lines fabricated using a red-dye concentration of 8 mM yields continuous conductive lines after 20 laser scans and 10 laser scans were enough to give continuous lines with a 16 mM solution. The width of the deposited line is approximately 50 μm. The size and the resolution of the conductive patterns being fabricated by the LDSP process is mainly controlled by the size of the heat affect area [14], which are determined by the process parameters such as the laser intensity, scanning speed, and heat transfer properties of the process solution. In addition, the size and morphology of the LDSP deposited patterns could also be controlled with the assistance of microfluidic dispensers [17] in conjunction with adjusted laser parameters. Figure 4 shows the electrical resistivity, where measurable, of these silver lines. The resistivity decreased to become stable with an increase in the number of laser scans or red-dye concentration. It should be noted that, for the 16 mM solution, silver lines with 20 laser scans and 30 laser scans yield compatible resistivity, which suggests that the effective thickness of the line did not increase significantly with more than 20 laser scans. A similar trend had also been observed earlier [10,14]. This effect is attributed to the fact that, as the thickness of the silver line increases, it begins to reflect the incident laser light, the energy is no longer absorbed by the process liquid, and the chemical reaction therefore stops.

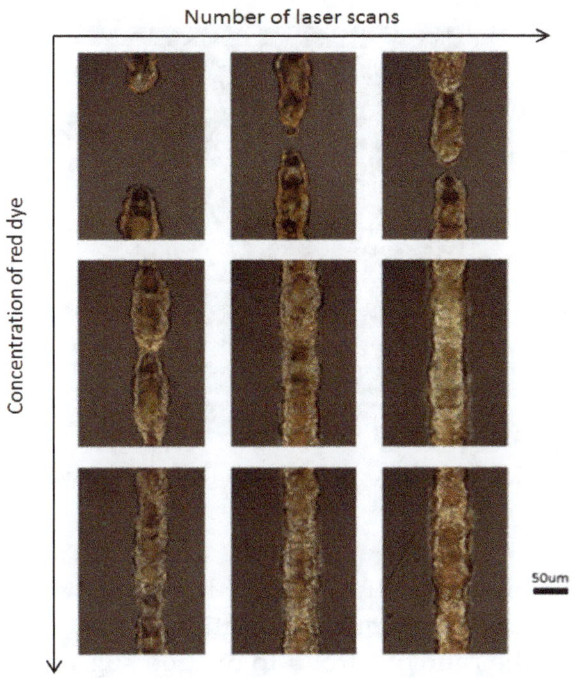

Figure 3. Silver lines fabricated by the LDSP process on the polyethylene terephthalate (PET) substrate with different numbers of laser scans: 10×, 20×, 30× and red-dye concentrations in the process solution of 4 mM, 8 mM, and 16 mM.

Figure 5 shows the silver line fabricated with the 16 mM process solution and 30 laser scans. The rough surface mainly results from solution flow and the possible sintering of silver nanoparticles that were produced during the laser scan [18], which could be improved by increasing the viscosity of the fluid [10]. The cross section of the silver line exhibits a concave shape. This surface morphology is one of the typical shapes found for patterns fabricated by laser chemical vapor deposition (LCVD), which is similar to the mechanism of LDSP. The formation of the morphology is closely related to the temperature distribution, the transport of reactants and products in the liquid (in this case, possibly silver nanoparticles being produced during the laser scan) and the surface diffusion of species [19]. The concave and round shape indicates strong impact from capillary-driven flow and thermophoresis, which are significant when the local temperature gradient in the fluid near the laser focal spot is large.

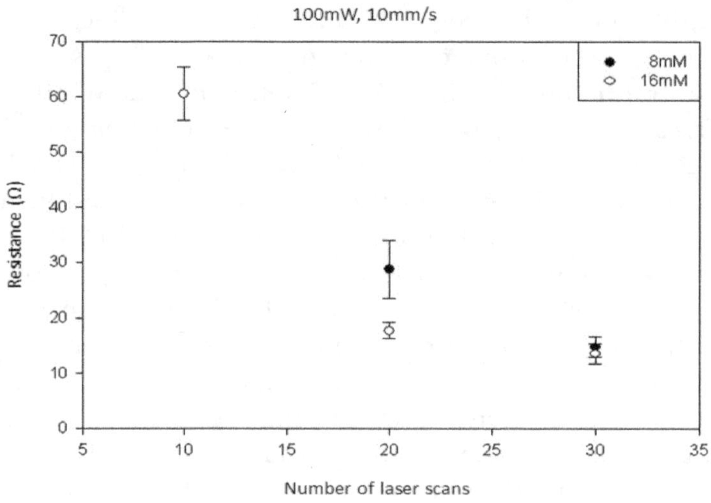

Figure 4. Electrical resistivity of silver lines with respect to the red-dye concentration and number of laser scans.

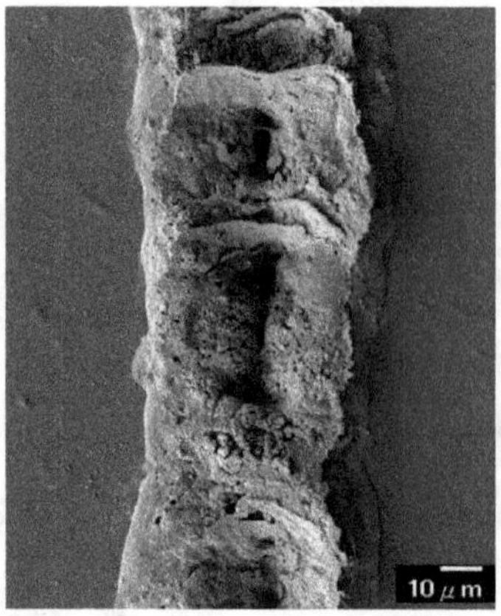

Figure 5. Scanning electron microscope (SEM) image of a silver line fabricated using a 16 mM process solution and 30 laser scans.

Figure 6 shows the results of energy dispersive X-ray analysis (EDX) analysis. It can be seen that the PET substrate after the LDSP in the process solution without silver acetate (but with red dye) contains no silver element, and the ratio of carbon to oxygen slightly increased compared with the original PET substrate. It can be assumed that very little red dye would be sintered and deposited on the PET substrate during the LDSP process. Silver lines fabricated with the process solution containing silver acetate shows dominant silver content and a small amount of carbon and oxygen. The carbon and oxygen could be from the red dye embedded in the silver line. These carbon and oxygen components in the conductive pattern are thought to be the main reason for its higher resistivity compared to pure silver.

Figure 7 shows the morphology of the cross section of the silver lines fabricated with LDSP (16 mM process solution, 30 laser scans). The measurements were carried out using a white light interferometer. The concave morphology can be clearly seen. The negative value at the center of the silver line could be attributed to slight melting of the PET substrate during the LDSP process. Using the average altitude profile shown in Figure 7 to calculate the cross-sectional area of the line, and assuming the canyon at the center under the plain substrate surface does not contribute to significant electrical conductivity, the electrical resistivity of the silver line was estimated to be 2.25×10^{-7} Ω·m, which is approximately 10 times that of pure silver. This can again be attributed to the fact that the composition of the conductive line contains a fraction of carbon and oxygen. Nevertheless, the conductivity of the line is sufficient to conduct electricity effectively as demonstrated in Figure 8. The durability test apparatus and the results are shown in Figure 9. After 10,000 repetitive bends, the drift in electrical resistance of the silver line was approximately 3 times higher than the initial value. Referring to the microstructures of the silver line showing in Figure 9b, there are silver micro-flakes that tend to be easily deformed and deteriorated. The variation in the electrical property is mainly attributed to the change in mechanical structure of the line. Further effort should be made to improve the mechanical robustness and the durability of the conductive patterns fabricated by this technique. It should also be noted that the processing speed could be increased by increasing the laser intensity, the absorptivity, and/or the reactivity of the process solution. The productivity of the LDSP process can thus be improved.

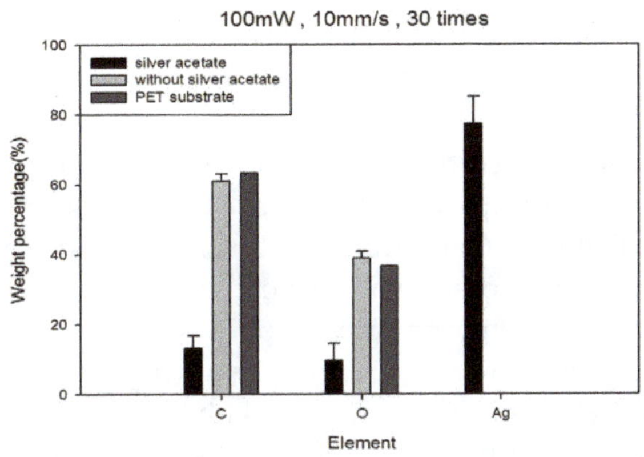

Figure 6. Energy dispersive X-Ray analysis (EDX) elemental analyses.

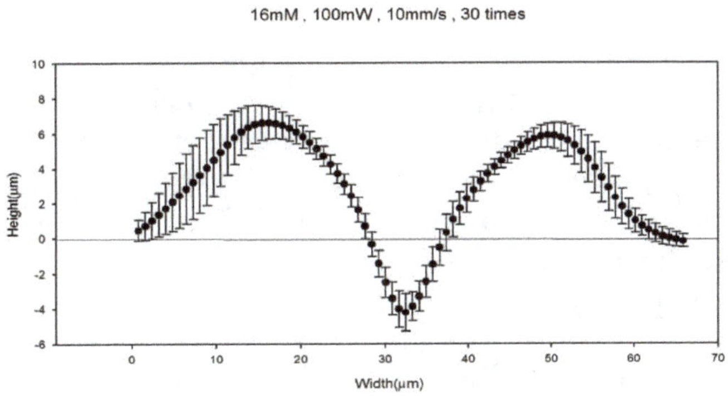

Figure 7. The morphology of the cross section of a silver line fabricated with LDSP (16 mM process solution, 30 laser scans).

Figure 8. Demonstration of conductive silver line fabricated by LDSP on the transparent PET substrate.

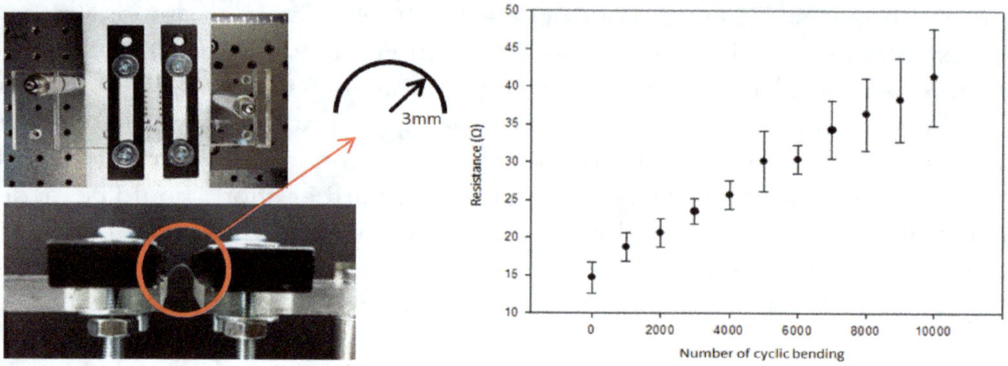

(a) (b)

Figure 9. (**a**) Bending test apparatus; (**b**) resistance drift.

Numerical simulations were carried out to further investigate the transport phenomena during the fabrication process. COMSOL Multiphysics® software (COMSOL Inc., Burlington, MA, USA) was applied in the current study. Details of the numerical model can be found in an earlier report [14] and in [19,20]. In comparison to our previous numerical study of transport phenomena in the LDSP process, the main difference in the current numerical model is the absorbing material. In earlier studies, the substrate (polyimide film) served as a laser light absorbing material, while in the current study the laser energy was absorbed by the process solution. The heat source Q_{abs} from the absorption of a Gaussian laser beam moving in the liquid can be determined by [3,20,21]:

$$Q_{abs} = (1 - R)\gamma I_{pk} \exp\left[-\left(\frac{x - Ut}{\omega}\right)^2 - \left(\frac{y}{\omega}\right)^2 - \gamma z\right] \quad (1)$$

where R is the reflectivity, γ is the absorption coefficient of the process solution estimated from Figure 2, I_{pk} is the peak laser intensity, and ω is determined from the size of the laser beam on the x-y plane of the PET substrate. In the current simulation, the surface reflectivity from the PET substrate and the interfaces of the PET and process solution were assumed to be negligible. Chemical reactions and liquid-vapor phase change were also neglected in this analysis considering that there was no significant boiling observed during the process.

Figure 10a shows the temperature profile of the cross section at the center of the laser beam in the scan direction. It can be seen that the highest temperature spot is inside the reaction fluid and is some tens of microns above the PET surface. In contrast to the previous studies where the laser beam energy was absorbed by the polymer substrate itself, so that the highest temperature in the reaction fluid was on the interface between the solid substrate and the fluid, the process solution absorbs the incident laser energy gradually along the laser path within the fluid, and the hottest spot is in the interior of the process fluid region. This, in turn, results in a strong in-plane flow right on the PET surface below the hottest spot. Strong in-plane flow, especially near the centerline (the symmetric line), could adversely affect the deposition and adhesion of the reduced silver (from the hottest spot) on the PET surface. Additionally, the symmetric swirling flow as shown in Figure 10b indicates that the reduced silver could precipitate on the shoulders along the laser scanning path, which also corresponds to the morphology of the silver line patterns shown in Figure 5. It should also be noted that the upward flow at the centerline due to buoyancy exists with an absorptive substrate (polyimide) or an absorptive fluid as in the current case. The morphology of the deposited pattern, however, is closely related to the intensity of the buoyancy flow, which generally increases with laser intensity, which determines the temperature of the focal point [19]. Other forces such as surface tension would be important for improving the surface morphology [22]. Furthermore, the roughness of the surface of the deposited patterns could yield a random scattering of the incident laser irradiation [23]. Thus, the reflectance to the laser light by the deposited patterns should be carefully considered. Continuous study in this regard is being carried out.

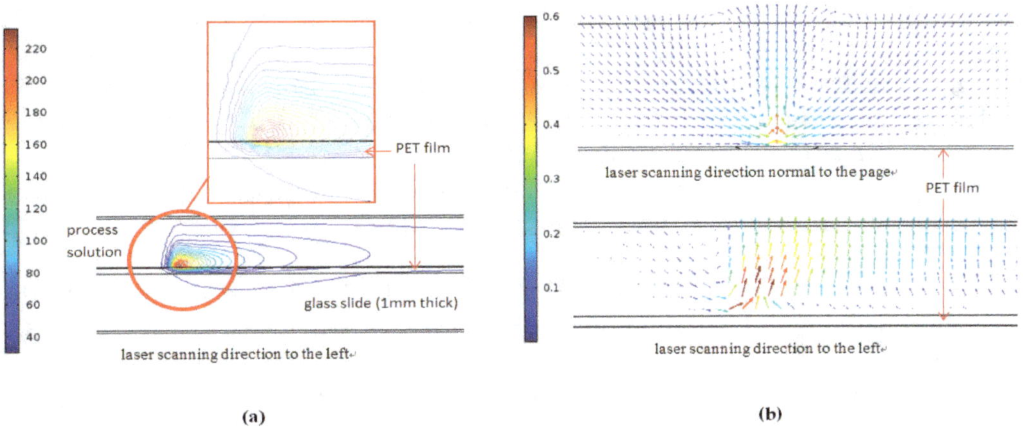

Figure 10. (**a**) Temperature (unit: °C) profile; and (**b**) velocity (unit: mm/s) profile near the laser focal point.

4. Conclusions

In the current study, a new and improved approach to the rapid and green fabrication of highly conductive nano-/micro-scale silver structures on low-cost transparent polyethylene terephthalate (PET) flexible substrate was successfully demonstrated. Predetermined silver pattern was formed by photothermochemical reduction of the silver ions induced by the focused CW laser beam in a processing solution containing a red azo dye as the absorbing material. The effects of the additive materials, concentration, and the number of laser scans were all investigated. The electrical resistivity of the silver line was estimated to be 2.25×10^{-7} Ω·m, which is approximately 10 times that of pure silver. Numerical simulation was also carried out on the heat and flow transport near the laser focal spot to provide insights of the impact on the resulting silver line structures from the temperature and velocity fields. The concave and round shape of the conductive lines indicates strong impact from capillary-driven flow, thermophoresis, and buoyancy flow in the current LDSP configuration. This study and the revised LDSP technology provides a potential approach for preparing nano/micro-scale circuitry on inexpensive, flexible, and transparent polymer substrates that are fast, environmentally benign, and cost-effective.

Acknowledgments: The authors would like to acknowledge the financial support for this study from the Ministry of Science and Technology, Taiwan R.O.C. Project Nos. 105-2221-E-005-041 and 105-2218-E-005-005.

Author Contributions: Chen-Jui Lan, Song-Ling Tsai, and Ming-Tsang Lee conceived and designed the experiments; Chen-Jui Lan and Song-Ling Tsai performed the experiments; Song-Ling Tsai performed the numerical simulations; Chen-Jui Lan and Ming-Tsang Lee analyzed the data; Ming-Tsang Lee wrote the paper.

References

1. Perelaer, J.; Smith, P.J.; Mager, D.; Soltman, D.; Volkman, S.K.; Subramanian, V.; Korvink, J.G.; Schubert, U.S. Printed electronics: The challenges involved in printing devices, interconnects, and contacts based on inorganic materials. *J. Mater. Chem.* **2010**, *20*, 8446–8453. [CrossRef]
2. Subramanian, V.; Lee, T. Nanotechnology-based flexible electronics. *Nanotechnology* **2012**, *23*, 340201. [CrossRef] [PubMed]
3. Lee, D.; Pan, H.; Sherry, A.; Ko, S.H.; Lee, M.T.; Kim, E.; Grigoropoulos, C.P. Large-area nanoimprinting on various substrates by reconfigurable maskless laser direct writing. *Nanotechnology* **2012**, *23*, 344012. [CrossRef] [PubMed]

4. Yu, X.; Mahajan, B.; Shou, W.; Pan, H. Materials, mechanics, and patterning techniques for elastomer-based stretchable conductors. *Micromachines* **2016**, *8*, 7. [CrossRef]
5. Kullmann, C.; Schirmer, N.C.; Lee, M.-T.; Ko, S.H.; Hotz, N.; Grigoropoulos, C.P.; Poulikakos, D. 3D micro-structures by piezoelectric inkjet printing of gold nanofluids. *J. Micromech. Microeng.* **2012**, *22*, 055022. [CrossRef]
6. Yeo, J.; Hong, S.; Lee, D.; Hotz, N.; Lee, M.T.; Grigoropoulos, C.P.; Ko, S.H. Next generation non-vacuum, maskless, low temperature nanoparticle ink laser digital direct metal patterning for a large area flexible electronics. *PLoS ONE* **2012**, *7*, e42315. [CrossRef] [PubMed]
7. Lee, M.-T.; Lee, D.; Sherry, A.; Grigoropoulos, C.P. Rapid selective metal patterning on polydimethylsiloxane (PDMS) fabricated by capillarity-assisted laser direct write. *J. Micromech. Microeng.* **2011**, *21*, 095018. [CrossRef]
8. Hong, S.; Lee, H.; Lee, J.; Kwon, J.; Han, S.; Suh, Y.D.; Cho, H.; Shin, J.; Yeo, J.; Ko, S.H. Highly stretchable and transparent metal nanowire heater for wearable electronics applications. *Adv. Mater.* **2015**, *27*, 4744–4751. [CrossRef] [PubMed]
9. Buffat, P.; Borel, J.P. Size effect on the melting temperature of gold particles. *Phys. Rev. A* **1976**, *13*, 2287–2298. [CrossRef]
10. Liu, Y.K.; Lee, M.T. Laser direct synthesis and patterning of silver nano/microstructures on a polymer substrate. *ACS Appl. Mater. Interfaces* **2014**, *6*, 14576–14582. [CrossRef] [PubMed]
11. Liu, Y.-K.; Sie, Y.-Y.; Liu, C.-A.; Lee, M.-T. A novel laser direct writing system integrated with A&F XXY alignment platform for rapid fabrication of flexible electronics. *Smart Sci.* **2015**, *3*, 87–91.
12. Yeo, J.; Hong, S.; Kim, G.; Lee, H.; Suh, Y.D.; Park, I.; Grigoropoulos, C.P.; Ko, S.H. Laser-induced hydrothermal growth of heterogeneous metal-oxide nanowire on flexible substrate by laser absorption layer design. *ACS Nano* **2015**, *9*, 6059–6068. [CrossRef] [PubMed]
13. Walker, S.B.; Lewis, J.A. Reactive silver inks for patterning high-conductivity features at mild temperatures. *J. Am. Chem. Soc.* **2012**, *134*, 1419–1421. [CrossRef] [PubMed]
14. Tsai, S.-L.; Liu, Y.-K.; Pan, H.; Liu, C.-H.; Lee, M.-T. The coupled photothermal reaction and transport in a laser additive metal nanolayer simultaneous synthesis and pattering for flexible electronics. *Nanomaterials* **2016**, *6*, 12. [CrossRef]
15. Siegel, R.; Howell, J. *Thermal Radiation Heat Transfer*, 4th ed.; Taylor & Francis: New York, NY, USA, 2002.
16. Xu, L.; Lee, B.J.; Hanson, W.L.; Han, B. Brownian motion induced dynamic near-field interaction between quantum dots and plasmonic nanoparticles in aqueous medium. *Appl. Phys. Lett.* **2010**, *96*, 174101. [CrossRef]
17. Pan, P.-T.; Hsieh, M.-C.; Liu, C.-Y.; Pan, C.-W.; Zhao, K.-W.; Hsieh, Y.-W.; Wang, A.-B. A precision capillary coating system and applications. *Smart Sci.* **2016**, *4*, 151–159. [CrossRef]
18. Zhang, R.; Lin, W.; Moon, K.S.; Wong, C.P. Fast preparation of printable highly conductive polymer nanocomposites by thermal decomposition of silver carboxylate and sintering of silver nanoparticles. *ACS Appl. Mater. Interfaces* **2010**, *2*, 2637–2645. [CrossRef] [PubMed]
19. Grigoropoulos, C.P. *Transport in Laser Microfabrication*; Cambridge University Express: Cambridge, UK, 2009.
20. Lee, D.G.; Kim, D.K.; Moon, Y.J.; Moon, S.J. Effect of laser-induced temperature field on the characteristics of laser-sintered silver nanoparticle ink. *Nanotechnology* **2013**, *24*, 265702. [CrossRef] [PubMed]
21. Paeng, D.; Yeo, J.; Lee, D.; Moon, S.-J.; Grigoropoulos, C.P. Laser wavelength effect on laser-induced photo-thermal sintering of silver nanoparticles. *Appl. Phys. A* **2015**, *120*, 1229–1240. [CrossRef]
22. Fukuda, K.; Sekine, T.; Kumaki, D.; Tokito, S. Profile control of inkjet printed silver electrodes and their application to organic transistors. *ACS Appl. Mater. Interfaces* **2013**, *5*, 3916–3920. [CrossRef] [PubMed]
23. Gu, M.-J.; Chen, Y.-B. Modeling bidirectional reflectance distribution function of one-dimensional random rough surfaces with the finite difference time domain method. *Smart Sci.* **2014**, *1*, 101–106. [CrossRef]

2

Stretchable Tattoo-Like Heater with On-Site Temperature Feedback Control

Andrew Stier [1], Eshan Halekote [1], Andrew Mark [1], Shutao Qiao [2], Shixuan Yang [2], Kenneth Diller [3] and Nanshu Lu [1,2,3,4,*]

1 Department of Electrical and Computer Engineering, University of Texas at Austin, Austin, TX 78712, USA; andrewcstier@gmail.com (A.S.); ehalekote@gmail.com (E.H.); andrewemark@gmail.com (A.M.)

2 Center for Mechanics of Solids, Structures and Materials, Department of Aerospace Engineering and Engineering Mechanics, University of Texas at Austin, Austin, TX 78712, USA; shutao2011@gmail.com (S.Q.); rock002008@gmail.com (S.Y.)

3 Department of Biomedical Engineering, University of Texas at Austin, Austin, TX 78712, USA; kdiller@mail.utexas.edu

4 Texas Materials Institute, the University of Texas at Austin, Austin, TX 78712, USA

* Correspondence: nanshulu@utexas.edu

Abstract: Wearable tissue heaters can play many important roles in the medical field. They may be used for heat therapy, perioperative warming and controlled transdermal drug delivery, among other applications. State-of-the-art heaters are too bulky, rigid, or difficult to control to be able to maintain long-term wearability and safety. Recently, there has been progress in the development of stretchable heaters that may be attached directly to the skin surface, but they often use expensive materials or processes and take significant time to fabricate. Moreover, they lack continuously active, on-site, unobstructive temperature feedback control, which is critical for accommodating the dynamic temperatures required for most medical applications. We have developed, fabricated and tested a cost-effective, large area, ultra-thin and ultra-soft tattoo-like heater that has autonomous proportional-integral-derivative (PID) temperature control. The device comprises a stretchable aluminum heater and a stretchable gold resistance temperature detector (RTD) on a soft medical tape as fabricated using the cost and time effective "cut-and-paste" method. It can be noninvasively laminated onto human skin and can follow skin deformation during flexure without imposing any constraint. We demonstrate the device's ability to maintain a target temperature typical of medical uses over extended durations of time and to accurately adjust to a new set point in process. The cost of the device is low enough to justify disposable use.

Keywords: epidermal electronics; wearable heater; temperature sensor; feedback control

---

1. Introduction

There exists a need for soft, stretchable electronic heating devices that can conform to human skin unobstructively and stay attached during long term use. Such devices can serve a variety of applications in the medical field. As some examples, heat is commonly used in physical therapy following exercise-induced delayed onset muscle soreness (DOMS) [1,2]. Heating injured joints can induce thermal expansion of the collagen tissue and thus reduce pain and stiffness [3–5]. When hypothermia occurs due to anesthesia [6], applying heat to the palms and soles of a patient with distended blood vessels can re-warm the body's core temperature [7,8]. Applying heat over a skin surface can accelerate the transdermal diffusion of chemicals from a drug patch [9,10].

Conventional heaters used to treat muscle pain or joint injuries include electric heat packs [11] and heat wraps [12]. Heat packs do not have very controllable temperature and are heavy and bulky.

Heat wraps are easier to control but are also heavy and their rigidity makes it difficult for them to be worn seamlessly [5]. These products' inability to conform well to skin [13–15] make them less comfortable and also present a more severe problem—lack of uniform and consistent adhesion to the skin surface could lead to air gaps which cause hotspots [16]. These hotspots could burn the skin if the heater is operated near the safety threshold of 43 °C [17]. This can severely limit the range and thus the effectiveness of the conventional heaters.

One heating method that can safely heat the body at temperatures close to 43 °C and the current gold standard for preventing the hypothermia caused by anesthesia, is forced air warming. Forced air warming heats air and pumps it into blankets covering large portions of the patient. While effective at raising the core body temperature, forced air warming has some disadvantages including bulkiness, obstructiveness to surgeries and high cost [18,19].

Recently there has been an expansion of the development of stretchable electronics [20–23]. Methods that have been used to produce these type of electronics include embedding carbon nanotubes (CNTs) in elastomers [24,25], depositing silver (Ag) nanoparticles in polyeruthane [5,26], chemically bonding Ag flakes to CNTs [27,28], combining Ag nanoparticles with elastomeric fibers [29], electrospinning Ag nanofibers onto a flexible substrate [30], constructing CuZr nanotrough networks that function as stretchable electrodes [31], constructing stretchable gold (Au) electrodes from multi-layers of Au nanosheets [32] and patterning metal thin films into serpentine [33,34] or fractal shapes [35,36] to minimize their strain during stretching. This last method has enabled the creation of epidermal electronics—ultrathin, ultrasoft electronics, physiological sensors, and electrical and thermal stimulators—that can adhere and conform to skin surfaces and bend and stretch without breaking, detaching, or imposing any mechanical constraint to the skin [36–38].

With the development of stretchable electronics, stretchable patch heaters have emerged in recent years. Examples include joule heating devices fabricated from soft Ag nanowire composites [5,30,39], copper (Cu) nanowire based fabric [40], stretchable copper zirconium electrodes [31], or stretchable Au serpentines [6,11,29,32]. Using a stretchable and conformable heater could solve the major disadvantages of conventional solid heaters. However, the existing stretchable heaters involve expensive nanomaterials or time-consuming procedures to produce. Moreover, most of them have no method of acquiring temperature feedback from the heater as they are not equipped with any temperature sensors. As a result, most of the reported stretchable heaters do not use temperature feedback to autonomously maintain a set temperature for the heater. One of the exceptions is a wearable fabric heater described by Cheng et al. [40]. This heater is loose enough that it is able to use an unspecified temperature sensor to monitor and control the heater. Of the more tightly conformable heaters, one with temperature feedback only has the functionality of turning the heater off if it gets too hot—aside from that, the temperature feedback is not used to actively control the heat of the heater [9]. The best existing example of a tightly conforming stretchable heater with continuous feedback control is the metal nanofiber heater developed by Jang et al. [30]. This heater uses a thermistor placed on the outer edge of the heater to detect the heater's temperature and uses an unspecified control algorithm to keep the heater at a specified temperature. This set-up relies on the assumption that the temperature at the outer edge of the heater is representative of the temperature of the heater overall but the heat profile across the heater is not actually uniform. Moreover, the thermistor appears to introduce disturbances in the uniformity of the heater.

In other stretchable heaters with temperature sensors, the temperature sensing element is the same as the heating element [9,15]. The biggest disadvantage of this setup is that the dual-purpose element is measuring its own temperature instead of the actual skin temperature. The closest example of a stretchable heater with separate, unobstructive temperature feedback is one developed for a prosthetic hand but not tested on human skin, where a multilayer heater and sensor array is laminated onto a prosthetic hand [41]. Without the use of effective temperature feedback, past stretchable heating devices have relied on the relationship between voltage and heat generated in order to maintain the heater at a desired temperature. However, heat transfer conditions vary from person to person and it is inaccurate to assume a consistent relationship between voltage and temperature if you wish to apply the same heater to multiple subjects. For example, changes in blood flow can cause changes in epidermal skin temperature [7].

Our group has developed a "cut-and-paste" method [42], in which stretchable patterns are cut out of ultrathin metal-polymer laminates and pasted to an adhesive substrate, allowing for cheaper, quicker and greener fabrication of tattoo-like sensors. This method also allows for easy integration of independent heaters and resistance temperature detectors (RTDs) on the same substrate. Using this method, we herein present an inexpensive, easy to fabricate and power-efficient programmable tattoo-like heating device which comprises a stretchable resistive heating element (RHE) of serpentine-shaped aluminum (Al) ribbons and a stretchable RTD of serpentine-shaped Au ribbons. The RTD is thin enough to not disturb the uniformity of the heat from the heating element. Included with this device is a customized proportional-integral-derivative (PID) control software which uses real time temperature feedback to control the heater and can maintain it at a target temperature over a large area of skin for extended periods of time.

2. Materials and Methods

The "cut-and-paste" manufacturing process of the stretchable tattoo-like heater is illustrated in Figure 1a, and a picture of the as-fabricated sample on 3M Tegaderm tape is offered in Figure 1b. As depicted in the first row of Figure 1a, the process began with placing a blanket 7 μm/13 μm Al/PET bilayer laminate (Neptco Inc., Pawtucket, RI, USA) smoothly on a thermal release tape (Semiconductor Equipment Corp., Moorpark, CA, USA) with Al facing up. A Silhouette mechanical cutter plotter was programmed to cut the designed seams on the bilayer within 3 min. Excessive Al/PET was removed once the thermal release tape (TRT) was heated and the remaining Al/PET ribbon was printed on a 3M Tegaderm tape with the Al side facing the Tegaderm and the bluish PET side facing outward. The 13-μm thick PET layer allows for increased mechanical integrity [43] and electrical insulation. The same process was repeated to cut and paste the stretchable RTD ribbon, which was made out of 100 nm/15 nm/13 μm Au/Cr/PET laminate, with the PET facing the Tegaderm and the Au facing outward, as illustrated by the second row of Figure 1a. This arrangement resulted in two layers of insulation between the Al RHE and the Au RTD at locations where the two intersected. Both the RHE and the RTD were cut into serpentine ribbons, which contributes to the stretchability and softness of the device. Specifically, the stretchability and softness of these serpentine ribbons can be maximized by fabricating their width to be as narrow as possible [37]. Due to the resolution of the Silhouette cutter, all ribbon widths were fixed to be 400 μm [42]. Although the resolution is far from photolithographic patterning technologies, the cost of time, materials and facilities is significantly reduced using the freeform cut-and-paste process because it does not require any chemicals, photomasks or cleanroom facilities. Moreover, while photolithographic process is limited to wafer scale, the patterning area of the cutter plotter can be as large as 30 cm. wide and a meter long.

Costs of previous epidermal electronic systems are dominated by fabrication processes such as spin coating, photolithography, wet and dry etching and transfer-printing. These methods require expensive clean-room fees and chemical purchases and they are also very time intensive. Using the "cut-and-paste" method allows for the fabrication of the presented device without any of those costs, making it significantly more cost effective than other similar epidermal electronics [38,42,44,45].

Finally, an ultrathin, ultrasoft double-sided tattoo adhesive was laid on top of the RHE and the RTD, providing a final layer of electrical insulation as well as increased adhesion between the skin and the patch. Snap button connectors were used to connect lead wires to both the RHE and the RTD (Supplementary Information, Figure S1).

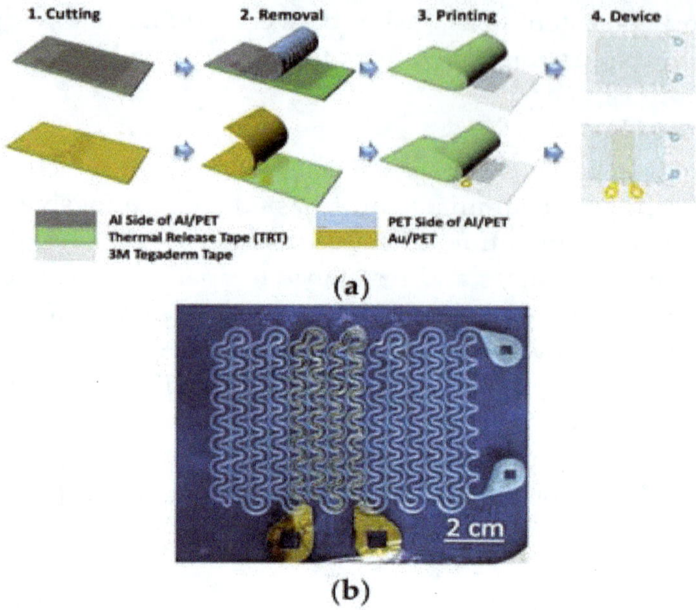

Figure 1. (**a**) Fabrication process used for heater and resistance temperature detector (RTD), shown for heater. Material is put on the thermal release tape (TRT) and cut with Silhouette cutter. TRT is heated, excess material is removed and remaining material is transferred to Tegaderm; (**b**) Complete device on tegaderm. Aluminum with blue polyimide backing forms the resistive heating element while Au/Cr 100/10 nm forms the resistance temperature detector.

The palm of a human subject's hand was chosen for the location to test the device on. The palms of the hand are glabrous skin surfaces, and heating them along with the soles of the feet can efficiently warm the body during anesthesia [7,8]. Presenting that the heater works effectively on the palms of the hand therefore demonstrates that perioperative warming is a feasible application of this device. When attached to the skin, this device conformed to the skin and deformed alongside it without mechanical resistance, as evidenced by Figure 2a,b. When a DC voltage of 5.1 V was applied across the Al RHE, it supplied an even amount of heat over the palm around the target temperature of 40 °C. There was minimum change in temperature during severe skin deformation such as hand clenching, as demonstrated by Figure 2c,d, which were taken by an infrared (IR) FLIR T620 camera (FLIR, Wilsonville, OR, USA). The University of Texas at Austin IRB protocol number for the human subject experiments was 2010-03-0050.

To calibrate the RTD, it was placed on an insulated hot plate and its resistance was compared against the temperature readings of two custom made type T thermocouples. The RTD exhibited the expected linear relationship [15] between resistance and temperature with a temperature coefficient of resistance (TCR) of 0.0025 $^{\circ}C^{-1}$ (Supplementary Information, Figure S2). To obtain the TCR under service condition, calibration of the RTD was also conducted on skin together with the RHE, and an IR camera was used for temperature measurement. In this set-up, the heat comes from the RHE instead of the hotplate and the RTD is in intimate contact with the actual heat sink—the skin. Due to these differences, we hypothesized that the TCR would be different from that measured on the hotplate. A schematic of the calibration set-up is depicted in (Figure 3a). The RHE was linked to a DC voltage supply (Mastech Linear Power Supply HY1803D, Pittsburgh, PA, USA) while the RTD was connected to a digital multimeter (DMM, NI Elvis II). Resistance readings were logged using the DMM and LabVIEW 2014. The device was covered with a fine layer of Johnson's Baby Powder to control its thermal radiation emissive properties [46]. The DC voltage supply was set to different voltages and the temperature and resistance of the RTD were measured simultaneously using the IR camera and the DMM, respectively. Temperature readings from the IR camera were logged using FLIR Tools+.

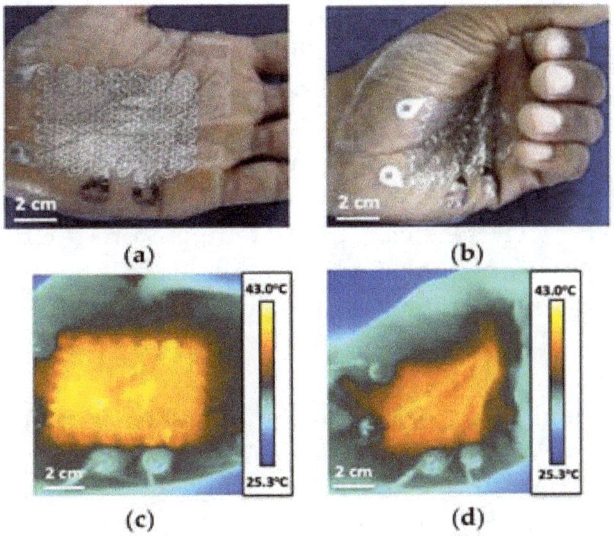

Figure 2. (**a**,**b**) Device conforms to hand and maintains its conformability during opening and closing; (**c**,**d**) Infrared (IR) images of the device powered with proportional-integral-derivative (PID) control as the hand is opened and closed. The PID controller automatically adjusts power output so the hand does not overheat when it closes.

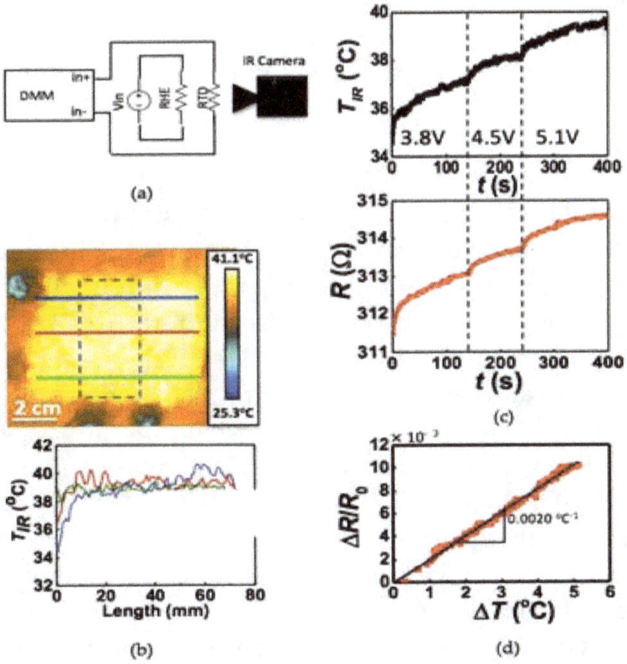

Figure 3. (**a**) Circuit diagram of set-up for calibration of RTD in situ. Heater is brought to different temperatures by adjusting Vin. Resistance and temperature are measured simultaneously using a digital multimeter (DMM) and IR camera, respectively; (**b**) Lateral heat distribution of heater. Blue, red and green lines on IR image mark the horizontal line across which temperature was measured for their respective red, blue and green plots. Temperature distribution is fairly uniform. Dotted purple line on IR image shows area that the IR camera calculated the average temperature for; (**c**) Average temperature of area marked by the dotted purple line in Figure 3b (top) and resistance of Au/Cr RTD as measured by DMM (bottom) each plotted across time as Vin was changed to 3.8 V, 4.5 V and finally 5.1 V; (**d**) The calibration curve for the RTD: $\Delta R/R_0$ of the RTD versus ΔT of the average temperature of the area around the RTD as marked by the dotted purple line in Figure 3b. The calibration constant, β, is marked and is equal to 0.000203.

For safety purposes, the RHE-RTD calibration was first carried out on a glass slide, which has thermal properties similar to those of human skin [17]. The TCR was measured to be 0.0022 °C (Supplementary Information, Figure S3), which is slightly lower than that measured by the hotplate calibration. After ensuring the RHE behavior, a similar RHE-RTD calibration was performed on human skin. Figure 3b upper frame shows the IR image of the heater on human palm. It is evident that the temperature across the heater is fairly uniform over an area of 60 mm × 45 mm. The dotted black box indicates where the RTD resides. The IR temperature for the RTD calibration used in Figure 3c,d was obtained by averaging the temperature within this boxed area. It is clear in Figure 3b that the existence of the RTD does not affect the RHE or the temperature distribution.

The three solid horizontal lines drawn across the heater mark the locations where the temperature is plotted as a function of distance along the lines in Figure 3b lower frame. Within the area covered by the RHE, temperature variation is between 38 °C and 40 °C. To continuously increase the temperature, the DC voltage was set to 3 increasing values: 3.8 V, 4.5 V and 5.1 V. Synchronously measured temperature and resistance versus time curves are provided in Figure 3c, which shows excellent alignment. Also visible in Figure 3c is the steady state average temperature of the RTD-area of the heater reaches when power is applied directly to the RHE. The average temperature to voltage ratio for the device was found to be 8.7 °C/V (Supplementary Information, Table S1). Plotting relative resistance change versus temperature change in Figure 3d, a linear fit with a TCR of 0.0020 $°C^{-1}$ can be obtained. As expected, this is lower than the TCR found with the hotplate calibration (0.0025 $°C^{-1}$) or the glass substrate calibration (0.0022 $°C^{-1}$) due to the fact that the RTD is well conformed to human skin, beneath which blood flow can help mitigate the heat.

To verify the experimental findings, we ran a COMSOL simulation of the device heating human skin. The skin was modeled as a multilayer substrate made up of epidermis (0.1 mm thick), papillary dermis (0.7 mm thick), reticular dermis (0.8 mm thick), fat (2 mm thick) and muscle (16.4 mm thick), each with different thermophysical properties taken from literature [47]. No blood perfusion effects were included in the model. Ambient radiation from the RHE and convective cooling between the RHE and the environment were taken into consideration. With the environment temperature set at 15 °C and the core temperature set at 37 °C, the skin surface temperature stabilized at 34.4 °C when the heater was off. The effective electrical conductivity of the RHE was calibrated by setting the maximum temperature to be 41.4 °C when the applied voltage was 5.1 V. Using a Joule heating model for the RHE and a heat transfer model for the other components of the device and the skin, the modeled temperature distribution across the skin was found under transient and equilibrium states. Figure 4a,b displays the top and 3D cross-sectional views of the temperature distribution within the skin under equilibrium while the heater was on. Figure 4c plots temperature distributions along the three lines drawn on the left frame of Figure 4a where the blue, red and green curves correspond to the blue, red and green lines, respectively.

The close agreement between Figures 3b and 4c validates the COMSOL model and gives more credit to the simulated equilibrium temperature distribution in the skin along the depth direction (as indicated by the black arrow in Figure 4a right frame), as plotted in Figure 4c. When the heater is off (dashed curve), skin surface temperature is 34.4 °C. As the depth increases, the curve approaches the core temperature of 37 °C. When the heater is on (solid curve), skin surface is heated to 41.4 °C. The temperature gradually decays to 37 °C as we go deep into the skin. The slight kinks in the curves are due to the change of the thermophysical properties of the different layers of human skin.

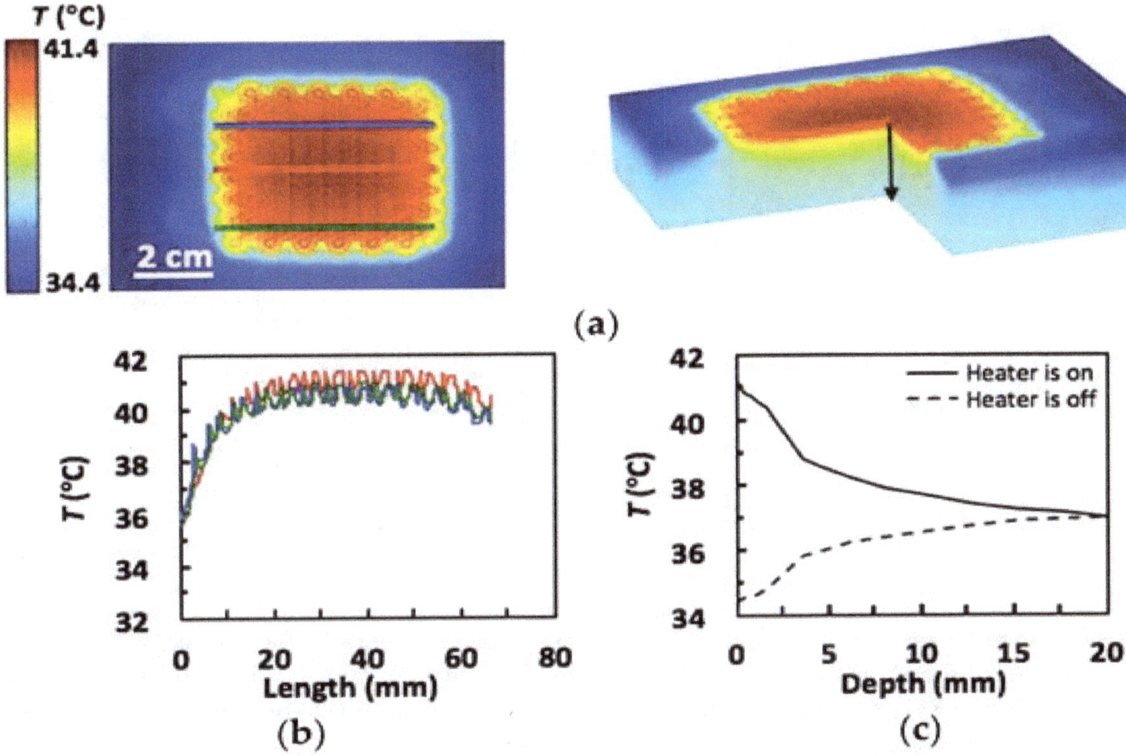

Figure 4. (**a**) COMSOL thermal simulation results (left: top view; right: 3D view); (**b**) Lateral heat distribution of heater. Blue, red and green lines on simulation image (left) mark the horizontal line across which temperature was collected for their respective red, blue and green plots; (**c**) Vertical heat distribution of skin from the black line on simulation image (right).

After calibrating the RTD and characterizing the RHE, we were able to establish a real time PID feedback control as illustrated by the diagram in Figure 5a. The purpose of this system was to demonstrate the functionality of the tattoo-like heater itself and was thus built with wires connecting the heater to a data acquisition (DAQ) unit and a PC. The system can be made to be wireless in the future by integrating it with a microcontroller unit (MCU), a Bluetooth low energy (BLE) chip and a rechargeable battery on a miniature printed circuit board (PCB). The DC power to the RHE was routed through an Omron DC-DC relay (G3CN) which was controlled by a computer using an output DAQ (NI USB-6009). The computer ran a LabVIEW program which controlled the temperature of the RHE using pulse width modulation (PWM). The RTD was connected to the DMM of an NI Elvis II, which measured the RTD's resistance and sent the readings to the LabVIEW program in real time. The LabVIEW program converted the resistance readings into temperature using the following equation:

$$T = T_0 + \frac{\Delta R}{0.002R_0}, \tag{1}$$

where the initial resistance R_0 was measured at the room temperature T_0 and the coefficient 0.0020 °C^{-1} was the TCR obtained from the calibration on human skin in Figure 3d. The PID program then used the real-time temperature feedback, along with a desired temperature set point, to determine how to control the relay and thereby the PWM of the RHE. This allowed the program to keep the heater at a set temperature or to adjust to a new temperature when demanded.

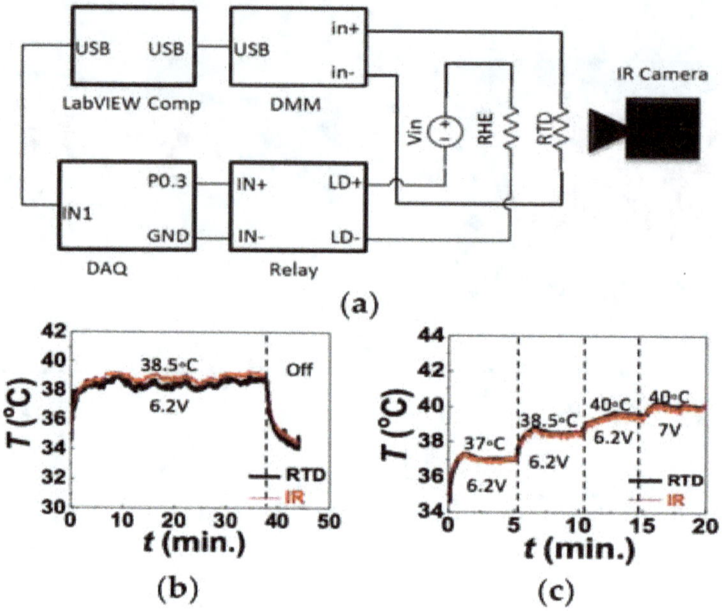

Figure 5. (**a**) Circuit diagram of set up for operating heater with PID control. DMM measures resistance of RTD and feeds it into a computer with LabVIEW, The LabVIEW program calculates the temperature of the RTD using the RTD's starting temperature, starting resistance and calibration constant. It then uses a PID algorithm to calculate the optimal duty cycle for PWM of the heater given the heater's current temperature and the set point temperature for the heater. The LabVIEW program then uses the data acquisition unit (DAQ) to switch the relay on and off with the determined duty cycle, thus controlling how much total power is fed to the heater; (**b**) Temperature of the heater versus time measured with both the RTD and the IR camera as the heater is turned on at a set point of 38.5 °C and then turned off. Heater is able to maintain set point temperature for an extended period of time; (**c**) Temperature of the heater versus time measured with both the RTD and the IR camera as the set point of the heater is changed while the voltage remains constant. At 40 °C, 6.2 V is not sufficient for the heater to reach the set point, so the voltage is increased to 7 V, at which point the heater is able to reach and maintain a temperature of 40 °C.

3. Results and Discussion

First, to test if the device could effectively maintain a target temperature the DC voltage supply was set to 6.2 V and the temperature was set to 38.5 °C. The device was able to maintain a constant temperature of 38.5 °C for over 30 min on human skin until the heater was completely turned off to finish the experiment as shown in Figure 5b. The heater reached the target temperature within 3 min and the error between the target temperature and the device's temperature never reached more than a degree (Supplementary Information, Figure S4). The temperature readings of the RTD (black curve) was also verified by the IR camera results (red curve).

To test if the device could self-adjust when set temperature changes, we conducted an experiment with multiple set temperatures (37 °C, 38.5 °C, 40 °C) while the voltage supply was kept constant at 6.2 V (Figure 5c). For the first two temperatures (Stages I and II), the device was able to reach the set temperatures and to maintain them at a steady state. In switching between these temperatures, no changes were made except changing set point in the LabVIEW program. When the voltage was kept at 6.2 V and the target temperature was set to 40 °C, which is marked as Stage III, the actual skin surface temperature was not able to reach 40 °C. It could only reach up to 39 °C. This indicates insufficient power supply even when the duty cycle of the PWM reached 100%. Therefore, the maximum steady state temperature a heater can reach at 6.2 V with this set-up and under these circumstances is 39 °C, demonstrating a temperature to voltage ratio of 6.19 V/°C. This is lower than what was observed when the power was applied directly to the RHE and not routed through the relay. This indicates that some power may be lost as the electricity passes through the relay and the wires thereto. We therefore increased the voltage to 7 V and the skin surface was then successfully heated to 40 °C, as in Stage IV. Again, the temperature measured by the RTD (black) and the IR camera (red) are well matched. This experiment demonstrates that when given a sufficient voltage supply, the stretchable tattoo-like heater can automatically reach, maintain and change between desired temperatures without any manual adjustment of the voltage.

Due to the negligible stiffness of serpentine ribbons [48,49], the mechanical stiffness of our tattoo-like heater is dominated by the supporting Tegaderm tape, whose Young's modulus was measured to be 7 MPa [42]. The effects of strain and skin deformation on our stretchable heater are discussed in Supplementary Information, Figures S5 and S6. Figure S5 indicates that our RTD can survive more than 70% tensile strain but its resistance is unfortunately slightly sensitive to strain. Due to such strain effects, Figure S6 indicates that the RTD temperature is not accurate when the hand closes but its measurement can recover when the hand restores its original configuration. With the PID control, the RHE may under heat the skin due to the falsely perceived increase of RTD temperature during hand closure, which will not cause any skin burn.

To evaluate the power consumption of the tattoo-like heater, the duty cycles at different set temperatures were investigated. The device was placed on a human palm with PID control. The top frame in Figure 6a plots the actual skin temperature measured by the RTD versus time and the labels are again voltage supply and set temperature. Some small overshooting of the temperature occasionally happens at the points where voltage is changed but they are small and quickly rectified by the controller. The device is not expected to undergo step changes in voltages during real life application but these experiments show that the controller can react appropriately to those step changes should they occur. At each set temperature, the steady state duty cycle was recorded. The middle frame of Figure 6a shows the duty cycle versus time plot. The numbers mark the plateaus where the device was considered to have reached steady state. The duty cycle for each set temperature was calculated as the average of the duty cycle readings at these plateaus. The following equation was then used for power calculation:

$$P = \frac{D}{100\%} \times \frac{V^2}{R}, \tag{2}$$

where D is the duty cycle, V is the voltage supplied to the RHE and R is the resistance of the RHE.

If we define power density to be the power delivered to the skin per unit area of the RHE, power density can be calculated through:

$$\text{Power Density} = P/A, \tag{3}$$

where A represents the total area of the heater, which is 38.7 cm^2 for our RHE. Plotting power density versus the corresponding temperature as red markers in the bottom frame of Figure 6a, a linear relation can be fitted. The slope of this linear curve is defined as the specific power flow (SPF), which represents power density normalized to the applied thermal driving potential, that is, temperature difference. The SPF of our stretchable tattoo-like heater is estimated to be 0.846 mW/(cm^2·°C), which means that to heat up a 1 cm^2 area of this specific human palm by 1 °C would consume a power of 0.846 mW.

Considering convection and radiation between the heater and the ambient environment, it is inaccurate to assume that all the heat generated by the RHE completely goes into the skin. To obtain a more accurate estimation of the specific power flow into the skin, the entire experiment of Figure 6a was repeated in Figure 6b but with insulation over the heater. A 4 cm thick layer of foam, which is a well-known heat insulator, was taken from a delivery package and applied over the heater on the palm to minimize heat loss into the environment. With this heat insulating foam, the SPF was found to be 0.784 mW/(cm^2·°C) as given in Figure 6b bottom frame, which is 7.33% lower compared with that of the exposed heater (0.846 mW/(cm^2·°C)). This result indicates that about 7.33% of the heat generated by the RHE was lost to the environment when the RHE was exposed to air.

To compare our stretchable tattoo-like heater with other stretchable heaters in the literature, we summarized their materials, substrates, power densities and SPFs in Table 1. In cases where power supplied without using PWM, the value of *D* in Equation (2) was set to 100%. It is evident that the SPF of our tattoo-like heater is the lowest among the stretchable heaters directly applied on human skin, which is an active heat sink in comparison with air, polydimethylsiloxane (PDMS) and glass. Moreover, our device is one of the first to implement real time feedback control for stretchable heaters on human skin.

If adopted for commercialization, our device could very feasibly be converted into a wireless portable device with a battery-operated microcontroller as has been demonstrated for other joule heating devices reported in the Table 1 [5,30,31].

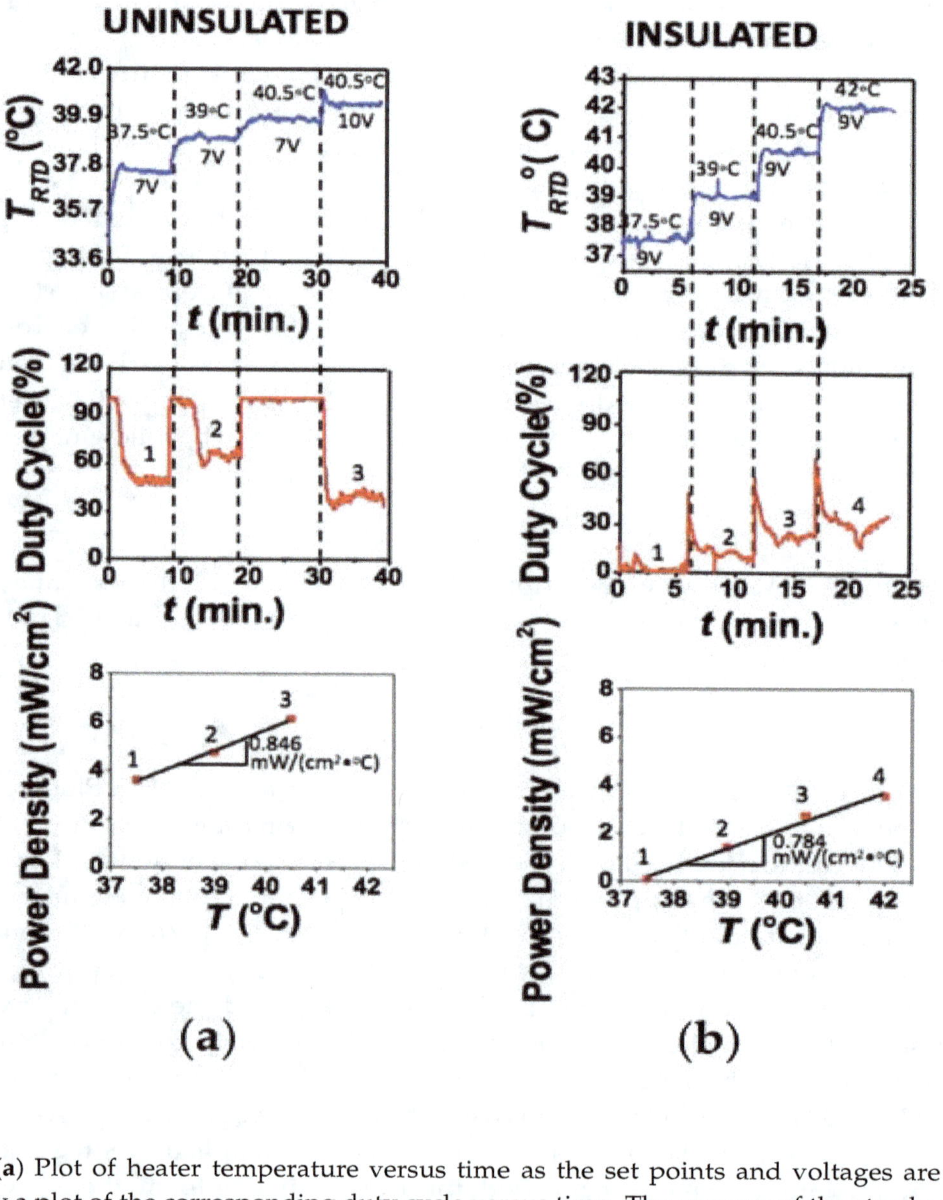

Figure 6. (**a**) Plot of heater temperature versus time as the set points and voltages are changed, followed by a plot of the corresponding duty cycle versus time. The average of the steady state duty cycles marked 1, 2 and 3 were used to calculate the power densities plotted below marked 1, 2 and 3, respectively, at different temperatures; (**b**) The same plots as figure A except the heater is insulated with a piece of foam.

Table 1. Compiled information about different stretchable heaters.

Ref.	T Sensor on Site	Feedback Control	Target T (°C)	Resistive Heating Element	Substrate	V (V)	R (Ω)	Area (cm^2)	P (W)	Power Density (mW/cm^2)	SPF $mW/(cm^2 \cdot °C)$
Our device	Yes	Yes	43	Al (9 μm)	Skin	10	17.2	38.7	2.38	61.44	0.78
[5]	No	No	43	Ag NW/SBS (18/82)	Skin	3.7	~2	91	~7	~80	0.9
[9]	Yes	Yes	40	Au (190 nm)	Pig Skin	12	95.9	<2.3	1.5	>652	>40 *
[15]	Yes	No	ΔT = 6	Au	Skin	–	–	0.64	0.01	20.31	3
[30]	Yes	Yes	250	Ag/Ethylene Glycol (50 wt %)	50 μm PI	4.5	0.75	–	27	650	3
[31]	No	No	40	CuZr nanotrough network	Skin	1.7	3.9	1.24	0.74	600	35
[39]	No	No	39	Ag NW /PDMS 132 mg m^{-2}	Air	4	50	38.5	0.32	8.31	0.6
[39]	No	No	56	Ag NW /PDMS 396 mg m^{-2}	Air	5	15	38.5	1.67	43.29	1
[41]	Yes	No	37	Cr/Au 7/70 nm	PDMS	4.4	550	9	0.04	3.91	0.8
[40]	Yes	Yes	31	PE Yarn with CuNW coating	Skin	1.4	1.66	30.16	1.18	39.12	5

* Assuming power for glass experiments was same as power for pig skin experiments.

4. Conclusions

A low cost, low power consumption stretchable tattoo-like heater was fabricated to reliably warm the skin surface to a target temperature. The device combines a stretchable RHE and a stretchable RTD into a single unit through the cost and time effective "cut-and-paste" fabrication method. The device is thin (60 μm thick) and soft (7.4 MPa modulus) so that it can conform to the complex 3-D surface of palms and can deform and remain attached during hand flexure without perceivable mechanical resistance. The RHE is able to reach set temperatures with relatively even distribution, including during hand movement. The RTD can monitor the real-time temperature of the palm accurately, as verified by simultaneous IR measurements. Through PID temperature feedback control, the device is able to maintain set temperatures for extended periods of time and can automatically adjust to a different temperature if the set point is changed on the controller. The SPF of our device is comparable with reported stretchable skin-mounted heaters. Its simple circuit and program can easily be downscaled to a battery powered printed circuit board (PCB) and microcontroller, giving it potential for point-of-care applications.

Acknowledgments: This work is supported by the Office of Naval Research Young Investigator Award under Contract N00014-16-1-2044 to N.L. and National Science Foundation Grant BET1250659 to K.R.D.

Author Contributions: A.S., E.H., S.Y. and N.L. conducted device design, fabrication, calibration and testing. A.M. constructed the initial PID control algorithm and A.S. later revised it for this project. S.Q. performed the COMSOL finite element modeling. N.L. and K.R.D. supervised and coordinated the project. A.S., N.L., S.Q. and K.R.D. wrote the paper.

References

1. Petrofsky, J.; Berk, L.; Bains, G.; Khowailed, I.A.; Hui, T.; Granado, M.; Laymon, M.; Lee, H. Moist heat or dry heat for delayed onset muscle soreness. *J. Clin. Med. Res.* **2013**, *5*, 416–425. [CrossRef] [PubMed]
2. Nadler, S.F.; Weingand, K.; Kruse, R.J. The physiologic basis and clinical applications of cryotherapy and thermotherapy for the pain practitioner. *Pain Physician* **2004**, *7*, 395–399. [PubMed]
3. Brosseau, L.; Yonge, K.; Welch, V.; Marchand, S.; Judd, M.; Wells, G.; Tugwell, P. Thermotherapy for treatment of osteoarthritis. *Cochrane Database Syst. Rev.* **2003**, *4*. [CrossRef]
4. Lehmann, J. *Therapeutic Heat and Cold*, 4th ed.; Williams & Wilkins: Baltimore, MD, USA, 1990.
5. Choi, S.; Park, J.; Hyun, W.; Kim, J.; Kim, J.; Lee, Y.B.; Song, C.; Hwang, H.J.; Kim, J.H.; Hyeon, T.; et al. Stretchable heater using ligand-exchanged silver nanowire nanocomposite for wearable articular thermotherapy. *ACS Nano* **2015**, *9*, 6626–6633. [CrossRef] [PubMed]
6. Sessler, D.I.; Rubinstein, E.H.; Moayeri, A. Physiologic responses to mild perianesthetic hypothermia in humans. *Anesthesiology* **1991**, *75*, 594–610. [CrossRef] [PubMed]
7. Diller, K.R. Heat transfer in health and healing. *J. Heat Transf.* **2015**, *137*, 1030011–10300112. [CrossRef] [PubMed]
8. Grahn, D.; Brock-Utne, J.G.; Watenpaugh, D.E.; Heller, H.C. Recovery from mild hypothermia can be accelerated by mechanically distending blood vessels in the hand. *J. Appl. Physiol.* **1998**, *85*, 1643–1648. [CrossRef] [PubMed]
9. Son, D.; Lee, J.; Qiao, S.; Ghaffari, R.; Kim, J.; Lee, J.E.; Song, C.; Kim, S.J.; Lee, D.J.; Jun, S.W.; et al. Multifunctional wearable devices for diagnosis and therapy of movement disorders. *Nat. Nanotechnol.* **2014**, *9*, 397–404. [CrossRef] [PubMed]
10. Bagherifard, S.; Tamayol, A.; Mostafalu, P.; Akbari, M.; Comotto, M.; Annabi, N.; Ghaderi, M.; Sonkusale, S.; Dokmeci, M.R.; Khademhosseini, A. Dermal patch with integrated flexible heater for on demand drug delivery. *Adv. Healthc. Mater.* **2016**, *5*, 175–184. [CrossRef] [PubMed]
11. Petrofsky, J.S.; Laymon, M.; Lee, H. Effect of heat and cold on tendon flexibility and force to flex the human knee. *Med. Sci. Monit. Int. Med. J. Exp. Clin. Res.* **2013**, *19*, 661–667. [CrossRef]
12. Michlovitz, S.; Hun, L.; Erasala, G.N.; Hengehold, D.A.; Weingand, K.W. Continuous low-level heat wrap therapy is effective for treating wrist pain. *Arch. Phys. Med. Rehabil.* **2004**, *85*, 1409–1416. [CrossRef] [PubMed]

13. Wang, S.; Li, M.; Wu, J.; Kim, D.-H.; Lu, N.; Su, Y.; Kang, Z.; Huang, Y.; Rogers, J.A. Mechanics of epidermal electronics. *J. Appl. Mech.* **2012**, *79*, 031022. [CrossRef]
14. Wang, L.; Lu, N. Conformability of a thin elastic membrane laminated on a soft substrate with slightly wavy surface. *J. Appl. Mech.* **2016**, *83*, 041007. [CrossRef]
15. Webb, R.C.; Bonifas, A.P.; Behnaz, A.; Zhang, Y.; Yu, K.J.; Cheng, H.; Shi, M.; Bian, Z.; Liu, Z.; Kim, Y.-S.; et al. Ultrathin conformal devices for precise and continuous thermal characterization of human skin. *Nat. Mater.* **2013**, *12*, 938–944. [CrossRef] [PubMed]
16. Cartmell, J.V.; DeRosa, J.F. Capacitively Coupled Indifferent Electrode. U.S. Patent 4,669,468, 2 June 1987.
17. Roselli, R.J.; Diller, K.R. *Biotransport: Principles and Applications*, 2011 ed.; Springer: New York, NY, USA, 2011; ISBN 978-1-4419-8118-9.
18. Tuckey, J. Forced-air warming blanket and surgical access. *Anaesthesia* **1999**, *54*, 97–98. [CrossRef] [PubMed]
19. Kimberger, O.; Held, C.; Stadelmann, K.; Mayer, N.; Hunkeler, C.; Sessler, D.I.; Kurz, A. Resistive polymer versus forced-air warming: Comparable heat transfer and core rewarming rates in volunteers. *Anesth. Analg.* **2008**, *107*, 1621–1626. [CrossRef] [PubMed]
20. Rogers, J.A.; Someya, T.; Huang, Y. Materials and mechanics for stretchable electronics. *Science* **2010**, *327*, 1603–1607. [CrossRef] [PubMed]
21. Kim, D.-H.; Ghaffari, R.; Lu, N.; Rogers, J.A. Flexible and stretchable electronics for biointegrated devices. *Annu. Rev. Biomed. Eng.* **2012**, *14*, 113–128. [CrossRef] [PubMed]
22. Suo, Z. Mechanics of stretchable electronics and soft machines. *MRS Bull.* **2012**, *37*, 218–225. [CrossRef]
23. Nassar, J.M.; Rojas, J.P.; Hussain, A.M.; Hussain, M.M. From stretchable to reconfigurable inorganic electronics. *Extreme Mech. Lett.* **2016**, *9*, 245–268. [CrossRef]
24. Sekitani, T.; Noguchi, Y.; Hata, K.; Fukushima, T.; Aida, T.; Someya, T. A rubberlike stretchable active matrix using elastic conductors. *Science* **2008**, *321*, 1468–1472. [CrossRef] [PubMed]
25. Jung, S.; Kim, J.H.; Kim, J.; Choi, S.; Lee, J.; Park, I.; Hyeon, T.; Kim, D.-H. Reverse-micelle-induced porous pressure-sensitive rubber for wearable human–machine interfaces. *Adv. Mater.* **2014**, *26*, 4825–4830. [CrossRef] [PubMed]
26. Kim, Y.; Zhu, J.; Yeom, B.; Di Prima, M.; Su, X.; Kim, J.-G.; Yoo, S.J.; Uher, C.; Kotov, N.A. Stretchable nanoparticle conductors with self-organized conductive pathways. *Nature* **2013**, *500*, 59–63. [CrossRef] [PubMed]
27. Chun, K.-Y.; Oh, Y.; Rho, J.; Ahn, J.-H.; Kim, Y.-J.; Choi, H.R.; Baik, S. Highly conductive, printable and stretchable composite films of carbon nanotubes and silver. *Nat. Nanotechnol.* **2010**, *5*, 853–857. [CrossRef] [PubMed]
28. Ma, R.; Lee, J.; Choi, D.; Moon, H.; Baik, S. Knitted fabrics made from highly conductive stretchable fibers. *Nano Lett.* **2014**, *14*, 1944–1951. [CrossRef] [PubMed]
29. Park, M.; Im, J.; Shin, M.; Min, Y.; Park, J.; Cho, H.; Park, S.; Shim, M.-B.; Jeon, S.; Chung, D.-Y.; et al. Highly stretchable electric circuits from a composite material of silver nanoparticles and elastomeric fibres. *Nat. Nanotechnol.* **2012**, *7*, 803–809. [CrossRef] [PubMed]
30. Jang, J.; Hyun, B.G.; Ji, S.; Cho, E.; An, B.W.; Cheong, W.H.; Park, J.-U. Rapid production of large-area, transparent and stretchable electrodes using metal nanofibers as wirelessly operated wearable heaters. *NPG Asia Mater.* **2017**, *9*, e432. [CrossRef]
31. An, B.W.; Gwak, E.-J.; Kim, K.; Kim, Y.-C.; Jang, J.; Kim, J.-Y.; Park, J.-U. Stretchable, transparent electrodes as wearable heaters using nanotrough networks of metallic glasses with superior mechanical properties and thermal stability. *Nano Lett.* **2016**, *16*, 471–478. [CrossRef] [PubMed]
32. Moon, G.D.; Lim, G.-H.; Song, J.H.; Shin, M.; Yu, T.; Lim, B.; Jeong, U. Highly stretchable patterned gold electrodes made of Au nanosheets. *Adv. Mater.* **2013**, *25*, 2707–2712. [CrossRef] [PubMed]
33. Gray, D.S.; Tien, J.; Chen, C.S. High-conductivity elastomeric electronics. *Adv. Mater.* **2004**, *16*, 393–397. [CrossRef]
34. Li, T.; Suo, Z.; Lacour, S.P.; Wagner, S. Compliant thin film patterns of stiff materials as platforms for stretchable electronics. *J. Mater. Res.* **2005**, *20*, 3274–3277. [CrossRef]
35. Xu, S.; Zhang, Y.; Cho, J.; Lee, J.; Huang, X.; Jia, L.; Fan, J.A.; Su, Y.; Su, J.; Zhang, H.; et al. Stretchable batteries with self-similar serpentine interconnects and integrated wireless recharging systems. *Nat. Commun.* **2013**, *4*, 1543. [CrossRef] [PubMed]

36. Fan, J.A.; Yeo, W.-H.; Su, Y.; Hattori, Y.; Lee, W.; Jung, S.-Y.; Zhang, Y.; Liu, Z.; Cheng, H.; Falgout, L.; et al. Fractal design concepts for stretchable electronics. *Nat. Commun.* **2014**, *5*, 3266. [CrossRef] [PubMed]
37. Yang, S.; Ng, E.; Lu, N. Indium Tin Oxide (ITO) serpentine ribbons on soft substrates stretched beyond 100%. *Extreme Mech. Lett.* **2015**, *2*, 37–45. [CrossRef]
38. Kim, D.-H.; Lu, N.; Ma, R.; Kim, Y.-S.; Kim, R.-H.; Wang, S.; Wu, J.; Won, S.M.; Tao, H.; Islam, A.; et al. Epidermal electronics. *Science* **2011**, *333*, 838–843. [CrossRef] [PubMed]
39. Hong, S.; Lee, H.; Lee, J.; Kwon, J.; Han, S.; Suh, Y.D.; Cho, H.; Shin, J.; Yeo, J.; Ko, S.H. Highly stretchable and transparent metal nanowire heater for wearable electronics applications. *Adv. Mater.* **2015**, *27*, 4744–4751. [CrossRef] [PubMed]
40. Cheng, Y.; Zhang, H.; Wang, R.; Wang, X.; Zhai, H.; Wang, T.; Jin, Q.; Sun, J. Highly stretchable and conductive copper nanowire based fibers with hierarchical structure for wearable heaters. *ACS Appl. Mater. Interfaces* **2016**, *8*, 32925–32933. [CrossRef] [PubMed]
41. Kim, J.; Lee, M.; Shim, H.J.; Ghaffari, R.; Cho, H.R.; Son, D.; Jung, Y.H.; Soh, M.; Choi, C.; Jung, S.; et al. Stretchable silicon nanoribbon electronics for skin prosthesis. *Nat. Commun.* **2014**, *5*, 5747. [CrossRef] [PubMed]
42. Yang, S.; Chen, Y.-C.; Nicolini, L.; Pasupathy, P.; Sacks, J.; Su, B.; Yang, R.; Sanchez, D.; Chang, Y.-F.; Wang, P.; et al. "Cut-and-Paste" Manufacture of multiparametric epidermal sensor systems. *Adv. Mater.* **2015**, *27*, 6423–6430. [CrossRef] [PubMed]
43. Lu, N.; Wang, X.; Suo, Z.; Vlassak, J. Metal films on polymer substrates stretched beyond 50%. *Appl. Phys. Lett.* **2007**, *91*, 221909. [CrossRef]
44. Lu, N.; Kim, D.-H. Flexible and stretchable electronics paving the way for soft robotics. *Soft Robot.* **2013**, *1*, 53–62. [CrossRef]
45. Kim, J.; Banks, A.; Cheng, H.; Xie, Z.; Xu, S.; Jang, K.I.; Lee, J.W.; Liu, Z.; Gutruf, P.; Huang, X.; et al. Epidermal electronics with advanced capabilities in near-field communication. *Small* **2015**, *11*, 906–912. [CrossRef] [PubMed]
46. Methods of Increasing Emissivity in the Infrared Spectrum. Available online: http://www.optotherm.com/emiss-increasing.htm (accessed on 2 May 2016).
47. Cetingül, M.P.; Herman, C. A heat transfer model of skin tissue for the detection of lesions: Sensitivity analysis. *Phys. Med. Biol.* **2010**, *55*, 5933–5951. [CrossRef] [PubMed]
48. Widlund, T.; Yang, S.; Hsu, Y.-Y.; Lu, N. Stretchability and compliance of freestanding serpentine-shaped ribbons. *Int. J. Solids Struct.* **2014**, *51*, 4026–4037. [CrossRef]
49. Yang, S.; Qiao, S.; Lu, N. Elasticity solutions to nonbuckling serpentine ribbons. *J. Appl. Mech.* **2017**, *84*, 021004. [CrossRef]

3

Suspended Graphene-Based Gas Sensor with 1-mW Energy Consumption

Jong-Hyun Kim [1], Qin Zhou [2] and Jiyoung Chang [1,*]

1 Department of Mechanical Engineering; University of Utah, Salt Lake City, UT 84112, USA; jonghyun.kim@utah.edu
2 Department of Mechanical Engineering; University of Nebraska–Lincoln, Lincoln, NE 68588, USA; zhou@unl.edu
* Correspondence: jy.chang@utah.edu

Academic Editors: Seung Hwan Ko, Daeho Lee and Zhigang Wu

Abstract: This paper presents NH_3 sensing with ultra-low energy consumption for fast recovery and a graphene sheet based on a suspended microheater. Sensitivity and repeatability are important characteristics of functional gas sensors embedded in mobile devices. Moreover, low energy consumption is an essential requirement in flexible and stretchable mobile electronics due to their small dimension and fluctuating resistivity during mechanical behavior. In this paper, we introduce a graphene-based ultra-low power gas detection device with integration of a suspended silicon heater. Dramatic power reduction is enabled by a duty cycle while not sacrificing sensitivity. The new oscillation method of heating improves the sensitivity of 0.049 ($\Delta R/R_0$) measured at a flow rate of 18.8 sccm $NH_{3(g)}$ for 70 s. Our experimental tests show that a 60% duty cycle does not sacrifice sensitivity or recovery by dropping the total power consumption from 1.76 mW to 1.05 mW. The aforementioned low energy consuming gas sensor platform not only attracts environmentally-related industries, but also has the potential to be applied to flexible and stretchable mobile electronic devices.

Keywords: graphene; sensor; heater; NH_3; oscillation; pulse; sensitivity; recovery; energy; flexible

---

1. Introduction

Graphene's unique and excellent electrical and mechanical properties make it a tremendous contribution to flexible and stretchable electronics [1–3]. In particular, gas detection at low temperatures has been one of the main topics explored in applications based on 2D materials [4–6]. Among many gases, ammonia (NH_3) is a one of major target compounds of toxic gases in the field of safety monitoring [2] Conventional toxic gas detection relies on catalytic reactions based on metal oxide, however recent study suggests that the sensitivity can be greatly enhanced by integration of atomically thin layered materials. There are multiple known advantages of graphene. First of all, graphene sheets can be used electronically in single electronic detectors operating at room temperature and in ultra-high sensitivity sensors with mechanical strain or magnetic fields [3]. In terms of mechanical aspects, graphene strength is an essential characteristic for wearable and embedded gas sensors in mobile devices, and graphene-based gas sensors are known to tolerate extreme sensitivity [4]. Thus, many attempts have been made recently to increase the performance of gas detection using chemical doping, array structures and UV-light exposure. The large arrays of sensors can raise the sensitivity due to broadening of the active detecting area [4]. However, there is a limit in size because small size and high sensitivity are required in order to match modern electronic devices.

Increased concentration of charge carrier in graphene induced by adsorbed gas molecules can be useful in highly sensitive sensors [5]. Metal oxide semiconductor and solid electrolyte sensors with a wide typical detection range at an operating temperature of a few hundred degrees Celsius are commercially available. Although these types of sensors are inexpensive and robust, they require high energy consumption and cannot be fabricated on flexible substrates [6]. Other research revealed that the UV light exposure shows much improved sensitivity [7], but there are critical limitations to the method, specifically with respect to its reliability, as the sensor eventually loses its performance due to UV exposure. This is because single-walled carbon nanotube, sensing probes, are gradually removed by the continuous UV light irradiation [8]. In addition, several researchers focused on reducing recovery time. According to the experiment by Schedin et al., adsorbates can be removed by annealing the device at 150 °C [3]. Additionally, another experiment showed that the low temperature in range of 50~150 °C also increasingly contributed to molecular desorption. Semantic et al. fabricated micro-scale suspended hot plate arrays and deposited metals onto the plate by post-chemical vapor deposition (CVD) method [9,10]. The post-CVD method for metal deposition can provide thermal shock effects on the device. In addition, the study also indicated that the manufacturing process of the suspended heater structure is a challenging task. As a separate study, Fowler et al. made a suspended nitride heater and deposit graphene dispersion using spin-coating [11]. However, those suspended heaters have large dimensions (100–500 μm^2), that are three-fold larger compared to the device in this paper. UV exposure for a short time offered an alternative to thermal annealing, however use of UV exposure is not recommended in general due to possible corrosion of carbon on the sensing sheet [8]. The temperature-programmed desorption curve using the Monte Carlo simulation methods and force-field parameters gives predictable data to grasp the fast recovery rate [12]. Therefore, there are clear needs for a device which can detect toxic gases at low power consumption without compromising sensitivity and recovery time. In this paper, a silicon-based microheater and single layered graphene are implemented to enable ultra-low energy consumption. The suspended structured heater is fabricated by surface micromachining followed by backside wet etching. Then, the graphene is deposited via poly-methyl-methacrylate (PMMA) transferring method. Then the sensing probe, covered with graphene, is formed by e-beam lithography. We have demonstrated ultra-low power consumption of up to 1.05 mW by implementing several factors including various input vibration parameters such as air-suspended microheaters, voltage, duty cycle and power input frequency. Multiphysics simulations are used to predict the temperature of a silicon heater and the results are used when designing a suspended heater.

2. Device and Methods

2.1. Structure

Figure 1a shows the working principle of low-power gas sensor. Gas molecules are absorbed on the graphene and sensing electrodes placed on top of silicon nitride layer. A structural side view of the sensor is shown in Figure 1b. The microheater layer is sandwiched between two low-stress silicon nitride (LSN) membranes and suspended in the air to minimize heat loss. Ti and Pt are patterned to be electrically connected to the graphene and the packaging. A single layer of graphene on the copper film synthesized by CVD is transferred onto the device surface and patterned by e-beam lithography as shown in Figure 1c. The graphene sheet, as an absorber, is located on the narrow bridge region of the microheater layer. Each device has four sensors which perform independently and they are positioned on top, left, right and bottom in Figure 1d. Each sensor has four yellow-colored Ti/Pt legs at the center of the device. Two square shaped electrical connecting pads in the middle are opened to the input power source. The other two pads on the outside are connected to the electrical measurement setup. In Figure 1e, the device is mounted on a packaging platform that can be connected by wire bonding.

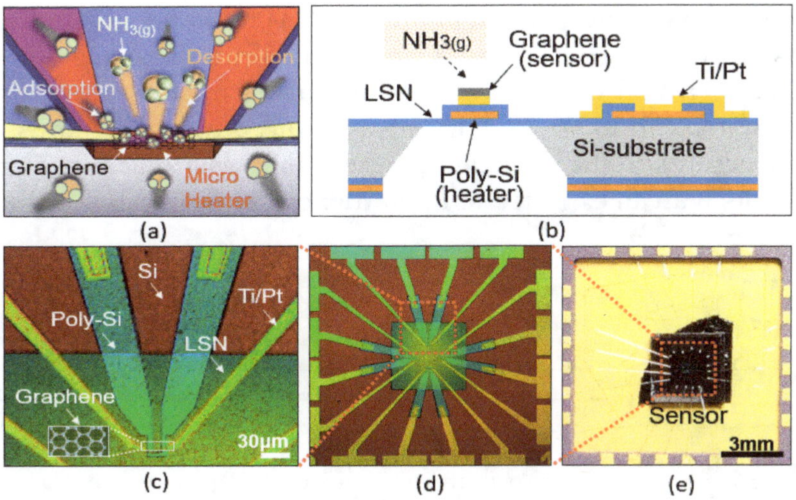

Figure 1. Schematic view of the sensor. (**a**) Concept of a microheater-based graphene gas sensor. Molecules are adsorbed on the graphene and fast recovery is achieved via heat-assisted desorbtion. (**b**) Cross-sectional view of the sensor. Heater is suspended in air to reduce heat loss. (**c**) The graphene is patterned on the bridge of the heater that generates high temperature due to narrow carrier pathway. (**d**) Top view of the device consists of four sensors on a single device. (**e**) A device is connected to a silicon-package by wirebonding. LSN: low-stress silicon-rich nitride.

2.2. Working Principle

A mechanism of adsorption and desorption of NH_3 molecules is shown in Figure 2. When the graphene surface is exposed to a stable environment composed of atmospheric pressure and room temperature, the graphene sheet exhibits an ohmic response before gas adsorption as shown in Figure 2a. When the graphene is exposed to NH_3 as shown in Figure 2b, molecules are adhered on the graphene surface. This $NH_{3(g)}$ sensing is based on changes in the resistivity due to molecular adsorption on the graphene sheet that act as donors [5]. Figure 2c shows desorption by air purification and annealing to accelerate the desorption rate. Molecular desorption reduces the electrical resistance by removing molecules from the graphene surface. The device is then purged with dry air to return graphene to its original resistance value [13]. In addition, the ultraviolet irradiation on graphene is known to increase the electrical resistance of the graphene [14]. Ultraviolet radiation catalyzes the sensitivity of graphene for sensing different types of gases, but graphene has weak light absorption and therefore light irradiation systems are difficult to apply to microelectronics [15]. The sensor's sensitivity has a strong relationship with the activation temperature, therefore, accurately knowing the temperature is important.

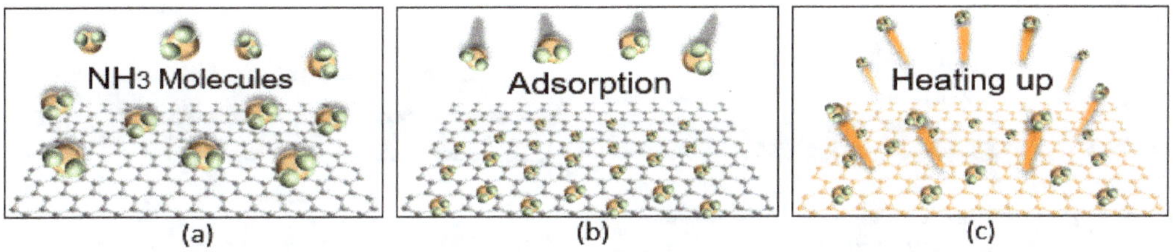

Figure 2. Conceptual process of interaction between molecules and graphene sheet. (**a**) The graphene has a steady condition without gas adsorption. (**b**) Molecules are adhered by interaction bonding. (**c**) Desorption occurs by air purging and annealing that accelerates desorption.

However, due to the size of the microheater, it is difficult to directly measure the temperature of the microheater. We have predicted the temperature of the heater to be about 200 °C by the 2 V input source using commercial finite element analysis (FEA) packaging. In addition, a silicon heater with the same dimensions was fabricated and tested with thermal couple on top of it to monitor the temperature variation along with voltage input. The heater's micro-scale bridge can be used over a wide temperature range of up to 800 °C, driven by input power. The emission of a 'U' shaped microheater (bridge structure) on the left side of Figure 3 shows a silicon heater that glows around an input voltage of 8 volts with an expected temperature of 900 °C.

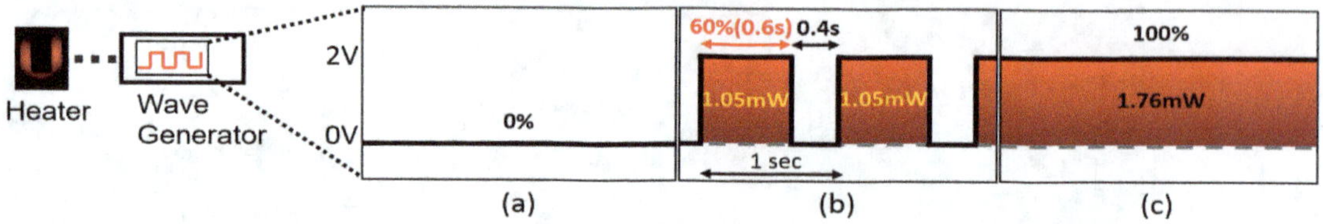

Figure 3. Schematic view of connection with microheater and power wave generator. (**a**) A 0% duty cycle means non-heating. (**b**) A 60% duty cycle of oscillation with 2 V shows 1.05-mW energy consumption. (**c**) Full duty cycle (continuous heating) generates 1.76 mW of energy consumption.

The zero duty cycle in Figure 3a has no energy consumption, and the 60% duty cycle with 2 V in Figure 3b means the generator releases 2 V power source for 0.6 s, then takes a rest for 0.4 s. Overall energy consumption during 1 s is equivalent to 1.05 mW of power consumption. In Figure 3c, the overall duty cycle shows a continuous power supply, reaching a higher energy consumption of 1.76 mW. Consolidating the duty cycle in the operation of the microheater greatly reduces power consumption.

2.3. Device Fabrication

The fabrication process of the sensor is shown in Figure 4. (a) First, a low-stress silicon-rich nitride (LSN) of 100 nm is deposited on 500-µm-thick silicon substrate using low-pressure chemical vapor deposition (LP-CVD). (b) A boron-doped poly-silicon is deposited by CVD. Then the substrate is annealed at 1050 °C. Micro heater structure is patterned by photolithography process followed by reactive ion etching (RIE) of silicon. (c) LSN on the top surface is removed by plasma etching. (d) The LSN is deposited to make a sandwich structure on both sides of the heater. (e) The LSN is patterned to open the electrical connection between the poly-silicon and Ti/Pt that is formed. (f) Ti/Pt, 10/90 nm, are deposited on the top surface and patterned for the electrical pathway using supplemental items including a graphene connector and wire bonding pads. The device is annealed at 350 °C for 1 h in a nitrogen environment. (g) The silicon substrate is etched from the bottom by KOH solution for the suspended microheater. (h) The PMMA-coated graphene sheet is transferred onto the top surface. (i) The PMMA is patterned to shape the graphene sheet as an absorber on the heater region using e-beam lithography. The graphene layer is etched to define the heater region. Single-layer graphene synthesis is as follows.

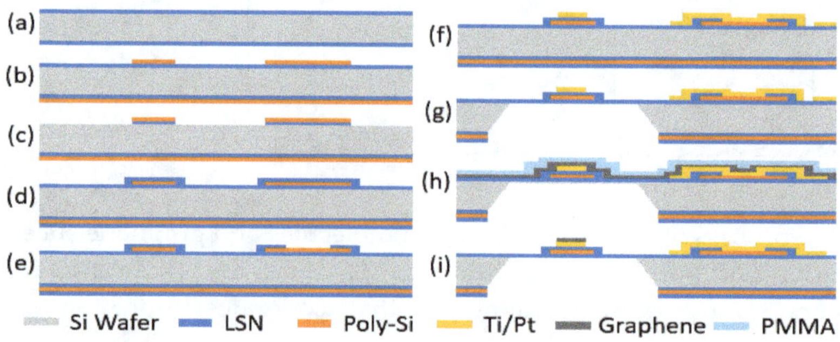

Figure 4. Fabrication process of suspended graphene gas sensor mounted on silicon based heater. Details of each process (**a~i**) are explained in above paragraph.

The single-layered graphene is grown using the CVD process. A copper film is used for the growth substrate. The film is exposed to H_2 for 100 sccm for 1 min and CH_4 for 30 sccm for 3 min in the 550 °C furnace. Then, 10 sccm of H_2 flows for 100 min in the 1050 °C furnace. After a pre-heating process, the film is exposed to both H_2 of 10 sccm and CH_4 22 sccm for 85 min simultaneously. The urnace is shut-off after 25 min in a growth process. All gas valves are closed after 85 min. The film is stored in the furnace for cooling to room temperature overnight. Based on the characterization by Raman spectroscopy, the grown film is identified by matching the characteristic wavelength of the single layer graphene [16].

2.4. Measurement

Figure 5 shows an experimental setup for the gas adsorption and desorption. The test is performed in ambient pressure and room temperature environment. Here, 18.8 sccm of air is supplied from the syringe on the left side of the schematic image. The liquid form of ammonium hydroxide $NH_{3(L)}$ evaporates and dissolves in the air. Then the air containing NH_3 molecules is pumped into the next flask containing drierites, which absorbs water vapor from a mixture of NH_3 and air. Then dry $NH_{3(g)}$ molecules penetrate a test chamber. For desorption of the molecules, a wave form generator (SDG1025, Siglent, Shenzhen, China) is used to generate power for the heating. Multiple sets of temperatures and input power are tested by varying input voltage (0~2.5V), duty cycle (45%~100%) and frequency (1100 Hz). As a reference, the same set of tests is performed using $argon_{(g)}$ gas. A multi-meter and a digital oscilloscope (TBS1052B, Tektronix, OR, USA) is used to measure the resistivity and wave form of the system. The detection capability is evaluated by the relative change in resistance per adsorption and desorption time $(R_1 - R_0)/R_0$. $NH_{3(g)}$ 70-s adsorption and air desorption 70 s was repeated four times. This is a cycle that repeats every 70 s.

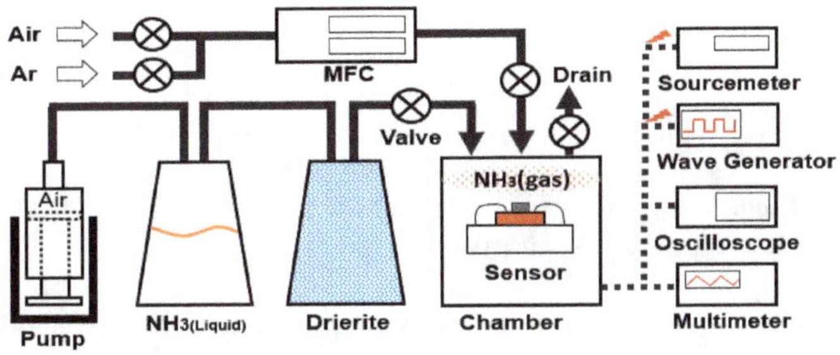

Figure 5. Experimental setup to flow target gas and measure the output from the gas sensor. MFC: mass flow controller.

3. Results

The graphene-based sensor is evaluated as a change in electrical resistance based on NH_3 gas and ultraviolet exposure. When the sensor is exposed to gas/UV, the resistance increases in both UV and NH_3. However, low temperature heating between room temperature and 100 °C does not significantly improve sensitivity/recovery of UV detection. On the other hand, the adsorption sensitivity of NH_3 increases by 40% when 0.5 V is applied compared to the same test conducted at room temperature. Argon (Ar) gas is supplied through a mass flow controller (MFC) in the gas tank. While heating from 100 to 200 °C, $\Delta R/R_0$ per 10 s recovery rate is monitored to be in the same range over all temperature ranges. The 200 sccm of Ar flow also provides 26% faster response than at 10 sccm. The air purging by uncovering with higher temperature gives a 6%~10% faster rate than 200 sccm Ar purging in detail. Using the air purging is the simplest method to refresh the sensor's surface. In addition, the air is able to be regularly utilized when the sensor is operating on the wearable device. Thus, NH_3 and the air worked in shifts for the evaluation of gas sensing and recovery. When the heater is heated to 100 °C or more by applying 1.5 V, the recovery speed is significantly reduced. In other words, the faster the recovery speed, the better the high voltage, but the constant power input of 2.0 V greatly reduces the sensitivity. Based on the tests, input power ranging between 1.5 and 2.0 V can serve as optimized refreshing input power, satisfying both reasonable recovery and sensitivity. As shown in Figure 6, increasing the temperature around 200 °C by 2 V input voltage with 60% duty cycle of oscillation improves the sensitivity and the recovery rate of $NH_{3(g)}$ sensing. It is noted that a test without heating (no input power source) clearly shows the poorest recovery for the gas. Consolidating the duty cycle to reduce power consumption significantly reduces power consumption over continuous heating without compromising sensitivity and recovery time. If the total input power is fixed at 0.6 mW through 40% of the duty cycle, the influence of the input frequency is not apparent. A 1-Hz input frequency improves performance if the total input power is fixed at 1.76 mW over 75% of duty cycle and 1.05 mW over 60%. For example, red (1 Hz) represents a higher absorption and desorption response than orange (100 Hz) with a total input power of 1 mW. However, the 45% and 75% of duty cycles with 1 Hz of frequency give low adsorption/desorption rates that are 0.041/0.041 and 0.038/0.047 ($\Delta R/R_0$) per 70 s, respectively, compared to the rates 0.049/0.052 ($\Delta R/R_0$) per 70 s at 60%. Next, the $NH_{3(g)}$ sensing device's dependence on input power frequency is tested and a small noticeable difference is observed as shown in Figure 7a,b. The 60% duty cycle at 1 and 100 Hz reaches the highest $\Delta R/R_0$ of sensitivity. A 1-Hz frequency shows a higher sensitivity/recovery rate than 100 Hz. Lastly, in an optimal test condition, ultra-low 1.05-mW energy consumption is achieved by a 2.0-V 1-Hz 60% duty cycle input power source that shows as high as 0.049 ($\Delta R/R_0$) sensitivity with 18.8 sccm of $NH_{3(g)}$ input flow.

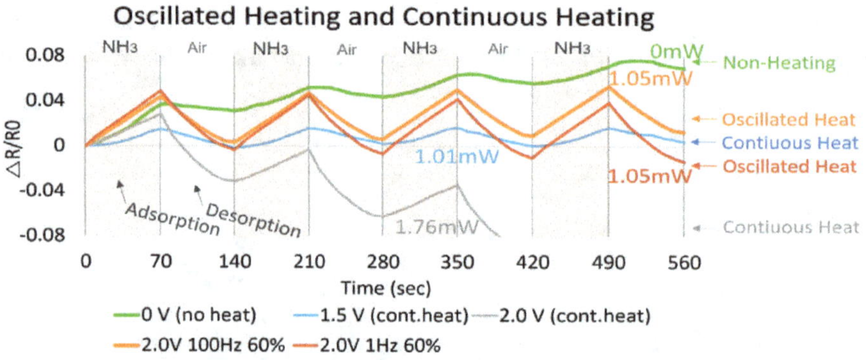

Figure 6. Comparison of sensing and recovery with varying duty cycles (0%, 60%, 100%). Adsorption cycle with $NH_{3(g)}$ 18.8 sccm for 70 s and desorption cycle with air for 70 s are repeated for four cycles each. Ultra-low 1.05-mW energy consumption by optimized heating mode (2.0 V, 1 Hz, 60% duty cycle, red color) shows significant adsorption rate 0.049 ($\Delta R/R_0$) per 70 s of sensitivity and desorption rate 0.052 ($\Delta R/R_0$) per 70s of recovery.

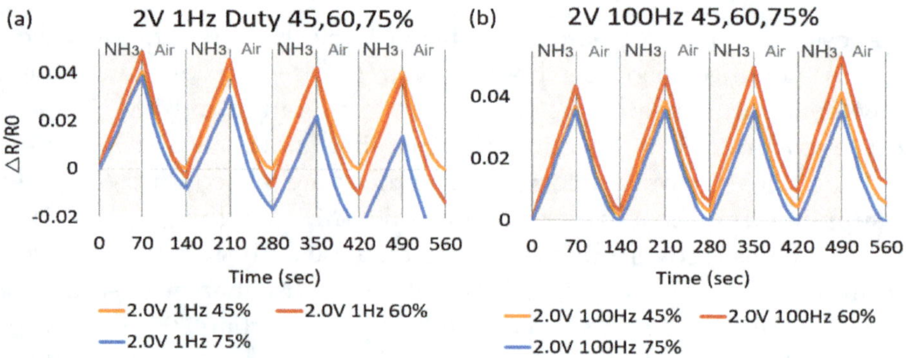

Figure 7. (**a**,**b**) shows a dependence of the frequency. A 60% Duty cycle achieves the highest 0.049 ($\Delta R/R_0$) of sensitivity. A 1-Hz frequency of all duty cycles shows higher sensitivity/recovery to gas. An optimized heating mode (2.0 V, 1 Hz, 60% duty cycle) shows the best performance.

4. Discussion

Various types of interaction, from weak van der Waals to strong covalent bonding, occur between adsorbed molecules and the graphene atoms during the experiment. This leads to a noticeable change in electrical conductivity of graphene [17]. Molecules adsorbed on the graphene sheet act as scattering centers, resulting in an increase in the resistivity of graphene [18]. Higher density of gas can decrease the greater conductivity. It can also be seen that, depending on the change in the emission of the microheater, the heater can generate various temperatures by changing the input power, as well as predicting the temperature in the low temperature range of 100 to 300 °C. The desorption is accelerated, especially with an input voltage of 1.5 V or more. This simulation shows that the 1.5 and 2.0 V input voltages produce approximately 100 and 200 °C respectively. Thus, the relationship between simulation and experimental data can accurately predict low heating temperatures. It is noted that when thermal energy is added, separation of molecules can be accelerated. This is because as the temperature increases, the hydrogen bond network is reorganized, and the NH_3 molecules are gradually desorbed from the surface by breaking of their intermolecular hydrogen bonds [12]. Moreover, there are various binding energies between molecules and graphene within one hexagon of graphene [19]. Thus, the bindings are increasingly broken as time goes on when the graphene is heated. Heating generally increases the sensitivity and at the same time increases the desorption rate. However, when the continuous heating is applied, the sensitivity of continuous heating does not show significant improvements compared to both non-heating and optimized oscillations. The optimized oscillation input is therefore suitable for overall performance and energy consumption. To further reduce energy consumption, the sensor can employ an oscillating power delivery scheme that uses a variety of duty cycles, frequencies, and voltages. According to various sources, high temperature and continuous power can gradually increase the supported recovery rate by weakening the interatomic coupling due to absorption of external atomic energy.

5. Conclusions

Ultra-low power consumption with graphene and micro-electro-mechanical-system (MEMS) based gas sensors have been successfully fabricated and demonstrated. A suspended microheater structure plays a key role in reducing power consumption. In addition, the duty cycle is integrated to further reduce power consumption to 1.05 mW. Although NH_3 gas is tested as one of the typical toxic gases, the platform is open to other types of gas detection through simple calibration. The sensor shows rapid recovery through heating of the silicon heater. As the temperature increases, the desorption gradually increases by reducing hydrogen bonding. With optimized configuration parameters, a 60% duty cycle of 2.0-V 1-Hz input power oscillation reduces energy consumption to 1.05 mW. The suspended microheater structure is suitable for preventing heat transfer which contributes to very low energy consumption. The pulsed input power may reduce damages of graphene and structures. The aforementioned low-energy consumption gas sensor platform will be an attractive platform for flexible and flexible mobile electronic devices.

Acknowledgments: The electron beam process and metal evaporation were conducted in the University of Utah NanoFab facility. This work was supported by the KIST Institutional Program (Project No. 2E27050)

Author Contributions: J.K. performed graphene fabrication, experiments and wrote the paper. Q.Z. and J.C. contributed design and analysis for fabrication of the device and wrote the paper.

References

1. Cho, B.; Yoon, J.; Hahm, M.G.; Kim, D.H.; Kim, A.R.; Kahng, Y.H.; Park, S.W.; Lee, Y.J.; Park, S.G.; Kwon, J.D.; et al. Graphene-based gas sensor: metal decoration effect and application to a flexible device. *J. Mater. Chem. C.* **2014**, 2, 5280. [CrossRef]

2. Lee, K.; Scardaci, V.; Kim, H.Y.; Hallam, T.; Nolan, H.; Bolf, B.E.; Maltbie, G.S.; Abbott, J.E.; Duesberg, G.S. Highly sensitive, transparent, and flexible gas sensors based on gold nanoparticle decorated carbon nanotubes. *Sens.Actuators B Chem.* **2013**, *188*, 571–575. [CrossRef]
3. Schedin, F.; Geim, A.K.; Morozov, S.V.; Hill, E.W.; Blake, P.; Katsnelson, M.I.; Novoselov, K.S. Detection of individual gas molecules adsorbed on graphene. *Nat. Mater.* **2007**, *6*, 652–655. [CrossRef] [PubMed]
4. Sheehan, P.E.; Whitman, L.J. Detection limits for nanoscale biosensors. *Nano Lett.* **2005**, *5*, 803–807. [CrossRef] [PubMed]
5. Leenaerts, O.; Partoens, B.; Peeters, F.M. Adsorption of H_2O, NH_3, CO, NO_2, and NO on graphene: A first-principles study. **2007**, *2*, 1–6.
6. Inoue, T.; Ohtsuka, K.; Yoshida, Y.; Matsuura, Y.; Kajiyama, Y. Metal oxide semiconductor NO_2 sensor. *Sens. Actuators B Chem.* **1995**, *25*, 388–391. [CrossRef]
7. Mitoma, N.; Nouchi, R.; Tanigaki, K. Enhanced sensing response of oxidized graphene formed by UV irradiation in water. *Nanotechnology* **2015**, *26*, 105701. [CrossRef] [PubMed]
8. Chen, g.; Paronyan, T.M.; Pigos, E.M.; Harutyunyan, A.R. Enhanced gas sensing in pristine carbon nanotubes under continuous ultraviolet light illumination. *Sci. Rep.* **2012**, *2*, 343. [CrossRef] [PubMed]
9. Semancik, S.; Cavicchi, R.E.; Wheeler, M.C.; Tiffany, J.E.; Poirier, G.E.; Walton, R.M.; Suehle, J.S.; Panchapakesan, B.; DeVoe, D.L. Microhotplate platforms for chemical sensor research. *Sens. Actuators B Chem.* **2001**, *77*, 579–591. [CrossRef]
10. Cavicchi, R.E.; Suehle, J.S.; Kreider, K.G.; Shomaker, B.L.; Small, J.A.; Gaitan, M.; Chaparala, P. Growth of SnO_2 films on micromachined hotplates. *Appl. Phys. Lett.* **1995**, *66*, 812–814. [CrossRef]
11. Fowler, J.D.; Allen, M.J.; Tung, V.C.; Yang, Y.; Kaner, R.B.; Weiller, B.H. Practical chemical sensors from chemically derived graphene. *ACS Nano.* **2009**, *3*, 301–306. [CrossRef] [PubMed]
12. Liu, L.; Zhao, L.; Sun, H. Simulation of NH_3 temperature-programmed desorption curves using an ab initio force field. *J. Phys. Chem. C.* **2009**, *113*, 16051–16057. [CrossRef]
13. Zhang, Z.; Zhang, X.; Luo, W.; Yang, H.; He, Y.; Liu, Y.; Zhang, X.; Peng, G. Study on adsorption and desorption of ammonia on graphene. *Nanoscale Res. Lett.* **2015**, *10*, 359. [CrossRef] [PubMed]
14. Akinwande, D.; Petrone, N.; Hone, J. Two-dimensional flexible nanoelectronics. *Nat. Commun.* **2014**, *5*, 5678. [CrossRef] [PubMed]
15. Novoselov, K.S.; Geim, A.K.; Morozov, S.V.; Jiang, D.; Zhang, Y.; Dubonos, S.V.; Grigorieva, I.V.; Firsov, A.A. Electric field effect in atomically thin carbon films. *Science* **2004**, *306*, 666–669. [CrossRef] [PubMed]
16. Liu, Y.; Liu, Z.; Lew, W.S.; Wang, Q.J. Temperature dependence of the electrical transport properties in few-layer graphene interconnects. *Nanoscale Res. Lett.* **2013**, *8*, 335. [CrossRef] [PubMed]
17. Silvestrelli, P.L. van der Waals interactions in density functional theory using wannier functions. *J. Phys. Chem. A.* **2009**, *113*, 5224–5234. [CrossRef] [PubMed]
18. Hwang, E.H.; Adam, S.; Das Sarma, S. Transport in chemically doped graphene in the presence of adsorbed molecules. *Phys. Rev. B* **2007**, *76*, 195421. [CrossRef]
19. Lin, X.; Ni, J.; Fang, C. Adsorption capacity of H_2O, NH_3, CO, and NO_2 on the pristine graphene. *J. Appl. Phys.* **2013**, *113*, 34306. [CrossRef]

4

Towards Flexible Transparent Electrodes Based on Carbon and Metallic Materials

Minghui Luo [1,2], Yanhua Liu [1,2,*], Wenbin Huang [1,2], Wen Qiao [1,2], Yun Zhou [1,2], Yan Ye [1,2] and Lin-Sen Chen [1,2,*]

1 College of Physics, Optoelectronics and Energy & Collaborative Innovation Center of Suzhou Nano Science and Technology, Soochow University, Suzhou 215006, China; lmhsuda@yeah.net (M.L.); wbhuang@suda.edu.cn (W.H.); wqiao@suda.edu.cn (W.Q.); zyun@suda.edu.cn (Y.Z.); yanye@suda.edu.cn (Y.Y.)

2 Key Laboratory of Advanced Optical Manufacturing Technologies of Jiangsu Province & Key Laboratory of Modern Optical Technologies of Education Ministry of China, Soochow University, Suzhou 215006, China

* Correspondence: yhliu@suda.edu.cn (Y.L.); lschen@suda.edu.cn (L.-S.C.)

Academic Editors: Seung Hwan Ko, Daeho Lee and Zhigang Wu

Abstract: Flexible transparent electrodes (FTEs) with high stability and scalability are in high demand for the extremely widespread applications in flexible optoelectronic devices. Traditionally, thin films of indium thin oxide (ITO) served the role of FTEs, but film brittleness and scarcity of materials limit its further application. This review provides a summary of recent advances in emerging transparent electrodes and related flexible devices (e.g., touch panels, organic light-emitting diodes, sensors, supercapacitors, and solar cells). Mainly focusing on the FTEs based on carbon nanomaterials (e.g., carbon nanotubes and graphene) and metal materials (e.g., metal grid and metal nanowires), we discuss the fabrication techniques, the performance improvement, and the representative applications of these highly transparent and flexible electrodes. Finally, the challenges and prospects of flexible transparent electrodes will be summarized.

Keywords: flexible transparent electrodes; flexible electronics; metal nanowire; metal grid; optoelectronics devices

---

1. Introduction

Flexible transparent electrodes (FTEs) are essential components for numerous flexible optoelectronic devices due to their excellent capacity for transparency and flexibility, including organic light-emitting diodes [1–6], solar cells [7–12], touch panels [13,14], and wearable devices [15]. Conventionally, transparent conductive oxides (TCO) such as the indium tin oxide films (ITO), fluorine doped tin oxide (FTO) [16], ZnO:Al (AZO) [17], and ZnO:Ga (GZO) [18,19] have governed the domain of optoelectronic devices for several decades. However, some innate drawbacks of the TCO films such as brittleness due to its ceramic nature [20] and high cost for the scarcity of materials such as indium limit the widespread use in flexible devices, where stretching, twisting, or bending are usually requested. Recently, potential alternative materials to TCO have been widely explored, including ultra-thin metallic film [21,22], carbon-based nanomaterials (e.g., carbon nanotubes (CNTs) [23,24], and graphene [25–28]), conducting polymer [29,30], and metallic materials (e.g., metal grid [7,31,32] and metal nanowires [33–37]). To replace conventional TCO films, new types of FTEs should have low processing costs and mechanical flexibility while maintaining a low sheet resistance (R_s) and high optical transparency (T). As for FTEs based on conducting polymer, although the flexibility is improved, low conductivity and T limit their optoelectronic performance, and stability in ambient atmosphere is not good enough. Commonly, CNTs with percolating networks cannot process a low R_s and a high T simultaneously. The low conductivity of the CNTs remains the main limiting factor of the overall conductivity. For example, to achieve an R_s of the CNTs based FTE less than 10 $\Omega\square^{-1}$, the T will decrease drastically because the required thickness of CNTs exceeds 100 nm [38]. There remain challenges to improve the electrical conductivity of CNT-based FTEs.

Monolayer graphene only absorbs 2.3% of visible light and can sustain 4% strain with negligible cracking [39,40]. The theoretical R_s of the graphene is as low as 30 $\Omega\square^{-1}$ [41]. However, the R_s of synthesized graphene usually exceeds several hundred $\Omega\square^{-1}$ using different synthesis methods (e.g., epitaxial grown graphene on silicon carbide, chemical vapor (CVD) deposited on Cu catalysts), due to lower quality graphene with polycrystalline structures and plenty of defects [42,43].

Metallic based FTEs constructed from random networks of nanowires or regular metal grids regarded as another potential alternative to TCO, due to their better R_s-T performance than other alternatives. The metal nanothough networks can obtain a FTE with R_s of 2 $\Omega\square^{-1}$ at a transmittance of 90% [44]. In addition, the fabrication of metal structures coincides with printing and roll-to-roll technology, which reduces the cost for mass-production of FTEs significantly. However, the stability of the metal-based FTEs needs to be investigated further.

In this review, we provide a summary of recent advances in emerging FTEs and related flexible optoelectronic devices, mainly focusing on works reported in the past three years. For early work on FTEs, readers can refer to the review by Hecht et al. [45] and Ellmer et al. [46]. Carbon-based nanomaterials and metallic nanomaterials are promising to replace the dominance of the TCO films due to their superior performance, which will be the focus of our review. Then, we discuss the fabrication techniques, the performance improvement, and the representative applications of these FTEs. The challenges and prospects of the FTEs will eventually be summarized.

2. Currently Emerging Materials

Among those emerging alternatives, carbon-based materials and metallic materials are considered promising candidates for next-generation FTEs, due to their high mechanical flexibility paralleling good optical transparency and electrical conductivity. The material properties, combined with low material costs and fabrication techniques, make these emerging materials very attractive for FTEs.

2.1. One-Deminsional CNT-Based Nanomaterials

Carbon nanotubes (CNTs) have been evaluated and verified as one of the future FTEs relying on their remarkable characteristics. In the past several years, transparent, conductive, and flexible CNT-based FTEs have been investigated widely, involving many applications (e.g., OLED and supercapacitor). They can be fabricated using different methods [47–49], including wet and dry processing, which could be further exploited by combination with the roll-to-roll process. Both single-walled CNTs (SWCNTs) and multiwalled CNT (MWCNT)-based FTEs have been fabricated via solution approaches [24,50,51]. In detail, the SWCNT powders made from the CVD process are centrifuged to remove large conglomerations, before blending into an SWCNT solution. After that, the FTEs can be fabricated by various methods with the solution, including spray coating, dip-coating, and infiltration. Wu et al. have reported a simple process for the fabrication of FTEs using pure SWCNTs [52]. Low sheet resistance of 30 $\Omega\square^{-1}$ was obtained with 90% transmittance. However, owing to the SWCNTs tendency of twining during the CVD process, only a small part of them can be efficiently used, which increases the production costs.

To explore cost-effective method, Feng et al. employed a straightforward roll-to-roll process to fabricate flexible and stretchable MWCNT films as FTEs, where the CNTs possess excellent performance with low sheet resistance (208 $\Omega\square^{-1}$) and high transmittance (90%) [47]. In the work, CNT arrays are obtained by batch growth. Lin et al. reported polyaniline composite films incorporated with aligned MWCNTs, which were fabricated using an easy electrodeposition process [48]. Additionally, Recep et al. reported FTEs based on SWCNTs (Figure 1a), which exhibited an optical transmittance of 82% with 0.02 mg SWCNTs (Figure 1b). The supercapacitors fabricated by the SWCNT FTEs showed good capacity retention (94%) upon cycling over 500 times (Figure 1c) [49].

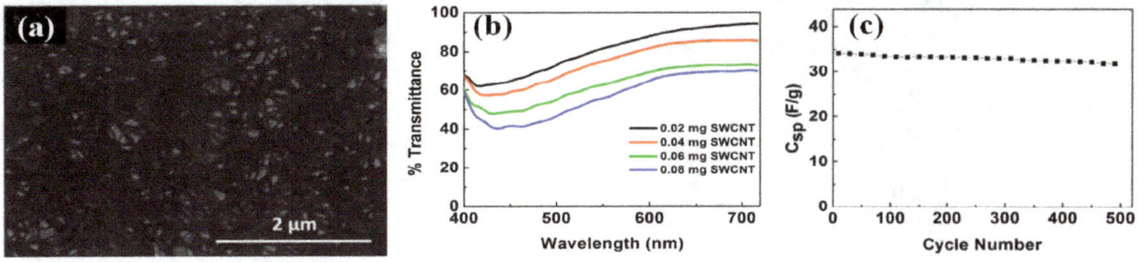

Figure 1. (**a**) SEM image of a SWCNT thin film with a sheet resistance of 75 $\Omega\square^{-1}$; (**b**) Optical transmittance characteristics of the full devices with respect to total SWCNT weight in both electrodes; (**c**) Cycle performance of flexible supercapacitors at a current density of 1.25 A/g. Reproduced with permission from Reference [49].

2.2. Two-Dimensional Graphene-Based Nanomaterials

Graphene, a two-dimensional carbon allotrope, becomes a popular material for transparent electrodes due to the excellent performance [25,53,54]. Generally, the R_s of synthesized graphene usually exceed several hundred $\Omega\square^{-1}$ [42,43]. To improve the photoelectric performance, combing graphene with other conductors is an effective method.

Liu et al. prepared the graphene/ITO flexible hybrid transparent electrode, which showed excellent mechanical and optoelectronic characteristics [55]. Figure 2a shows the optical image of the graphene/ITO hybrid electrode. The SEM image shows the surface of ITO bridges on left and right sides as well as the graphene film upon them. Figure 2b is the transmittance spectra of the samples on the PET substrate. Besides, combination of graphene with metal materials can significantly improve the optoelectronic performance. The metal materials and the graphene offer extra conductive paths for each other. Qiu et al. reported the combination of the metal grid with graphene oxide films via a facile, green, and room-temperature method [56]. The excellent photoelectric properties with the sheet resistance of 18 $\Omega\square^{-1}$ and the transmittance of 80% can be obtained. Hong et al. demonstrated omnidirectionally stretchable and transparent graphene electrodes with good mechanical durability and performance reliability (Figure 2c) [57]. The multilayered graphene based FTE exhibits a transmittance of 87.1% at 550 nm (Figure 2d). Figure 2e,f show that sheet resistances of the graphene/PDMS film systematically monitored under various bending and stretching conditions, respectively. The textured graphene/PDMS film sustains its electrical properties when the film is folded with a bending radius (r) as small as r = 0.5 mm or stretched up to 30% of the tensile strain. Deng et al. developed a full roll-to-roll production of FTEs based on metal NWs (AgNWs and CuNWs) encapsulated by a monolayer graphene film, as shown in Figure 3a–c [26]. The R_s of the encapsulated NWs film show a 60%–90% decrease relative to the pure one (Figure 3d), which is due to the graphene conduction channel. Meanwhile, the encapsulated film processes lower R_s than pristine graphene because metal NWs provide additional conduction channels.

In addition to the graphene-based hybrid FTEs, doped graphene-based FTEs also exhibit good performance [41,58]. Bae et al. demonstrated roll-to-roll production and wet-chemical doping of predominantly monolayer 30-inch graphene films grown by CVD [41]. The doped graphene based FTEs have sheet resistances as low as ~125 $\Omega\square^{-1}$ with 97.4% optical transmittance. Park et al. found that $AuCl_3$ doping on graphene could improve the conductivity and shift the work function of the graphene FTEs, which results in improved power conversion efficiency of the OPV devices [58].

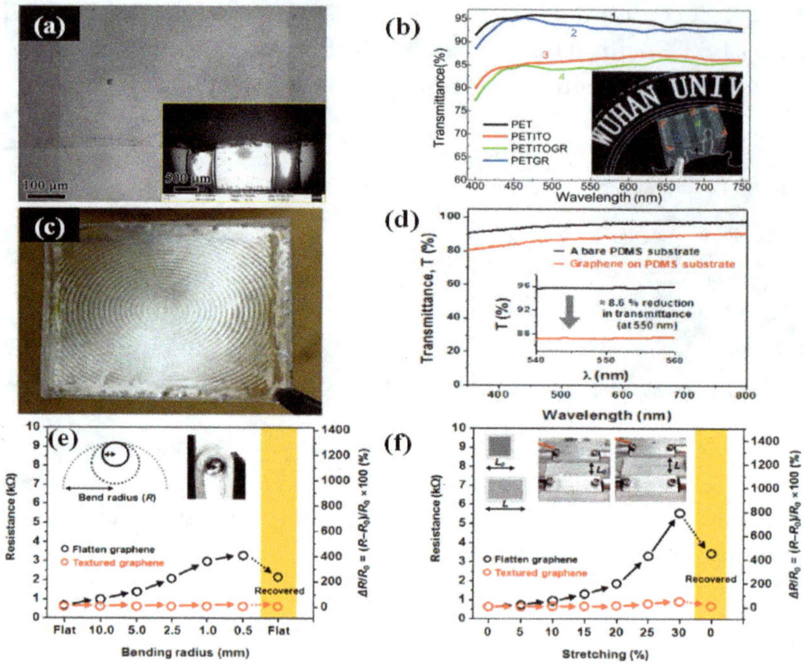

Figure 2. (**a**) Optical image of graphene/ITO hybrid electrode; (**b**) Transmittance spectra of the samples on PET substrate. (**a**,**b**) Reproduced with permission from Ref. [55]. Copyright 2016, Springer Nature; (**c**) Photograph of a graphene/PDMS free-standing film; (**d**) Transmission spectra of a flat PDMS substrate (black) and multilayered graphene on a flat PDMS substrate (red); (**e**) bending radius and (**f**) tensile strains (insets: actual test images for a graphene/PDMS film); (**c**–**f**) Reproduced with permission from Reference [57].

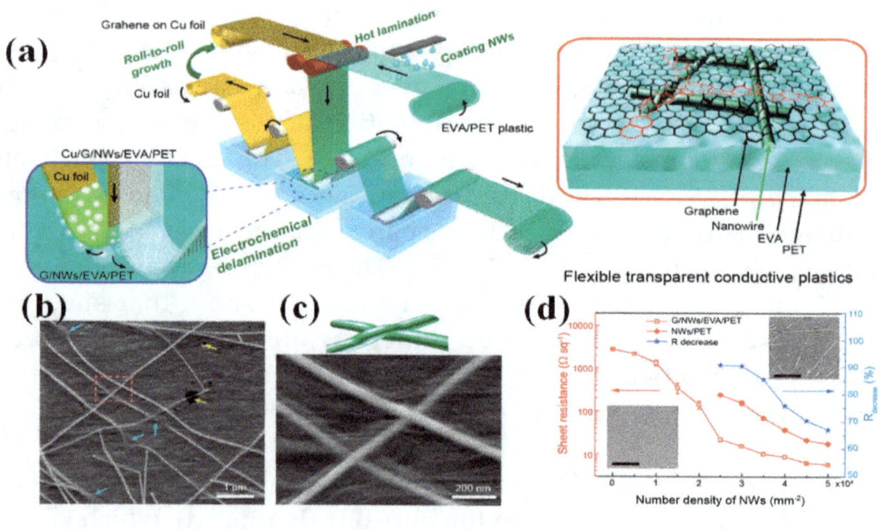

Figure 3. (**a**) Schematic diagram of the fabrication process includes coating of metal nanowires on polymer substrate (EVA/PET), hot-press lamination with graphene/Cu foil, delamination of graphene and Cu foil by electrochemical bubbling method, and the reuse of Cu foil to grow graphene by a continuous chemical vapor deposition system. The detailed structural schematic of the hybrid film labeled in the red cycle shows that nanowires are partly embedded into the EVA substrate and fully encapsulated by monolayer graphene film; (**b**) SEM image of the monolayer graphene; (**c**) Enlarged side-view SEM image of the hybrid film of graphene and AgNWs; (**d**) Sheet resistance versus number density of AgNWs for pure AgNW film and graphene/AgNW hybrid film. Reproduced with permission from Reference [26].

2.3. Metallic Materials

As mentioned above, the progress in carbon-based FTEs is exciting, but the R_s is still not low enough. For the realization of the next generation of flexible and large-area electronics, further improvement in R_s-T performance is required. Metal-based FTEs have gained prominence in both academic and industrial use in flexible electronics. Metal-based FTEs can be constructed from random distribution metallic NWs works [33–37] to regular metal grids [7,31,32], which have demonstrated great potential in optical transparency, electrical conductivity, and mechanical flexibility.

2.3.1. Random Distribution Metallic NWs as FTEs

Most metals have excellent electrical conductivity due to their high free-electron density. To utilize metal materials in FTEs, high transparency should be satisfied. Among common metallic nanomaterials, silver nanowires (AgNWs) are commerciogenic due to the excellent electrical conductivity and high transmittance in a broad wavelength range [33–37].

The fabrication of AgNW FTEs has been explored by many researchers in various fabrication processes [59], which can achieve a low R_s (8–50 $\Omega\square^{-1}$) with a high transmittance (80%–98%). However, metal NW-based FTEs suffer from degradation of the R_s-T performance because of high junction resistances, which have been systematical investigated by several groups [60,61]. Mutiso et al. demonstrated that a higher NW aspect ratio could decrease the R_s while keeping the T a constant [61]. To obtain $R_s \leq 10\ \Omega\square^{-1}$ at T = 90%, the aspect ratio should exceed 800 at a given junction resistance of 2 KΩ. Commonly, the strategies to reduce junction resistance involve bulk heating [62], plasmonic treatment [63], and chemical modifications [64], which tend to damage the plastic substrate and are not suitable for flexible applications. Therefore, other additional processes are introduced to produce hybrid structure electrodes. For example, Xiong et al. exhibited electroless-welding of an AgNW network coated with conductive ion gel that significantly reduces junction resistance between AgNWs, while its effect on the T of the FTEs (R_s of 8.4 $\Omega\square^{-1}$ at T = 86%) is negligible [65].

In addition, the material of AgNWs exhibits the limitation of high surface roughness, which leads to low carrier transport. Currently, AgNW-based composite materials have become promising materials as FTEs, exhibiting excellent optical and electric properties [34,66]. Lee et al. reported a percolating network of AgNWs with densities above the percolation threshold integrated into graphene as hybrid FTEs [67]. The hybrid structure can reduce R_s down to 33 $\Omega\square^{-1}$ with a transmittance of 94%. Other research groups demonstrated the combination of an AgNW network with graphene or graphene oxide could enhance electrical conductivity and decrease surface roughness [68–70]. Xu et al. mixed Ag NWs with graphene oxide and obtained Ag-doped graphene fibers, whose electrical conductivity increased from 4.1×10^4 S/m to 9.3×10^4 S/m, exhibiting a 330% enhancement factor [71]. Meanwhile, chemical and thermal stabilities also can be further improved. It is obvious that the composite materials enhance electrical properties of FTEs by decreasing junction resistance or providing new conductive pathways.

Not merely AgNWs, copper NWs (CuNWs) and gold NWs (AuNWs) are promising candidates for FTEs, due to the similarly excellent conductivity (Figure 4a–c) [34,72,73]. Cui et al. have reported a new synthetic approach for obtaining ultrathin high-quality CuNWs (Figure 4b), where oleylamine is used as a coordinating ligand [72]. Maurer et al. have demonstrated that ultrathin AuNWs can be synthesized at room temperature [73]. Triisopropylsilance (TIPS, 2 mL) is added into the solution after the gold salt was completely dissolved. The AuNWs will format standing for two days until the color turns from yellow into dark-red (Figure 4c). A high transparency of 79% can be achieved and maintained over 80 stretching cycles.

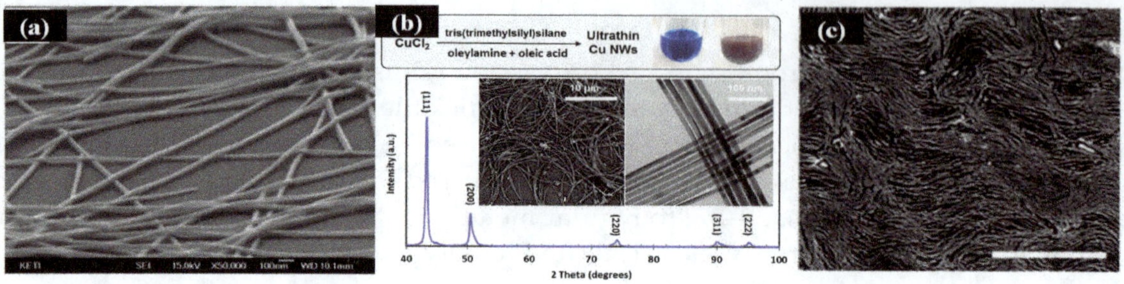

Figure 4. (**a**) SEM image of AgNWs. Reproduced with permission from Reference [34]. Copyright 2016, American Chemical Society; (**b**) Synthesis of ultrathin copper nanowires. Reproduced with permission from Reference [72]. Copyright 2015, American Chemical Society; (**c**) SEM image of monolayer AuNWs on PDMS substrate. Reproduced with permission from Reference [73].

2.3.2. Metal Grids

The randomly dispersed network in the whole scale cannot be used immediately as prepared. Subsequently, patterning processes are necessarily adopted depending on different device architectures. In particular, metal grid FTEs offer advantages over other NW FTEs due to the nature of artificial design. For example, the electrical and optical properties can be managed by modulating the grid pitch, width, and thickness.

Uniform metal grids as FTEs fabricated by many processes, including nano-patterning techniques such as photolithography, crackle, electroless plating, nano-transfer printing, and electrospinning [74–78]. Almost all of these processes involve deposition of a metal film onto a template to form inter-connected network excluding junction resistance. In contrast to the limited material choices for NWs, a variety of metallic materials can be exploited in different applications. Cui et al. employed electrospun and metal deposited technique to generate nanotrough networks. Correspondingly, the width and thickness of the networks are 420 nm and 80–100 nm, respectively. The new kind FTEs exhibited both superior optoelectronic performance (R_s of 2 $\Omega\square^{-1}$ at 90% transmission) (Figure 5a) and remarkable mechanical flexibility (Figure 5b,c) [44]. The R_s-T performance achieved is much better than that of TCO, carbon-based materials, or solution processed mental NWs networks.

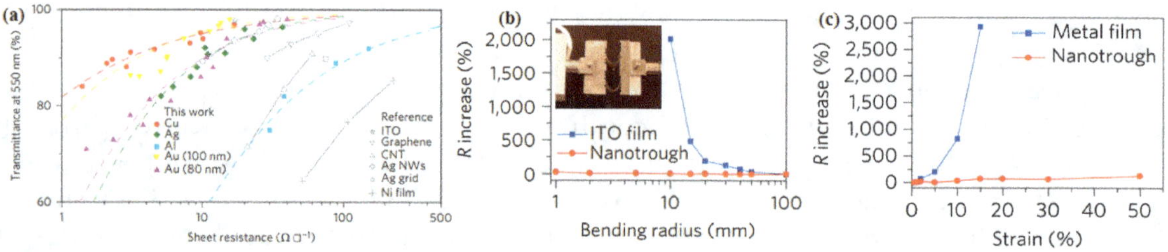

Figure 5. (**a**) Sheet resistance versus optical transmission (at 550 nm) for copper, gold, silver, and aluminum nanotrough networks, described by percolation theory. The performances of device-grade ITO, CNTs, graphene, silver nanowires (NWs), silver grid, and nickel thin films are shown for comparison; (**b**) R_s versus bending radius for bendable transparent electrodes consisting of gold nanotrough networks or ITO films on 178-μm-thick PET substrates; (**c**) R_s versus uniaxial strain for a stretchable transparent electrode consisting of gold nanotrough networks on 0.5-mm-thick PDMS substrate. Reproduced with permission from Ref. [44].

In another work, Han et al. proposed the cracked TiO_2 gel film as a template to make Ag networks [75]. The Ag networks with diameters of 1–2 μm, and widths of 4–100 μm, exhibited good electro-optical properties. The transmittance ranges from 82% to 45%, and correspondingly the R_s ranges from 4.2 $\Omega\square^{-1}$ to 0.5 $\Omega\square^{-1}$. To further improve the R_s-T performance of the NWs based FTEs,

Hsu et al. introduced a mesoscale metal-wire concept in conjunction with NWs [78]. The mesoscale metal-wire networks show the extraordinary R_S-T performance with the R_s being 0.36 and the T being 92%. However, fabricating metal grid FTEs usually employs the physical deposition of metal materials involving thermal evaporation or sputtering, which requires expensive vacuum-based processing. Therefore, the fabrication process is not simple and cost-effective.

Recently, electrohydrodynamic (EHD) jet printing employed to fabricate metal grid FTEs [31,79]. Lee et al. fabricated FTEs by thermal pressing of metal lines, which was provided by EHD jet printing [31]. The excellent properties have been demonstrated with the sheet resistance of 0.5 $\Omega\square^{-1}$ at the transmittance of 80%. Seong et al. have reported a method of fabricated FTEs by EHD jet printing. Sheet resistance of 1.49 $\Omega\square^{-1}$ can be achieved by printing the Ag mesh on the convex glass [79]. Another route utilized to fabricate FTE by using inject printing [80]. Mohl et al. obtained grid meshes by inject printing and subsequent chemical copper plating [32]. The printed metal grids from reactive ink plated with copper can create high performance FTEs. The achieved R_s is 10 $\Omega\square^{-1}$ and transmittance is 80%.

A novel fabrication method via hybrid printing technique has been demonstrated for FTEs with embedded metal grids [81]. Cui et al. developed high-resolution metal mesh as FTEs by nanoimprinting technology [14]. The silver nanoparticle inks embedded into the metal mesh, where the metal mesh was fabricated by the roll-to-roll progress. Low-cost fabrication is the most important advantage of this new approach. Moreover, high performance could be obtained with the low sheet resistance of 0.69 $\Omega\square^{-1}$ at 88% transparency. Most important is that the fabrication cost can be further decreased by exploiting their compatibility with printing technologies, attributing to the efficient use of material, a simple fabrication process, and easy scalability to large scale. In addition, Kiruthika et al. employed the roll and spray coating methods to fabricate the FTEs by a simple solution process using crackle lithography [82]. A transmittance of 78% and sheet resistance of ~20 $\Omega\square^{-1}$ can be obtained (Figure 6a). Figure 6b shows the optical micrographs of Ag meshes with different widths on PET substrate. Figure 6c–f exhibits the mechanical stability of the metallized Ag in the crackle network. The mesh was subjected to 500 bending cycles with 1.5 mm radius, the change in the resistance remained within 5%.

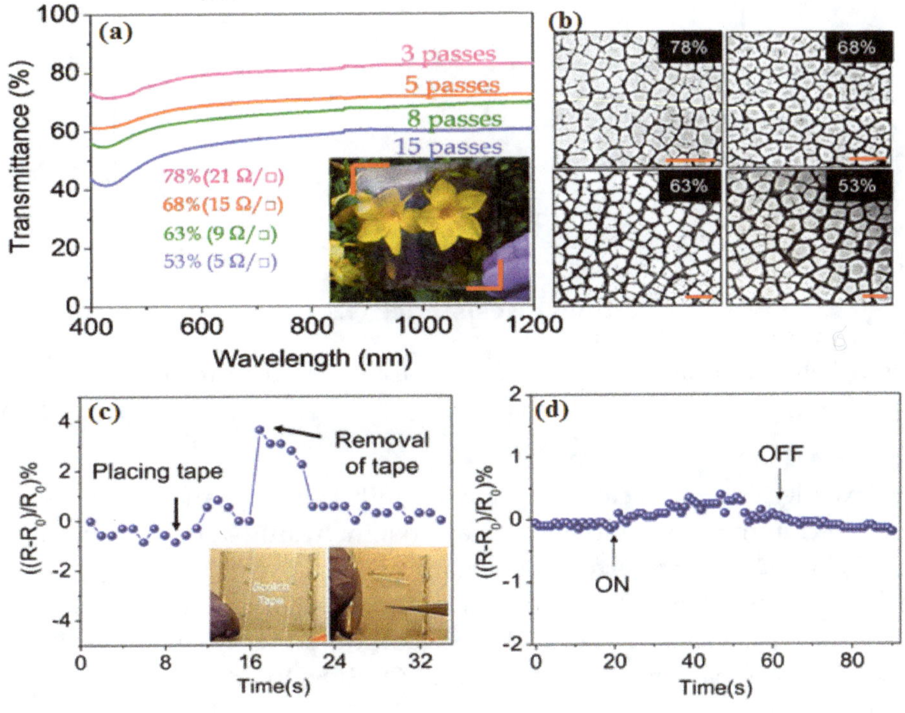

Figure 6. *Cont.*

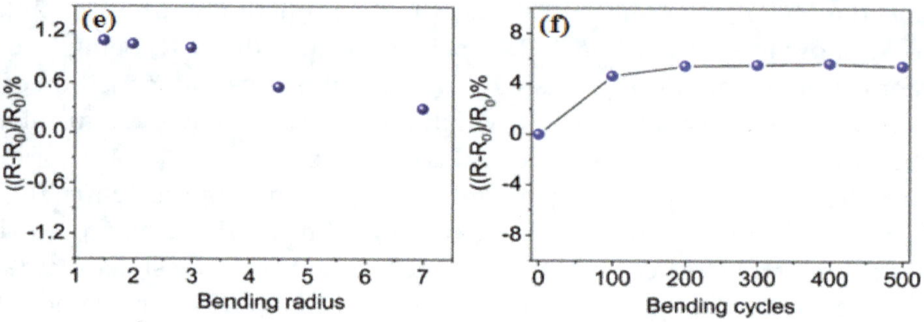

Figure 6. (**a**) Specular optical transmittance of Ag mesh prepared from crackles of different widths. The sheet resistances of the samples are mentioned in parentheses. Photograph of the prepared TCE with 78% transmittance at 550 nm is shown as an inset. (**b**) Optical micrographs of Ag mesh of different widths. Scale bar is 1 mm. Relative variations in the resistance of the Ag mesh during (**c**) the scotch tape adhesion test, (**d**) sonication test, (**e**) bending to different radii, and (**f**) 500 bending cycles with a radius of 1.5 mm. The photographs in the inset in (**a**) show the scotch tape pasted over the Ag mesh and while peeling off. Reproduced with permission from Reference [82].

Finally, some typical values of sheet resistance and transmission of the novel FTEs up to date are summarized in Figure 7. Great progress has made in FTEs towards lower sheet resistance and high transmittance.

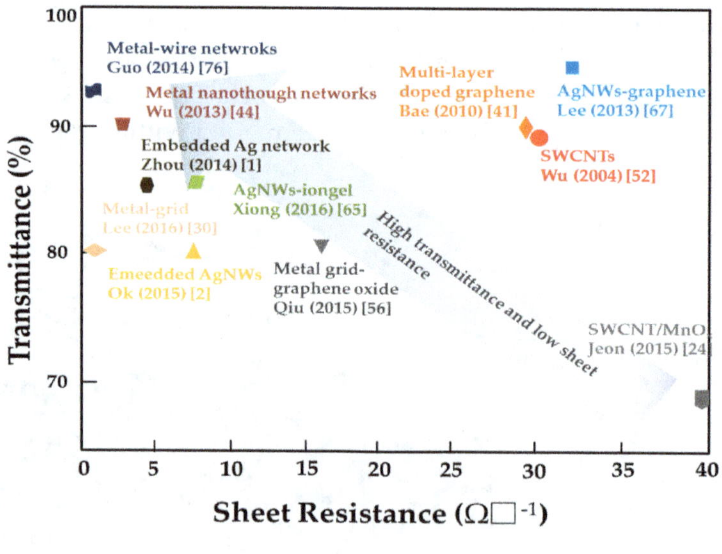

Figure 7. Sheet resistance/Transmission values of FTEs reported in recent works.

3. Applications

Based on the excellent mechanical compliance, high transmittance, and electric conductivity, the FTEs can be utilized in many essential applications, including supercapacitor, OLED, solar cell, and touch panel [45,46]. Different performances are needed for different applications, due to the requirement of technical indexes [60]. To satisfy the requirement of touch screens, the R_s of FTE should range from 100 $\Omega\square^{-1}$ to 1000 $\Omega\square^{-1}$, a T exceeds 85% combined with low haze. The FTEs must have the R_s ~10 $\Omega\square^{-1}$ for T > 90% to provide solar cells and OLEDs. In addition, for use of FTEs in flexible optoelectronics, several requirements must be met, including the reduced haze value for the optical transparency, and ultra-smooth surface of the electrode to avoid the disconnection problem. Finally, a simple, low-cost, and large-scale process in fabricating FTEs is a necessity for commercial applications.

To function as an FTE in OLEDs, the metal-based electrodes must alleviate the leakage current by planarizing its surface. Zhou et al. investigated embedded Ag-grid FTEs as the anode of OLED, which exhibits a power efficiency of 106 $lm \cdot W^{-1}$ at 1000 $cd \cdot m^{-2}$ (Figure 8) [1].

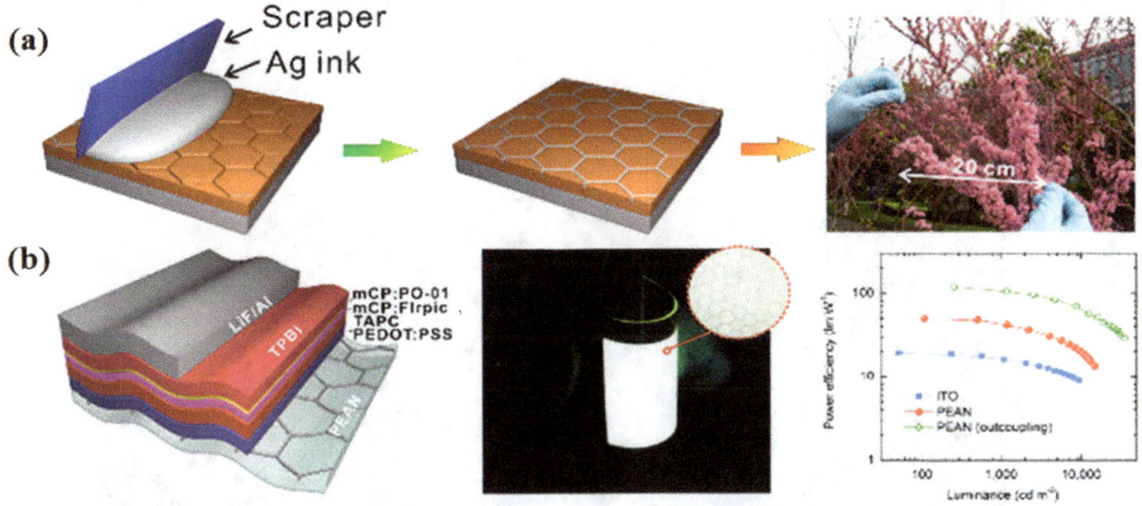

Figure 8. (**a**) Schematic illustration of the fabrication process of an embedded Ag network on PET (PEAN) and its optical image; (**b**) Device structure and performance of flexible white OLED using PEAN anode. Reproduced with permission from Reference [1].

Liu et al. reported a composite electrode developed from poly (3,4-ethylenedioxythiophene): poly (styrenesulfonate) (PEDOT: PSS) and AgNWs with a template stripping method [3]. The efficiency of the resultant OLEDs improved by 25% compared to that of traditional PEDOT: PSS FTEs, attributing to the reduction of surface roughness and the improvement of electric conductivity. Ok and co-worker developed an ultra-thin and smooth FTE by embedding AgNWs in a colorless polymide (cPI) [2]. The AgNWs-cPI composite electrode exhibited a $T > 80\%$, a low R_s of 8 $\Omega\square^{-1}$, and ultra-smooth surfaces comparable to glass (4.1 nm), as shown in Figure 9a. The OLEDs fabricated from such composite electrodes showed a stable performance with a luminance reduction of <3% after 10 repeated bendings at a radius of 30 μm (Figure 9b).

For applications in display devices and touch screens, high visibility is needed. Haze property is a vital value to the display industry. Ag networks that are fabricated by cracked template and AgNWs solution processed methods [75,83] can survive multiple finger touchings. The low R_s (<10 $\Omega\square^{-1}$) and the large area fabrication are rapidly improving in the large format touch panels market. Currently, iV-touch can offer large projected capacitive touch panels from 21.5″ to 55″ with stunning touch performance [84].

Similar to the case of OLEDs, the application of NW networks and metal grids as the TCEs employed in functional organic solar cells. In order to achieve the required R_s (<50 $\Omega\square^{-1}$) and T (>90%), constructing hybrid structured FTEs is a promising technology. Singh et al. demonstrated solution processed solar cells employing AgNW film binding with a 40 nm thin overlayer of sputtered ZnO as FTEs [85]. The ZnO overlayer is used to increase adhesion between AgNW and ZnO/buffer layer and to interconnect the AgNWs junctions. More recently, Li et al. integrated embedded Ag grids and conducting polymer hybrid electrodes into a perovskite-based photovoltaic cell and demonstrated an ultrathin flexible device delivering a power conversion efficiency of 14.0% [7]. Wu et al. demonstrated AgNW grids with multi-length scaled structures as FTEs for an organic solar cell [86]. A power conversion efficiency of 9% was achieved for the organic solar cell devices.

(a)

(b)

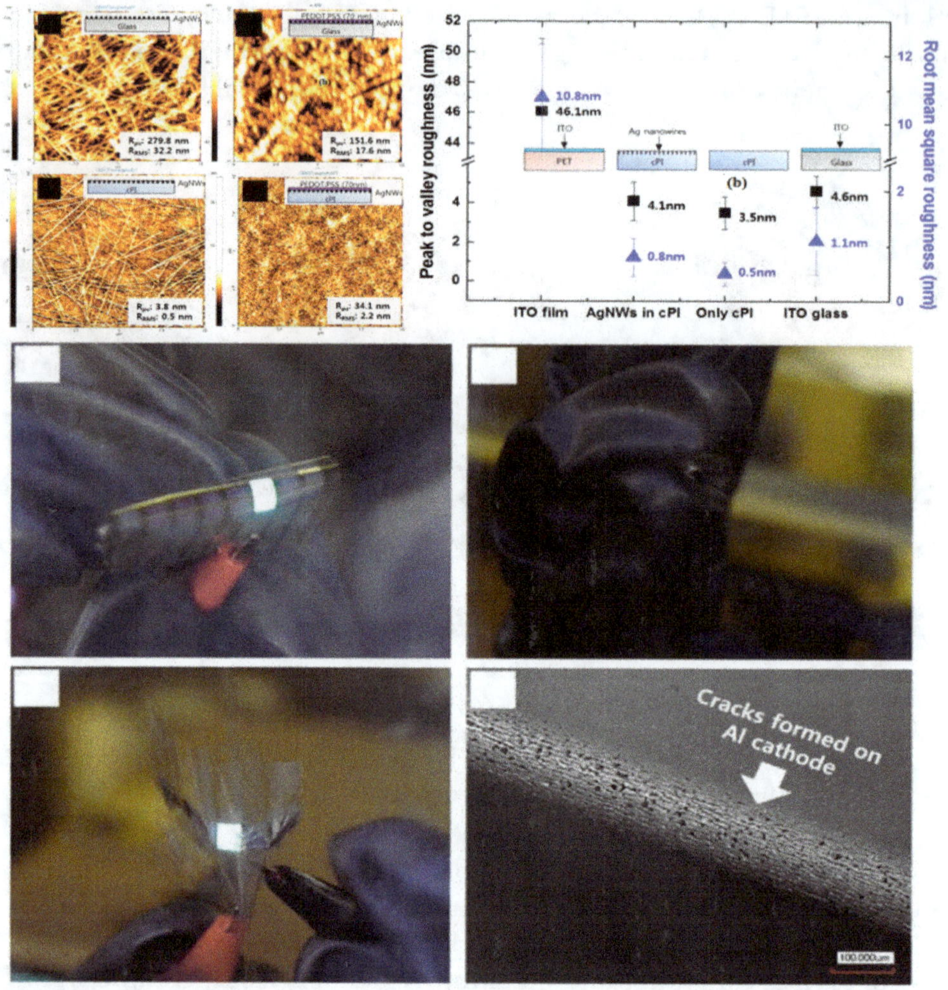

Figure 9. (**a**) Comparison of morphologies by AFM analyses and a diagram comparing the roughness of various samples; (**b**) Bending stability of the flexible OLEDs based on AgNWs-cPI composite electrodes. Reproduced with permission from Reference [2].

Transparent supercapacitors have been proposed in the past decade, which function both as a current collector to transport the electrons, and an active material to store electrochemical energy. For the application in supercapacitors, Lin et al. has synthesized polyaniline composite film incorporated with aligned MWCNTs through an easy electrodeposition process [48]. The conductive films are sufficiently used to fabricate transparent, flexible, and efficient supercapacitors with a maximum specific capacitance of 233 F/g at a current density of 1 A/g. Cai et al. demonstrated that the electrochemical stability of Ag grid FTEs can be enhanced by coating one layer of PEDOT:PSS [87]. The film sustained an optical modulation and a specific capacitance of 87.7% and 67.2% at 10 $A{\cdot}g^{-1}$, respectively. Cheng et al. reported hybrid electrodes composed of PEDOT:PSS and Ag-grids prepared via inject printing [88]. The hybrid structures not only compensated for the shortcomings of the single materials but also fully combined their advantages. The comparatively high Ct (capacitance of the supercapacitor) and Csc (capacitance of the electrode) of the supercapacitor based on PEDOT: PSS (three layers)/Ag grid electrodes were 1.13 $mF{\cdot}cm^{-2}$ and 4.52 $mF{\cdot}cm^{-2}$, respectively. In addition, Lee et al. have introduced a highly flexible and transparent supercapacitor based on electrochemically stable Ag-Au core-shell (AACS) nanowire percolation network electrode (Figure 10) [89]. Figure 10c confirms that the AACS NW networks possess superior optical transparency, exceeding 85% in the entire range of the visible wavelength. Figure 10d shows strain-dependent electrical resistance of Ag–Au core–shell NW electrode on an elastic PDMS substrate.

The various applications make carbon-based and metal-based FTE one of the most promising materials in optoelectronic devices. Considering the progress and the ongoing efforts on FTEs, there is no doubt that performance of these FTE-based flexible devices (OLED, OSC, supercapacitor, touch screen) can be further improved, and the new applications can served to an amazing degree.

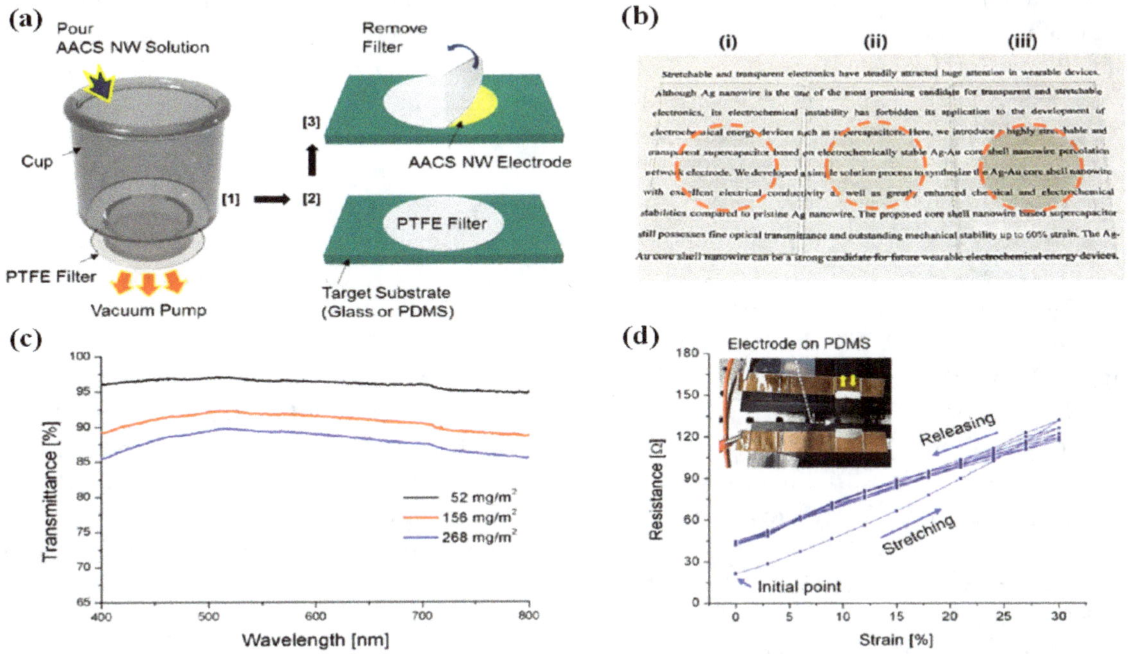

Figure 10. Fabrication and characterization of Ag–Au core–shell NW-based electrodes. (**a**) Schematic illustration of the electrode fabrication through vacuum filtration and transfer method; (**b**) Digital image of as-prepared transparent Ag–Au core–shell NW network electrodes at various sheet resistances; (**c**) Optical transmittance of the transparent electrodes at various areal NW densities; (**d**) Strain-dependent electrical resistance of Ag–Au core–shell NW electrode on an elastic PDMS substrate. Reproduced with permission from Reference [89].

4. Conclusions and Challenges

In conclusion, a comprehensive overview on recent developments and achievements in carbon-based and metal-based FTEs and related flexible optoelectronic devices is provided in this review. Carbon-based materials and metallic materials are promising to dominate the FTEs, especially hybrid FTEs exhibiting superior properties. Although outstanding performances—including low sheet resistance, high transmittance, and good mechanical properties—have been demonstrated, there are still many restrictions hindering large-scale fabrication. In addition, more exploration in the microcosmic aspect and the fundamental theories are needed. The present FTEs already play a significant role in many optical and electric applications, which urges more studies to improve the performance (stretchability, flexibility, transmittance, stability, and electric conductivity) from the material synthesis to device fabrication.

Acknowledgments: This work is supported by the National Natural Science Foundation of China (NSFC) (61405133, 61505131, 91323303), and by the Specialized Research Fund for the Doctoral Program of Higher Education (No. 20133201120027), and by the Natural Science Foundation of Jiangsu Province (No. BK20140348, BK20150309), and by the China Postdoctoral Science Foundation (Grant No. 2015M571816), and by a project funded by the Priority Academic Program Development (PAPD) of Jiangsu Higher Education Institutions.

Author Contributions: Minghui Luo gathered information and wrote the first draft of this review. Yanhua Liu supervised and amended this review. Wenbin Huang, Wen Qiao, Yun Zhou, and Yan Ye were auxiliary for perfecting this work. Lin-Sen Chen supervised and supported the work. All authors reviewed the manuscript.

References

1. Zhou, L.; Xiang, H.Y.; Shen, S.; Li, Y.Q.; Chen, J.D.; Xie, H.J.; Goldthorpe, I.A.; Chen, L.S.; Lee, S.T.; Tang, J.X. High-performance flexible organic light-emitting diodes using embedded silver network transparent electrodes. *ACS Nano* **2014**, *8*, 12796–12805. [CrossRef] [PubMed]
2. Ok, K.H.; Kim, J.; Park, S.R.; Kim, Y.; Lee, C.J.; Hong, S.J.; Kwak, M.G.; Kim, N.; Han, C.J.; Kim, J.W. Ultra-thin and smooth transparent electrode for flexible and leakage-free organic light-emitting diodes. *Sci. Rep.* **2015**, *5*, 9464. [CrossRef] [PubMed]
3. Liu, Y.S.; Feng, J.; Ou, X.L.; Cui, H.F.; Xu, M.; Sun, H.B. Ultrasmooth, highly conductive and transparent PEDOT:PSS/silver nanowire composite electrode for flexible organic light-emitting devices. *Org. Electron.* **2016**, *31*, 247–252. [CrossRef]
4. Lee, H.; Lee, D.; Ahn, Y.; Lee, E.W.; Park, L.S.; Lee, Y. Highly efficient and low voltage silver nanowire-based OLEDs employing a n-type hole injection layer. *Nanoscale* **2014**, *6*, 8565–8570. [CrossRef] [PubMed]
5. Yun, S.O.; Hwang, Y.; Park, J.; Jeong, Y.; Kim, S.H.; Noh, B.I.; Jung, H.S.; Jang, H.S.; Hyun, Y.; Choa, S.H. Sticker-Type Alq 3-Based OLEDs Based on Printable Ultrathin Substrates in Periodically Anchored and Suspended Configurations. *Adv. Mater.* **2013**, *25*, 5626–5631. [CrossRef] [PubMed]
6. Meng, H.; Luo, J.; Wang, W.; Shi, Z.; Niu, Q.; Dai, L.; Qin, G. Top-Emission Organic Light-Emitting Diode with a Novel Copper/Graphene Composite Anode. *Adv. Funct. Mater.* **2013**, *23*, 3324–3328. [CrossRef]
7. Li, Y.; Meng, L.; Yang, Y.; Xu, G.; Hong, Z.; Chen, Q.; You, J.; Li, G.; Yang, Y.; Li, Y. High-efficiency robust perovskite solar cells on ultrathin flexible substrates. *Nat. Commun.* **2016**, *7*. [CrossRef] [PubMed]
8. Kim, B.J.; Dong, H.K.; Lee, Y.Y.; Shin, H.W.; Sang, G.; Han, G.S.; Sug, J.; Mahmood, K.; Ahn, T.K.; Joo, Y.C. Highly efficient and bending durable perovskite solar cells: Toward a wearable power source. *Energy Environ. Sci.* **2014**, *8*, 916–921. [CrossRef]
9. Shin, S.S.; Yang, W.S.; Noh, J.H.; Suk, J.H.; Jeon, N.J.; Park, J.H.; Ju, S.K.; Seong, W.M.; Sang, I.S. High-performance flexible perovskite solar cells exploiting Zn_2SnO_4 prepared in solution below 100 °C. *Nat. Commun.* **2015**, *6*. [CrossRef] [PubMed]
10. Kranz, L.; Gretener, C.; Perrenoud, J.; Schmitt, R.; Pianezzi, F.; La, M.F.; Blösch, P.; Cheah, E.; Chirilă, A.; Fella, C.M. Doping of polycrystalline CdTe for high-efficiency solar cells on flexible metal foil. *Nat. Commun.* **2013**, *4*, 1431–1442. [CrossRef] [PubMed]
11. Li, Z.; Kulkarni, S.A.; Boix, P.P.; Shi, E.; Cao, A.; Fu, K.; Batabyal, S.K.; Zhang, J.; Xiong, Q.; Wong, L.H. Laminated Carbon Nanotube Networks for Metal Electrode-Free Efficient Perovskite Solar Cells. *ACS Nano* **2014**, *8*, 6797–6804. [CrossRef] [PubMed]
12. Han, J.; Yuan, S.; Liu, L.; Qiu, X.; Gong, H.; Yang, X.; Li, C.; Hao, Y.; Cao, B. Fully indium-free flexible Ag nanowires/ZnO:F composite transparent conductive electrodes with high haze. *J. Mater. Chem. A* **2015**, *3*, 5375–5384. [CrossRef]
13. Lee, J.; Lee, P.; Lee, H.B.; Hong, S.; Lee, I.; Yeo, J.; Lee, S.S.; Kim, T.S.; Lee, D.; Ko, S.H. Room-Temperature Nanosoldering of a Very Long Metal Nanowire Network by Conducting-Polymer-Assisted Joining for a Flexible Touch-Panel Application. *Adv. Funct. Mater.* **2013**, *23*, 4171–4176. [CrossRef]
14. Cui, Z.; Gao, Y. 27.5L: Late-News Paper: Hybrid Printing of High Resolution Metal Mesh as A Transparent Conductor for Touch Panels and OLED Displays. *SID Symp. Dig. Tech. Pap.* **2015**, *46*, 398–400. [CrossRef]
15. Wu, C.; Kim, T.W.; Li, F.; Guo, T. Wearable Electricity Generators Fabricated Utilizing Transparent Electronic Textiles Based on Polyester/Ag Nanowires/Graphene Core-Shell Nanocomposites. *ACS Nano* **2016**, *10*, 6449–6457. [CrossRef] [PubMed]
16. Baek, W.H.; Choi, M.; Yoon, T.S.; Lee, H.H.; Kim, Y.S. Use of fluorine-doped tin oxide instead of indium tin oxide in highly efficient air-fabricated inverted polymer solar cells. *Appl. Phys. Lett.* **2010**, *96*, 133506. [CrossRef]
17. Pei, Z.L.; Zhang, X.B.; Zhang, G.P.; Gong, J.; Sun, C.; Huang, R.F.; Wen, L.S. Transparent conductive ZnO:Al thin films deposited on flexible substrates prepared by direct current magnetron sputtering. *Thin Solid Films* **2005**, *68*, 20–23. [CrossRef]
18. Assuncao, V.; Fortunato, E.; Marques, A.; Aguas, H.; Ferreira, I.; Costa, M.; Martins, R. Influence of the deposition pressure on the properties of transparent and conductive ZnO:Ga thin-film produced by r.f. sputtering at room temperature. *Thin Solid Films* **2003**, *427*, 401–405. [CrossRef]

19. Abduev, A.; Akmedov, A.; Asvarov, A.; Chiolerio, A. A revised growth model for transparent conducting Ga doped ZnO films: Improving crystallinity by means of buffer layers. *Plasma Process. Polym.* **2015**, *12*, 725–733. [CrossRef]
20. Gutruf, P.; Shah, C.M.; Walia, S.; Nili, H.; Zoolfakar, A.S.; Karnutsch, C.; Kalantarzadeh, K.; Sriram, S.; Bhaskaran, M. Transparent functional oxide stretchable electronics: Micro-tectonics enabled high strain electrodes. *NPG Asia Mater.* **2013**, *5*, 759–760. [CrossRef]
21. Zhang, C.; Zhao, D.; Gu, D.; Kim, H.; Ling, T.; Wu, Y.K.R.; Guo, L.J. An Ultrathin, Smooth, and Low-Loss Al-Doped Ag Film and Its Application as a Transparent Electrode in Organic Photovoltaics. *Adv. Mater.* **2014**, *26*, 5696–5701. [CrossRef] [PubMed]
22. Vj, L.; Kobayashi, N.P.; Islam, M.S.; Wu, W.; Chaturvedi, P.; Fang, N.X.; Wang, S.Y.; Williams, R.S. Ultrasmooth silver thin films deposited with a germanium nucleation layer. *Nano Lett.* **2009**, *9*, 178–182.
23. Wang, X.; Li, Z.; Xu, W.; Kulkarni, S.A.; Batabyal, S.K.; Zhang, S.; Cao, A.; Wong, L.H. TiO_2 nanotube arrays based flexible perovskite solar cells with transparent carbon nanotube electrode. *Nano Energy* **2015**, *11*, 728–735. [CrossRef]
24. Jeon, I.; Cui, K.; Chiba, T.; Anisimov, A.; Nasibulin, A.G.; Kauppinen, E.I.; Maruyama, S.; Matsuo, Y. Direct and Dry Deposited Single-Walled Carbon Nanotube Films Doped with MoO(x) as Electron-Blocking Transparent Electrodes for Flexible Organic Solar Cells. *J. Am. Chem. Soc.* **2015**, *137*, 7982–7985. [CrossRef] [PubMed]
25. Chen, Y.Z.; Medina, H.; Tsai, H.W.; Wang, Y.C.; Yen, Y.T.; Manikandan, A.; Chueh, Y.L. Low Temperature Growth of Graphene on Glass by Carbon-Enclosed Chemical Vapor Deposition Process and Its Application as Transparent Electrode. *Chem. Mater.* **2015**, *27*, 1636–1655. [CrossRef]
26. Deng, B.; Hsu, P.C.; Chen, G.; Chandrashekar, B.N.; Liao, L.; Ayitimuda, Z.; Wu, J.; Guo, Y.; Lin, L.; Zhou, Y. Roll-to-Roll Encapsulation of Metal Nanowires between Graphene and Plastic Substrate for High-Performance Flexible Transparent Electrodes. *Nano Lett.* **2015**, *15*, 4206–4213. [CrossRef] [PubMed]
27. Li, N.; Yang, G.; Sun, Y.; Song, H.; Cui, H.; Yang, G.; Wang, C. Free-Standing and Transparent Graphene Membrane of Polyhedron Box-Shaped Basic Building Units Directly Grown Using a NaCl Template for Flexible Transparent and Stretchable Solid-State Supercapacitors. *Nano Lett.* **2015**, *15*, 3139–3203. [CrossRef] [PubMed]
28. Liu, Z.; Parvez, K.; Li, R.; Dong, R.; Feng, X.; Müllen, K. Transparent Conductive Electrodes from Graphene/PEDOT:PSS Hybrid Inks for Ultrathin Organic Photodetectors. *Adv. Mater.* **2015**, *27*, 669–675. [CrossRef] [PubMed]
29. Lipomi, D.J.; Lee, J.A.; Vosgueritchian, M.; Tee, C.K.; Bolander, J.A.; Bao, Z. Electronic Properties of Transparent Conductive Films of PEDOT:PSS on Stretchable Substrates. *Chem. Mater.* **2012**, *24*, 373–382. [CrossRef]
30. Jung, S.; Lim, K.; Kang, J.W.; Kim, J.K.; Oh, S.I.; Eun, K.; Kim, D.G.; Choa, S.H. Electromechanical properties of indium-tin-oxide/poly(3,4-ethylenedioxythiophene): Poly(styrenesulfonate) hybrid electrodes for flexible transparent electrodes. *Thin Solid Films* **2014**, *550*, 435–443. [CrossRef]
31. Lee, Y.; Jin, W.Y.; Cho, K.Y.; Kang, J.W.; Kim, J. Thermal pressing of a metal-grid transparent electrode into a plastic substrate for flexible electronic devices. *J. Mater. Chem. C* **2016**, *4*, 4577–4583. [CrossRef]
32. Mohl, M.; Dombovari, A.; Vajtai, R.; Ajayan, P.M.; Kordas, K. Self-assembled large scale metal alloy grid patterns as flexible transparent conductive layers. *Sci. Rep.* **2015**, *5*, 13710. [CrossRef] [PubMed]
33. Kim, D.H.; Yu, K.C.; Kim, Y.; Kim, J.W. Highly Stretchable and Mechanically Stable Transparent Electrode Based on Composite of Silver Nanowires and Polyurethane-Urea. *ACS Appl. Mater. Interfaces* **2015**, *7*, 15214–15222. [CrossRef] [PubMed]
34. Sciacca, B.; Groep, J.V.D.; Polman, A.; Garnett, E.C. Nanowires: Solution-Grown Silver Nanowire Ordered Arrays as Transparent Electrodes. *Adv. Mater.* **2016**, *28*, 976. [CrossRef]
35. Guo, F.; Azimi, H.; Hou, Y.; Przybilla, T.; Hu, M.; Bronnbauer, C.; Langner, S.; Spiecker, E.; Forberich, K.; Brabec, C.J. High-performance semitransparent perovskite solar cells with solution-processed silver nanowires as top electrodes. *Nanoscale* **2015**, *7*, 1642–1649. [CrossRef] [PubMed]
36. José, A.L.; Fe, M.M.; Gómez, D.; Luisa, M.A.; Bristow, N.; Paul, K.J.; Menéndez, A.; Ruiz, B. Rapid synthesis of ultra-long silver nanowires for tailor-made transparent conductive electrodes: Proof of concept in organic solar cells. *Nanotechnology* **2015**, *26*, 26.

37. Li, B.; Ye, S.; Stewart, I.E.; Alvarez, S.; Wiley, B.J. Synthesis and Purification of Silver Nanowires To Make Conducting Films with a Transmittance of 99. *Nano Lett.* **2015**, *15*, 6722–6726. [CrossRef] [PubMed]
38. Kumar, A.; Zhou, C. The race to replace tin-doped indium oxide: Which material will win? *ACS Nano* **2010**, *4*, 11–14. [CrossRef] [PubMed]
39. Nair, R.R.; Blake, P.; Grigorenko, A.N.; Novoselov, K.S.; Booth, T.J.; Stauber, T.; Peres, N.M.; Geim, A.K. Fine structure constant defines visual transparency of graphene. *Science* **2008**, *320*, 1308. [CrossRef] [PubMed]
40. Chen, J.H.; Jang, C.; Xiao, S.; Ishigami, M.; Fuhrer, M.S. Intrinsic and extrinsic performance limits of graphene devices on SiO2. *Nat. Nanotechnol.* **2008**, *3*, 206–209. [CrossRef] [PubMed]
41. Mcdonald, C.; Salter, D.M.; Chetty, U.; Anderson, T.J. Roll-to-roll production of 30-inch graphene films for transparent electrodes. *Nat. Nanotechnol.* **2010**, *5*, 574–578.
42. Park, J.U.; Nam, S.W.; Lee, M.S.; Lieber, C.M. Synthesis of monolithic graphene–graphite integrated electronics. *Nat. Mater.* **2011**, *11*, 120–125. [CrossRef] [PubMed]
43. Li, X.; Cai, W.; An, J.; Kim, S.; Nah, J.; Yang, D.; Piner, R.; Velamakanni, A.; Jung, I.; Tutuc, E. Large-area synthesis of high-quality and uniform graphene films on copper foils. *Science* **2009**, *324*, 1312–1314. [CrossRef] [PubMed]
44. Wu, H.; Kong, D.; Ruan, Z.; Hsu, P.C.; Wang, S.; Yu, Z.; Carney, T.J.; Hu, L.; Fan, S.; Cui, Y. A transparent electrode based on a metal nanotrough network. *Nat. Nanotechnol.* **2013**, *8*, 421–425. [CrossRef] [PubMed]
45. Hecht, D.S.; Hu, L.; Irvin, G. Emerging transparent electrodes based on thin films of carbon nanotubes, graphene, and metallic nanostructures. *Adv. Mater.* **2011**, *23*, 1482–1513. [CrossRef] [PubMed]
46. Ellmer, K. Past achievements and future challenges in the development of optically transparent electrodes. *Nat. Photonics* **2012**, *6*, 809–817. [CrossRef]
47. Feng, C.; Liu, K.; Wu, J.S.; Liu, L.; Cheng, J.S.; Zhang, Y.; Sun, Y.; Li, Q.; Fan, S.; Jiang, K. Flexible, Stretchable, Transparent Conducting Films Made from Superaligned Carbon Nanotubes. *Adv. Funct. Mater.* **2010**, *20*, 885–891. [CrossRef]
48. Lin, H.; Li, L.; Ren, J.; Cai, Z.; Qiu, L.; Yang, Z.; Peng, H. Conducting polymer composite film incorporated with aligned carbon nanotubes for transparent, flexible and efficient supercapacitor. *Sci. Rep.* **2013**, *3*, 1353. [CrossRef] [PubMed]
49. Yuksel, R.; Sarioba, Z.; Cirpan, A.; Hiralal, P.; Unalan, H.E. Transparent and Flexible Supercapacitors with Single Walled Carbon Nanotube Thin Film Electrodes. *ACS Appl. Mater. Interfaces* **2014**, *6*, 15434–15439. [CrossRef] [PubMed]
50. Ren, L.; Pint, C.L.; Arikawa, T.; Takeya, K.; Kawayama, I.; Tonouchi, M.; Hauge, R.H.; Kono, J. Broadband Terahertz Polarizers with Ideal Performance Based on Aligned Carbon Nanotube Stacks. *Nano Lett.* **2012**, *12*, 787–790. [CrossRef] [PubMed]
51. Kim, B.J.; Park, J.S.; Hwang, Y.J.; Park, J.S. Characteristics of silver meshes coated with carbon nanotubes via spray-coating and electrophoretic deposition for touch screen panels. *Thin Solid Films* **2015**, *596*, 68–71. [CrossRef]
52. Wu, Z.; Chen, Z.; Du, X.; Logan, J.M.; Sippel, J.; Nikolou, M.; Kamaras, K.; Reynolds, J.R.; Tanner, D.B.; Hebard, A.F. Transparent, conductive carbon nanotube films. *Science* **2004**, *305*, 1273–1276. [CrossRef] [PubMed]
53. Eda, G.; Fanchini, G.; Chhowalla, M. Large-area ultrathin films of reduced graphene oxide as a transparent and flexible electronic material. *Nat. Nanotechnol.* **2008**, *3*, 270–274. [CrossRef] [PubMed]
54. Hu, W.; Peng, C.; Luo, W.; Lv, M.; Li, X.; Li, D.; Huang, Q.; Fan, C. Graphene-based antibacterial paper. *ACS Nano* **2010**, *4*, 4317–4323. [CrossRef] [PubMed]
55. Liu, J.; Yi, Y.; Zhou, Y.; Cai, H. Highly Stretchable and Flexible Graphene/ITO Hybrid Transparent Electrode. *Nanoscale Res. Lett.* **2016**, *11*, 1–7. [CrossRef] [PubMed]
56. Qiu, T.; Luo, B.; Liang, M.; Ning, J.; Wang, B.; Li, X.; Zhi, L. Hydrogen reduced graphene oxide/metal grid hybrid film: Towards high performance transparent conductive electrode for flexible electrochromic devices. *Carbon* **2015**, *81*, 232–238. [CrossRef]
57. Hong, J.Y.; Kim, W.; Choi, D.; Kong, J.; Park, H.S. Omni-Directionally Stretchable and Transparent Graphene Electrodes. *ACS Nano* **2016**, *10*, 9446–9455. [CrossRef] [PubMed]
58. Park, H.; Rowehl, J.A.; Kim, K.K.; Bulovic, V.; Kong, J. Doped graphene electrodes for organic solar cells. *Nanotechnology* **2010**, *21*, 505204. [CrossRef] [PubMed]

59. Groep, J.V.D.; Spinelli, P.; Polman, A. Transparent Conducting Silver Nanowire Networks. *Nano Lett.* **2012**, *12*, 3138–3144. [CrossRef] [PubMed]
60. Bergin, S.M.; Chen, Y.H.; Rathmell, A.R.; Charbonneau, P.; Li, Z.Y.; Wiley, B.J. The effect of nanowire length and diameter on the properties of transparent, conducting nanowire films. *Nanoscale* **2012**, *4*, 1996–2004. [CrossRef] [PubMed]
61. Mutiso, R.M.; Sherrott, M.C.; Rathmell, A.R.; Wiley, B.J.; Winey, K.I. Integrating simulations and experiments to predict sheet resistance and optical transmittance in nanowire films for transparent conductors. *ACS Nano* **2013**, *7*, 7654–7663. [CrossRef] [PubMed]
62. Jiu, J.; Nogi, M.; Sugahara, T.; Tokuno, T.; Araki, T.; Komoda, N.; Suganuma, K.; Uchida, H.; Shinozaki, K. Strongly adhesive and flexible transparent silver nanowire conductive films fabricated with a high-intensity pulsed light technique. *J. Mater. Chem.* **2012**, *22*, 23561–23567. [CrossRef]
63. Garnett, E.C.; Cai, W.; Cha, J.J.; Mahmood, F.; Connor, S.T.; Christoforo, M.G.; Cui, Y.; Mcgehee, M.D.; Brongersma, M.L. Self-limited plasmonic welding of silver nanowire junctions. *Nat. Mater.* **2012**, *11*, 241–249. [CrossRef] [PubMed]
64. Chung, C.H.; Song, T.B.; Bob, B.; Zhu, R.; Yang, Y. Solution-processed flexible transparent conductors composed of silver nanowire networks embedded in indium tin oxide nanoparticle matrices. *Nano Res.* **2012**, *5*, 805–814. [CrossRef]
65. Xiong, W.; Liu, H.; Chen, Y.; Zheng, M.; Zhao, Y.; Kong, X.; Wang, Y.; Zhang, X.; Kong, X.; Wang, P. Highly Conductive, Air-Stable Silver Nanowire@Iongel Composite Films toward Flexible Transparent Electrodes. *Adv. Mater.* **2016**, *28*. [CrossRef] [PubMed]
66. Menamparambath, M.M.; Ajmal, C.M.; Kim, K.H.; Yang, D.; Roh, J.; Park, H.C.; Chan, K.; Choi, J.Y.; Baik, S. Silver nanowires decorated with silver nanoparticles for low-haze flexible transparent conductive films. *Sci. Rep.* **2015**, *5*, 16371. [CrossRef] [PubMed]
67. Lee, M.S.; Lee, K.; Kim, S.Y.; Lee, H.; Park, J.; Choi, K.H.; Kim, H.K.; Kim, D.G.; Lee, D.Y.; Nam, S. High-performance, transparent, and stretchable electrodes using graphene-metal nanowire hybrid structures. *Nano Lett.* **2013**, *13*, 2814–2821. [CrossRef] [PubMed]
68. Liang, J.; Li, L.; Tong, K.; Ren, Z.; Hu, W.; Niu, X.; Chen, Y.; Pei, Q. Silver Nanowire Percolation Network Soldered with Graphene Oxide at Room Temperature and Its Application for Fully Stretchable Polymer Light-Emitting Diodes. *ACS Nano* **2014**, *8*, 1590–1600. [CrossRef] [PubMed]
69. Moon, I.K.; Kim, J.I.; Lee, H.; Hur, K.; Kim, W.C.; Lee, H. 2D Graphene Oxide Nanosheets as an Adhesive Over-Coating Layer for Flexible Transparent Conductive Electrodes. *Sci. Rep.* **2012**, *3*, 1112. [CrossRef]
70. Liu, Y.; Chang, Q.; Huang, L. Transparent, flexible conducting graphene hybrid films with a subpercolating network of silver nanowires. *J. Mater. Chem. C* **2013**, *1*, 2970–2974. [CrossRef]
71. Xu, Z.; Liu, Z.; Sun, H.; Gao, C. Highly Electrically Conductive Ag-Doped Graphene Fibers as Stretchable Conductors. *Adv. Mater.* **2013**, *25*, 3249–3253. [CrossRef] [PubMed]
72. Cui, F.; Yu, Y.; Dou, L.; Sun, J.; Yang, Q.; Schildknecht, C.; Schierle-Arndt, K.; Yang, P. Synthesis of Ultrathin Copper Nanowires Using Tris(trimethylsilyl)silane for High-Performance and Low-Haze Transparent Conductors. *Nano Lett.* **2015**, *15*, 7610–7615. [CrossRef] [PubMed]
73. Maurer, J.H.; Gonzalez-Garcia, L.; Reiser, B.; Kanelidis, I.; Kraus, T. Templated Self-Assembly of Ultrathin Gold Nanowires by Nanoimprinting for Transparent Flexible Electronics. *Nano Lett.* **2016**, *16*, 2921–2925. [CrossRef] [PubMed]
74. An, B.W.; Gwak, E.J.; Kim, K.; Kim, Y.C.; Jang, J.; Kim, J.Y.; Park, J.U. Stretchable, Transparent Electrodes as Wearable Heaters Using Nanotrough Networks of Metallic Glasses with Superior Mechanical Properties and Thermal Stability. *Nano Lett.* **2015**, *16*, 471–478. [CrossRef] [PubMed]
75. Han, B.; Pei, K.; Huang, Y.; Zhang, X.; Rong, Q.; Lin, Q.; Guo, Y.; Sun, T.; Guo, C.; Carnahan, D. Uniform Self-Forming Metallic Network as a High-Performance Transparent Conductive Electrode. *Adv. Mater.* **2014**, *26*, 873–877. [CrossRef] [PubMed]
76. Guo, C.F.; Sun, T.; Liu, Q.; Suo, Z.; Ren, Z. Highly stretchable and transparent nanomesh electrodes made by grain boundary lithography. *Nat. Commun.* **2014**, *5*, 155–164. [CrossRef] [PubMed]
77. Khan, A.; Lee, S.; Jang, T.; Xiong, Z.; Zhang, C.; Tang, J.; Guo, L.J.; Li, W.D. High-Performance Flexible Transparent Electrode with an Embedded Metal Mesh Fabricated by Cost-Effective Solution Process. *Small* **2016**, *12*, 3021–3030. [CrossRef] [PubMed]

78. Hsu, P.C.; Wang, S.; Wu, H.; Narasimhan, V.K.; Kong, D.; Ryoung, L.H.; Cui, Y. Performance enhancement of metal nanowire transparent conducting electrodes by mesoscale metal wires. *Nat. Commun.* **2013**, *4*, 2522. [CrossRef] [PubMed]
79. Seong, B.; Yoo, H.; Dat Nguyen, V.; Jang, Y.; Ryu, C.; Byun, D. Metal-mesh based transparent electrode on a 3-D curved surface by electrohydrodynamic jet printing. *J. Micromech. Microeng.* **2014**, *24*, 9. [CrossRef]
80. Bao, B.; Jiang, J.; Li, F.; Zhang, P.; Chen, S.; Yang, Q.; Wang, S.; Su, B.; Jiang, L.; Song, Y. Fabrication of Patterned Concave Microstructures by Inkjet Imprinting. *Adv. Funct. Mater.* **2015**, *25*, 3286–3294. [CrossRef]
81. Liu, Y.; Shen, S.; Hu, J.; Chen, L. Embedded Ag mesh electrodes for polymer dispersed liquid crystal devices on flexible substrate. *Opt. Express* **2016**, *24*, 25774–25784. [CrossRef] [PubMed]
82. Kiruthika, S.; Gupta, R.; Anand, A.; Kumar, A.; Kulkarni, G.U. Fabrication of Oxidation Resistant Metal Wire Network Based Transparent Electrodes by a Spray-Roll Coating Process. *ACS Appl. Mater. Interfaces* **2015**, *7*, 27215–27222. [CrossRef] [PubMed]
83. Lee, J.; Lee, P.; Lee, H.; Lee, D.; Lee, S.S.; Ko, S.H. Very long Ag nanowire synthesis and its application in a highly transparent, conductive and flexible metal electrode touch panel. *Nanoscale* **2012**, *4*, 6408–6414. [CrossRef] [PubMed]
84. SVG Optronics. Available online: http://www.svgoptronics.com/en/product.asp?id=29 (accessed on 29 December 2016).
85. Singh, M.; Rana, T.R.; Kim, S.Y.; Kim, K.; Yun, J.H.; Kim, J.H. Silver Nanowires Binding with Sputtered ZnO to Fabricate Highly Conductive and Thermally Stable Transparent Electrode for Solar Cell Applications. *ACS Appl. Mater. Interfaces* **2016**, *8*, 12764–12771. [CrossRef] [PubMed]
86. Wu, J.; Que, X.; Hu, Q.; Luo, D.; Liu, T.; Liu, F.; Russell, T.P.; Zhu, R.; Gong, Q. Organic Solar Cells: Multi-Length Scaled Silver Nanowire Grid for Application in Efficient Organic Solar Cells. *Adv. Funct. Mater.* **2016**, *26*, 4822–4828. [CrossRef]
87. Cai, G.; Darmawan, P.; Cui, M.; Wang, J.; Chen, J.; Magdassi, S.; Lee, P.S. Highly Stable Transparent Conductive Silver Grid/PEDOT:PSS Electrodes for Integrated Bifunctional Flexible Electrochromic Supercapacitors. *Adv. Energy Mater.* **2015**, *6*. [CrossRef]
88. Cheng, T.; Zhang, Y.; Yi, J.; Yang, L.; Zhang, J.D.; Lai, W.Y.; Huang, W. Inkjet Printed Flexible, Transparent and Aesthetic Energy Storage Devices based on PEDOT:PSS/Ag Grids Electrodes. *J. Mater. Chem. A* **2016**, *4*, 13754–13763. [CrossRef]
89. Lee, H.; Hong, S.; Lee, J.; Suh, Y.D.; Kwon, J.; Moon, H.; Kim, H.; Yeo, J.; Ko, S.H. Highly Stretchable and Transparent Supercapacitor by Ag–Au Core–Shell Nanowire Network with High Electrochemical Stability. *ACS Appl. Mater. Interfaces* **2016**, *8*, 15449–15458. [CrossRef] [PubMed]

5

Rapid Fabrication of Epidermal Paper-Based Electronic Devices using Razor Printing

Behnam Sadri [1], Debkalpa Goswami [1] and Ramses V. Martinez [1,2,*]

1 School of Industrial Engineering, Purdue University, 315 N. Grant Street, West Lafayette, IN 47907, USA; bsadri@purdue.edu (B.S.); dgoswami@purdue.edu (D.G.)

2 Weldon School of Biomedical Engineering, Purdue University, 206 S. Martin Jischke Drive, West Lafayette, IN 47907, USA

* Correspondence: rmartinez@purdue.edu

Abstract: This work describes the use of a benchtop razor printer to fabricate epidermal paper-based electronic devices (EPEDs). This fabrication technique is simple, low-cost, and compatible with scalable manufacturing processes. EPEDs are fabricated using paper substrates rendered omniphobic by their cost-effective silanization with fluoroalkyl trichlorosilanes, making them inexpensive, water-resistant, and mechanically compliant with human skin. The highly conductive inks or thin films attached to one of the sides of the omniphobic paper makes EPEDs compatible with wearable applications involving wireless power transfer. The omniphobic cellulose fibers of the EPED provide a moisture-independent mechanical reinforcement to the conductive layer. EPEDs accurately monitor physiological signals such as ECG (electrocardiogram), EMG (electromyogram), and EOG (electro-oculogram) even in high moisture environments. Additionally, EPEDs can be used for the fast mapping of temperature over the skin and to apply localized thermotherapy. Our results demonstrate the merits of EPEDs as a low-cost platform for personalized medicine applications.

Keywords: epidermal sensors; stretchable electronics; wireless power; hydrophobic paper; wearable stimulators; paper electronics; low-cost manufacture

1. Introduction

The ever-growing demand for wearable technologies capable of monitoring key physiological signals have a predicted market growth from $15b in 2015 to $150b in 2027 [1]. Wearable healthcare devices fabricated using conventional rigid platforms and silicon-based technologies have been demonstrated to be useful in the continuous collection of clinically relevant personalized information for diseases such as heart failure [2], erythema [3], and diabetes [4,5]. Unfortunately, these rigid or semi-rigid wearable devices can lead to inconsistent measurements due to their limited conformability to human skin during motion and they are often perceived by patients as uncomfortable, thus hindering their adoption.

A new class of thin, flexible, and stretchable electronics, known as epidermal electronic systems [6], has emerged as wearable healthcare tools capable of efficiently monitoring a variety of physiological signals [7] and stimulating different tissues [8]. The stretchability and low mechanical impedance (similar to human skin) make epidermal electronics suitable for continuous health monitoring using wearable devices, due to their intimate contact with the skin and their compliance to its natural moves.

Thin films of ductile metals such as gold or platinum have been extensively used in the fabrication of contact electrodes for epidermal electronics due to their chemical stability and low resistivity [9]. Several electrode designs such as serpentine patterns [10], fractal designs [11], and self-similar buckles [12] have been explored to match the mechanical impedance of human skin.

The manufacture of epidermal electronics with conformable designs often require clean room processes, such as photolithography [13], wet etching [9,14], and physical vapor deposition [7] to create conductive electrodes capable of conforming to the skin. A variety of epidermal electronic systems assembled on flexible substrates have demonstrated excellent measuring performances even during stretching or severe bending, with a resolution comparable to that of advanced CMOS technologies [15]. Unfortunately, the high cost of the materials and the complexity of the fabrication processes (incompatible with large scale manufacturing) required to manufacture epidermal electronics make them unsuitable for personalized medicine applications.

Paper has become a popular substrate for flexible electronics due to its printability, low cost, light weight, and disposability [16]. Conductive inks [17,18], semiconductors [19,20], and insulators [21] can be introduced to papers to tailor the electrical properties of the final device using simple manufacturing processes such as inkjet printing [19,22], spin coating [23], or screen-printing [24] to make devices such as transistors [25], batteries [26], solar cells [27], light-emitting diodes [28], triboelectric generators [29], and antennas [30]. The limited stretchability (below 5%) of paper, however, makes paper-based electronics unsuitable for epidermal applications. Moreover, the performance of electronic devices printed on conventional paper is sensitive to relative humidity and temperature. The development of a simple method to fabricate a variety of paper-based epidermal electronic devices will be desirable to significantly reduce their cost and manufacturing time.

Several approaches have been proposed to change the wetting properties of paper to improve its electrical stability and mechanical integrity in high humidity environments [31]. Infusing hydrophobic materials such as wax or photoresists has enabled the selective modification of the wettability of the paper, enabling the fabrication of low-cost microfluidic devices [32,33]. Recently, our group introduced the selective functionalization of the cellulose fibers of paper using fluoroalkyl trichlorosilanes (R^F) as a fast and simple way to render paper omniphobic, limiting its wettability by aqueous solutions and non-polar solvents [34]. Omniphobic paper reduces the consumption of conductive ink during printing processes, reducing the price of printed electronics, and avoids the degradation of the mechanical properties of the paper due to environmental humidity. The moisture insensitivity and light weight of omniphobic paper has promoted its use as a low-cost substrate for applications in MEMS [35], microfluidics [36], and portable analytical devices [37]. Additionally, omniphobic paper devices can be easily prototyped using a variety of scalable tools such as engravers, laser cutters, and razor printers [38].

Our previous work on paper-based microfluidics demonstrated the low-cost fabrication of analytical systems by employing a thin cutting blade [6]. Here, we demonstrate the use of razor printing to rapidly fabricate epidermal paper-based electronic devices (EPEDs). EPEDs described in this study comprise a layer of a conductive material (metallic thin film or microparticle-based ink) attached to a layer of cellulose paper rendered omniphobic through silanization. Razor printed EPEDs can be rapidly fabricated at a low cost and offer several advantages as follows: (i) They are lightweight, thin, flexible, and even more stretchable than human skin; (ii) They are capable of real-time monitoring of bio-signals such as electrocardiogram (ECG), electro-oculogram (EOG), and electromyogram (EMG) with high precision, independently of environmental moisture or sweating of the wearer; (iii) The wide variety of thin metallic films and conductive inks compatible with razor printing provides an ample range of conductive agents to tailor the functionality of the EPED; (iv) EPEDs fabricated with flexible conductive inks can be used to monitor temperature and provide localized heat therapy, and (v) The omniphobic fibers of the paper reinforce the conductive layer of the EPEDs, preserving the electrical conductivity of the device upon stretching and making these devices compatible with wireless power transfer applications.

2. Materials and Methods

2.1. Choice of Materials

We purchased Whatman#1 paper (GE Healthcare Inc., Philadelphia, PA, USA) and thin paper (70-µm-thick, Elements 300, amazon.com) to serve as substrates for the EPEDs. Thin copper foils (20-µm-thick, Kraftex Products, Gloucestershire, UK) and Ag/AgCl ink (AGCL-675, Applied Ink Solutions, Westborough, MA, USA) were employed as conductive layers. We used a solution of a long-chain fluorinated organosilane (Diisopropyl(3,3,4,4,5,5,6,6,7,7,8,8,9,9,10,10,10-heptadecafluorodecyl)silane, Sigma-Aldrich Corp., St. Louis, MO, USA) to render the paper substrates omniphobic.

2.2. Fabrication of EPEDs by Razor Printing

We functionalized the paper substrates by spraying the organosilane solution at ambient conditions and letting it dry in a desiccator at 36 Torr for 20 min [6]. The open mesh serpentine layout of the EPEDs was designed using Adobe Illustrator CC (Adobe Systems Inc., San Jose, CA, USA) according to geometries previously reported in [10,11]. The minimum line width of the serpentine layout was kept at 200 µm (Figures 1b and 2b), the minimum resolution of our programmable razor printer (Silhouette Cameo^{TM}, Silhouette America Inc., Lindon, UT, USA), which uses a 100-µm-thick blade as the cutting tool. Prior to shaping the serpentine layout of the EPEDs, we attached adhesive copper tape (for copper-based EPEDs, Figure 1) or stencil printed Ag/AgCl ink (for Ag/AgCl-based EPEDs) on the functionalized paper. These functionalized paper substrates covered with a conductive layer (thickness of the composite ranging 70–190 µm) were then attached to a water-soluble tape (Aquasol Corp., North Tonawanda, NY, USA), which acted as the transfer layer to mount the EPEDs on skin (Figure 1c–f). Prior to the placement of EPEDs on skin, we sprayed medical glue (Medique products, Fort Myers, FL, USA) over the skin to maintain the conformal contact of the EPEDs on stretching. The transfer layer was then dissolved under a stream of running water (Figure 1e,f).

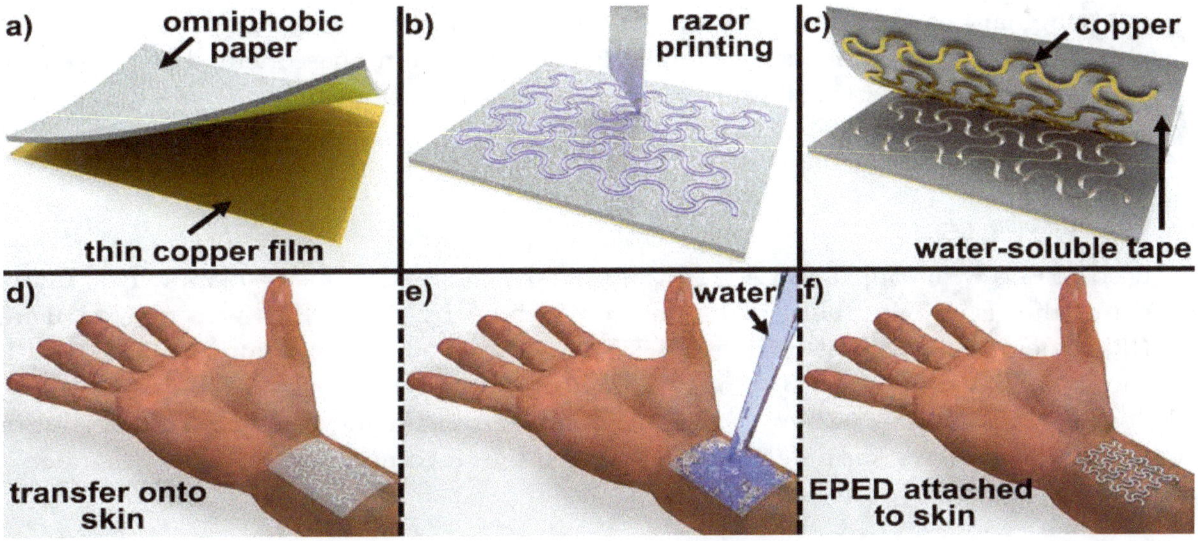

Figure 1. Fabrication of epidermal paper-based electronic devices (EPEDs) using razor printing: (**a**) A layer of omniphobic paper is glued to a thin metallic film that serves as a conductive layer (alternatively Ag/AgCl ink can be directly deposited on omniphobic paper); (**b**) A 100-µm-thick razor blade shapes the ensemble into a serpentine pattern; (**c**) A water-soluble tape, attached to the paper side of the EPED, is used as a temporary substrate for transfer onto skin; (**d**) The EPED is transferred onto skin previously sprayed with medical glue; (**e**) Placing the EPED under a stream of running water dissolves the temporary substrate; (**f**) EPED conformally attached to the skin.

2.3. Physiological Signal Measurement with EPEDs

We recorded ECG, EMG, and EOG signals using a three-electrode configuration [7]. The physiological signals were amplified, filtered, and displayed using a commercial electrophysiological recorder (Backyard Brains, Ann Arbor, MI, USA) coupled to a portable open-source microcontroller (UNO, Arduino Inc.). The thickness of the medical glue layer deposited on the skin is <2 μm [9], minimally affecting the performance of the EPED while recording physiological signals.

We attached external cables (28 AWG) directly onto the conductive layer of the EPEDs (over the contact pad area) using a small amount of low melting point soldering paste (SMD291AX, Chip Quik Inc., Niagara Falls, NY, USA). To perform underwater experiments, an extra layer of medical glue was deposited over the skin to encapsulate the flat connection between the EPEDs and the cables. Any excess of medical glue sprayed on the skin accumulated along the lateral walls of the EPED, preventing those exposed areas of the conductive layer from short-circuiting while under water.

To compare the performance of EPEDs with conventional electrodes, we ran parallel experiments using EPEDs and commercially available foam electrodes (Medline Industries Inc., Northfield, IL, USA). We coated the surface of the foam electrodes in contact with the skin with a conductive electrode gel (SPECTRA® 360, Parker Laboratories Inc., Fairfield, NJ, USA) to ensure a good electrical contact.

2.4. Characterization of Wirelessly Powered EPEDs

To wirelessly power functional components, such as LEDs, we attached a miniaturized half-wave rectifier fabricated using SMD components (Table S3) to the EPED antennas. We studied the wireless power transfer capabilities of EPEDs by performing a frequency-dependent characterization using a vector network analyzer (E5071B ENA, Agilent Technologies, Santa Clara, CA, USA). We used a copper coil (18 AWG wire, 6 turns, 5 cm diameter) connected to the network analyzer through an SMA connector (Digi-Key Electronics, Thief River Falls, MN, USA) to transfer wireless power to the EPEDs. All EPEDs were characterized passively at a distance of 15 cm from the center of the coil, in an orientation perpendicular to the axis of the coil. The network analyzer was programmed to record the real and imaginary parts of the impedance at 1601 frequency points linearly spaced in the range 1–20 MHz, finding the resonant frequency of the EPED using the min-phase method [39]. To enable the wireless powering of EPEDs, the coil was excited at the resonant frequency with a sinusoidal signal generated by a waveform generator (DG4062 Series, RIGOL Technologies Inc., Beaverton, OR, USA).

2.5. Heat Therapy Using EPEDs

We used EPEDs with copper and Ag/AgCl ink as the conductive layers to apply heat uniformly to the skin of the user. The thermal distribution created by the EPEDs was imaged using an infrared (IR) camera (FLIR E8, Wilsonville, OR, USA). We used a DC power supply (DP832A, RIGOL Technologies Inc., Beaverton, OR, USA) to generate heat through the resistive EPED, applying power levels below FCC guidelines (<2 W). To ensure the accuracy of the real-time monitoring of the temperature of the skin, we kept the distance between the IR camera and the EPED fixed at 20 cm during all the experiments.

2.6. Scanning Electron Microscopy (SEM)

We used a scanning electron microscope (Nova NanoSEM 200, FEI, Hillsboro, OR, USA) to examine the structure of the fabricated EPEDs. Before imaging, we used a sputter coater (208HR, Cressington, UK) to create a uniform conductive coating of ~10 nm platinum, using a D.C. current of 40 mA for 60 s. SEM images of the samples were captured at an electron accelerating potential of 5 kV, spot size 3, and working distance of 5 mm using an Everhart-Thornley detector (ETD).

2.7. Mechanical Characterization of EPEDs

We obtained stress-strain characteristics of bare paper substrates as well as fabricated EPEDs using a universal testing machine (MTS insight 10, MTS Systems Corp., Eden Prairie, MN, USA) equipped with a 100 N load cell (model 661.18.F01) according to ASTM D828-16 specifications. For the bare paper substrates, we fixed the gage length at 50 mm and applied a loading rate of 10 mm/min; while for the EPEDs, we had a gage length of 10 mm (comparable to the size of the device) and a loading rate of 5 mm/min.

3. Results and Discussion

3.1. Working Principle of EPEDs

Figure 2a shows the two layers of the EPEDs fabricated using razor printing (fabrication steps detailed in Figure 1): a conductive 20-μm-thick copper film in contact with the skin of the user and a silanized paper support (thickness ranging from 70 to 180 μm) that exhibits a static contact angle of 156°. The silane used to render paper omniphobic (both hydrophobic and oleophobic) prevents the EPEDs from being wetted by aqueous solutions and organic liquids with surface tension as low as 28 mN m^{-1} [6]. The covalent bonds generated between the alkyl trichlorosilanes and the cellulose fibers of the paper during the functionalization process are stable both in ambient conditions and under water for temperatures up to 150 °C [34]. Moreover, the chemical modification of the cellulose fibers of the paper do not affect its porosity (Figure 2c inset), preserving the gas permeability of the paper. The razor printing method used to fabricate the EPED enables the fabrication of flexible electrodes with a linewidth of 200 μm (Figure 2b) and a thickness of 78 μm when using Ag/AgCl ink as the conductive layer (70 μm is the thickness of the paper and 8 μm is the average thickness of the Ag/AgCl ink, see Figure 2c). The low thickness of the EPEDs fabricated using razor printing ensure their conformability to skin even when it wrinkles due to compression forces (Figure 2d) [15]. After dissolving the water-soluble transfer layer, 70-μm-thick EPEDs adhere to the skin solely by van der Waals and capillary forces, without requiring the spray-on medical glue. However, for a more robust adhesion of the EPEDs, we sprayed the skin with medical glue in all cases, regardless of the thickness of the paper used as a substrate. Since the thickness of the sprayed layer of glue is very small (<2 μm; [9]), its use does not adversely affect the functionality of the EPEDs or increase experimental noise in any significant way. The solvent of the medical glue sprayed on the skin of the user prior to placing the EPED does not affect the wetting properties of the paper substrate, which remains omniphobic after the medical glue dries. The omniphobic cellulose fibers provide a mechanical reinforcement to the thin film metals and conductive inks used in the conductive layer of the EPED, allowing them to withstand accidental stresses up to 2.5 MPa without tearing (Figure 2e). Despite the limited stretchability of unpatterned paper (~4%), the serpentine pattern used in the design of the EPED electrodes, enables these epidermal devices to endure stretching up to ~58% before failure. As a comparison, the maximum strain of human skin is ~30% [40].

3.2. Realtime Monitoring of Cardiac Activity

We recorded ECG signals from a human subject by attaching copper-based EPED electrodes on the wrist (measurement and ground) and the back of the hand (reference), as shown in Figure 3a. Figure 3b (top) shows the ECG signals recorded with the EPED electrodes. The silanization of the paper substrate to render EPEDs omniphobic allows us to capture high quality ECG signals even with the EPEDs completely immersed in water (Figure 3b bottom). We compared the signal to noise ratio (SNR) of EPED electrodes to conventional foam electrodes by placing them on the same locations of the hand (Figure 3c top). ECG measurements acquired by razor printed EPED electrodes ($SNR_{ECG\text{-}EPED,air}$ = 12.20 dB, $SNR_{ECG\text{-}EPED,water}$ = 10.37 dB; Table S2) exhibit no significant difference from conventional foam electrodes in air (Figure 3c bottom, $SNR_{ECG\text{-}foam,air}$ = 11.28 dB).

Conventional electrodes, though, cannot reliably capture ECG signals under water due to the swelling of their hydrogel terminals and their subsequent delamination and short-circuit.

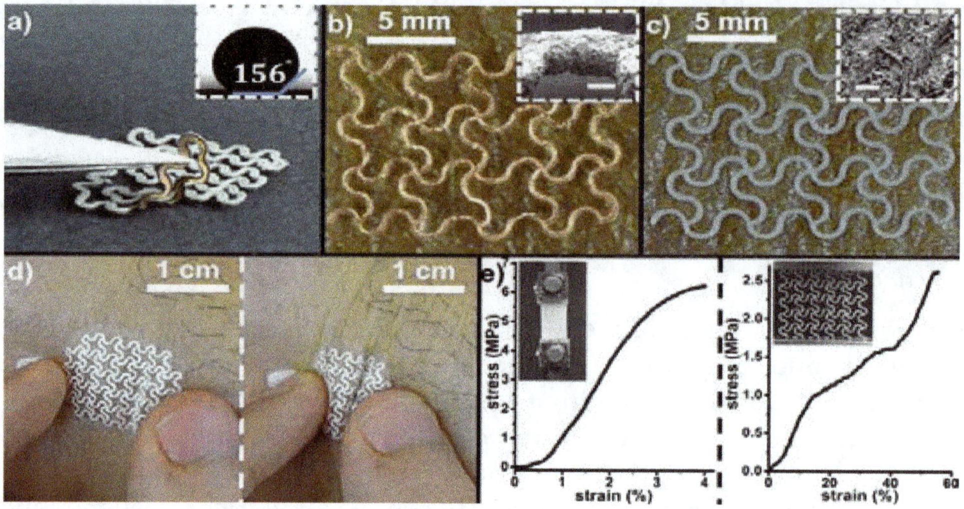

Figure 2. Razor printed EPEDs: (**a**) Omniphobic EPEDs comprising a thin conducting layer and a silanized paper serving as a back support. Inset shows the apparent contact angle of a 10 μL water droplet on top of the silanized paper substrate; (**b**) EPED on top of skin, with 20-μm-thick copper film as a conducting layer (copper facing up). The inset shows an SEM image of the 70-μm-thick patterned paper substrate (scale bar is 50 μm); (**c**) EPED with an 8-μm-thick layer of Ag/AgCl ink deposited on top of the omniphobic paper (Ag/AgCl facing up). The inset shows an SEM image of the Ag/AgCl/paper electrode, demonstrating that neither the functionalization of the paper nor the subsequent deposition of Ag/AgCl ink clogged the porous structure of the paper substrate (scale bar is 100 μm); (**d**) Conforming of EPEDs to skin bending and buckling due to severe compression; (**e**) Left: Representative stress-strain curve of an unpatterned paper substrate. Inset shows the experimental set up used. Right: Stress-strain curve of an Ag/AgCl EPED showing how the razor patterning of the EPED improves its stretching when compared to unpatterned paper. Inset shows the mechanical characterization of a representative EPED sample.

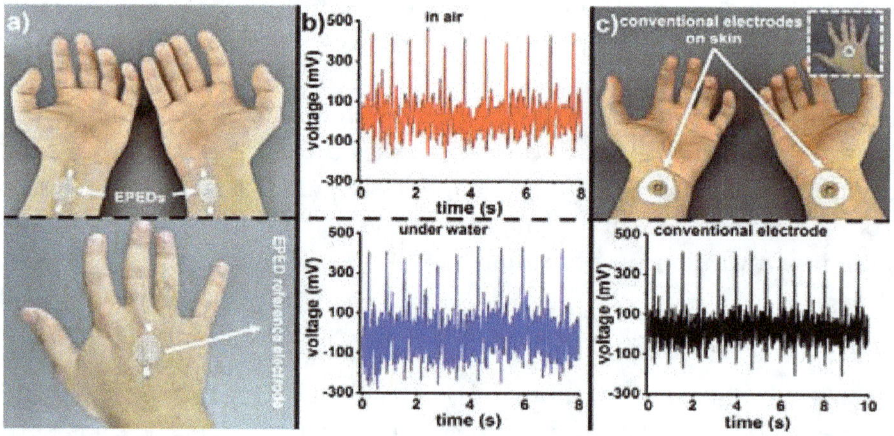

Figure 3. Comparison between the performance of razor printed copper-based EPEDs and conventional foam electrodes to record ECG signals: (**a**) Top: EPED measurement and ground electrodes used to record ECG signals from the wrist of a subject. Bottom: Reference EPED electrode; (**b**) Top: ECG signals recorded in air using razor printed EPEDs. Bottom: ECG signals recorded with both hands under water; (**c**) Top: Conventional foam electrodes placed at the same locations of the wrist as (**a**). The inset shows the location of the reference foam electrode on the back of the hand. Bottom: ECG signals recorded in air using conventional foam electrodes.

3.3. Real-Time Monitoring of Muscle Activity

We used copper-based EPEDs to record EMG signals from the forearm by placing the measurement and ground electrodes along the flexor muscle and the reference electrode on the back of the hand (Figure 4a). Figure 4b (top) shows the EMG signals recorded with EPEDs while lifting a 4.5 kg dumbbell, holding it for 5 s, and resting for 10 s. The omniphobic character of the EPED provided by the silanization of the paper substrate, enables EMG signals to be captured in high moisture environments without significant experimental noise or short circuiting the measuring electrodes. To demonstrate the moisture-independent collection of EMG signals, we repeated the measurements while keeping the arm in a water bath. We observed no significant difference in performance between the razor printed EPEDs ($SNR_{EMG\text{-}EPED,air}$ = 31.79 dB, $SNR_{EMG\text{-}EPED,water}$ = 30.16 dB; Table S2) and conventional foam electrodes ($SNR_{EMG\text{-}foam,air}$ = 26.58 dB) to record EMG signals in air. Conventional electrodes, however, are not capable to record EMG signals under water due to the short-circuit of their terminals.

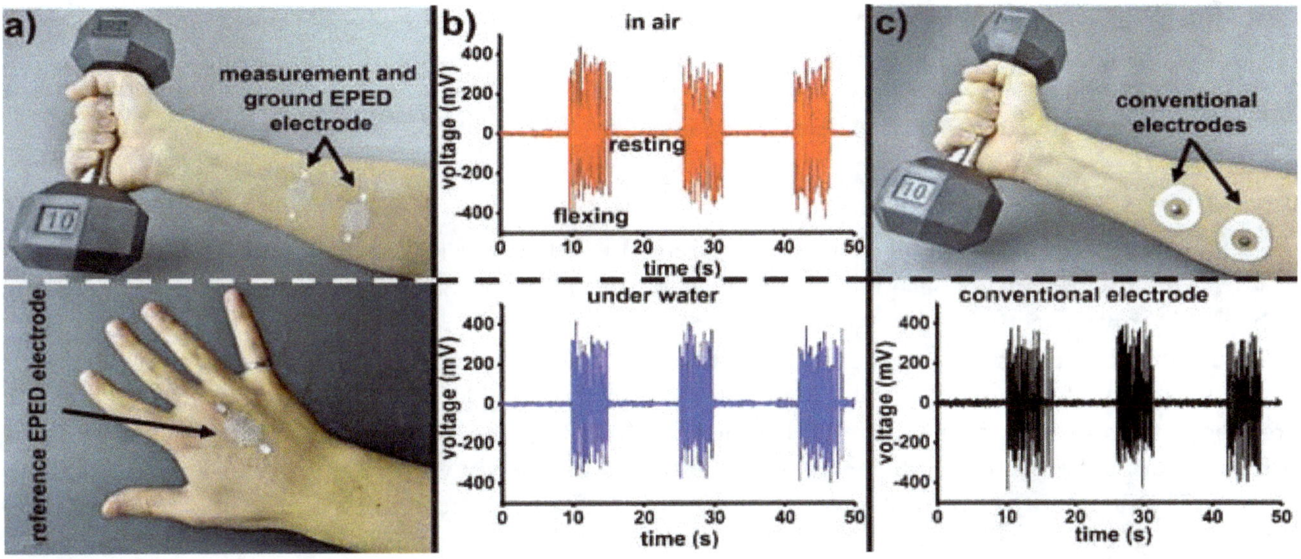

Figure 4. Comparison between the performance of razor printed EPEDs and conventional foam electrodes to monitor EMG signals: (**a**) Top: EPED measurement and ground electrodes used to record EMG signals from the forearm of an exercising subject. Bottom: Reference EPED electrode; (**b**) Top: EMG signals recorded in air using razor printed EPEDs. Bottom: EMG signals recorded under water; (**c**) Top: conventional foam electrodes placed at the same locations of the forearm as (**a**). Bottom: EMG signals recorded in air using conventional foam electrodes.

3.4. Monitoring Eye Motion

We used three copper-based EPED electrodes placed on the cheekbone (ground), forehead (measurement), and neck (reference) to capture EOG signals and to monitor the movement of the eye (Figure 5a). The mechanical conformability of EPEDs enabled the identification of the movement of the eyes (up and down) as well as blinking events with minimal experimental noise ($SNR_{EOG\text{-}EPED,air}$ = 33.77 dB; Figure 5b, Table S2). When compared with conventional foam electrodes ($SNR_{EOG\text{-}foam,air}$ = 31.17 dB), EPEDs exhibit better performance upon the natural moves of the user (Figure 5c).

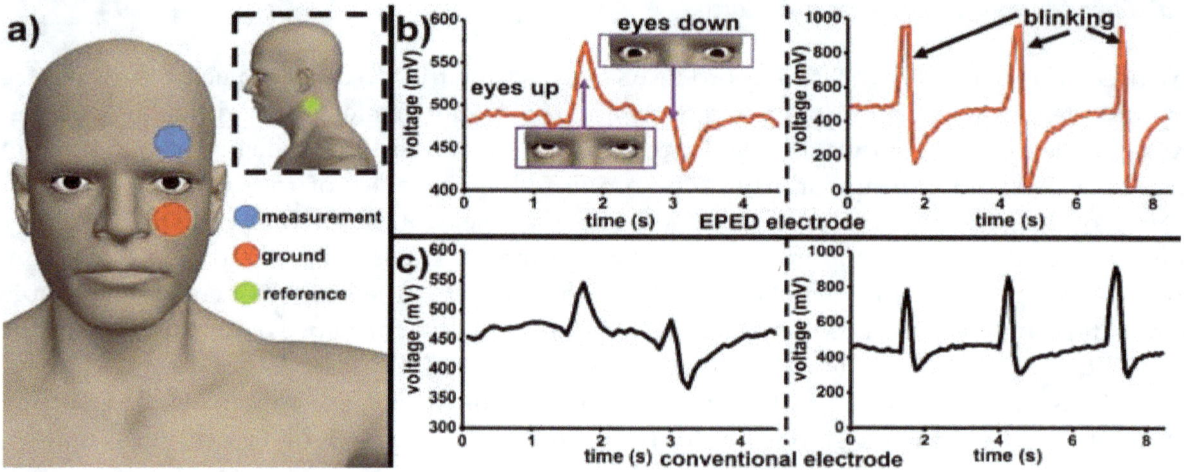

Figure 5. Monitoring eye motion using razor printed EPEDs: (**a**) Location of the measuring, ground, and reference copper-based EPED electrodes; (**b**) Left: EOG signal identifying eye movements (up and down) using EPEDs; Right: Identification of blinking events using EPEDs; (**c**) Left: Identification of eye movements (up and down) using conventional foam electrodes located as shown in (**a**); Right: Identification of blinking events using conventional foam electrodes.

3.5. Wireless Powering of EPEDs

The low resistivity (~20 nΩ m for copper-based EPEDs) of EPED antennas make them suitable for wireless power transfer based applications (Figure 6). Figure 6a,b shows a 10 mm EPED antenna with an LED and a rectifier circuit mounted on skin, being powered using far-field electromagnetic waves emitted from a primary coil placed 15 cm away. The geometry of the EPED antenna was chosen to match previously reported wireless epidermal stimulators [11]. This square pancake coil has only three loops to minimize their shaping with the razor printer, since we experimentally found that coils with three loops were able to efficiently power the LED wirelessly via inductive coupling at a distance of 15 cm. The line width of the antenna was made to match the minimum resolution of our razor printer (~200 μm). After the coil was shaped with the razor printer, we folded the external end of the coil towards the center to make both ends of the coil to rest flat at a distance of 2 mm and soldered the SMD components (Table S3) between the ends of the coil (Figure 6a,b). We analyzed the frequency-dependent electrical characteristics of this EPED using methods described in Section 2.4 (Figure 6c). We recorded the real and imaginary components of the impedance, $Z = R + jX$, $|Z| = (R^2 + X^2)^{1/2}$, where the real part, R, is the resistance, and the imaginary part, X, is the reactance. We used the recorded components of the impedance as a function of frequency to calculate electrical characteristics of the EPEDs such as inductance $L = X/2\pi f$, phase $\theta = \tan^{-1}(X/R)$, and quality factor $Q = X/R$. The resonant frequency f_0 of the EPED is determined by the min-phase method [39,41]: the frequency at which the θ response is minimized is taken as the resonant frequency of the EPED when coupled with the primary coil. Since the primary coil is connected to the network analyzer for a one-port measurement, only the S_{11} parameter is recorded (Figure 6d). The wireless power transfer efficiency is calculated as $\eta = (1 - |S_{11}|^2) \times 100\%$, where $|S_{11}|^2$ is defined as the reflectance. We found that the omniphobic functionalization of the paper does not significantly modify the resonant frequency or the power transfer efficiency of the EPED (Figure 6d,e). Modifying the values of the capacitors used to rectify the wireless signal, the wireless power transfer efficiency of the EPED can be easily optimized for a given frequency following the impedance-matching optimization method for magnetic resonance coupling systems [42].

Multiple EPEDs placed in close proximity can be selectively powered, if these EPEDs have different resonant frequency peaks (due to their different geometry or rectifying circuit). Figure 7 summarizes the passive electrical characteristics of a system of 2 EPEDs, one 8 mm and another 10 mm, placed side by side, but not in contact with each other. The different resonant frequencies of the EPEDs (9.0 MHz for 8 mm side EPED; 10.4 MHz for 10 mm side EPED) enable their selective activation, independently or at the same time, depending on the desired application (Figure 7d).

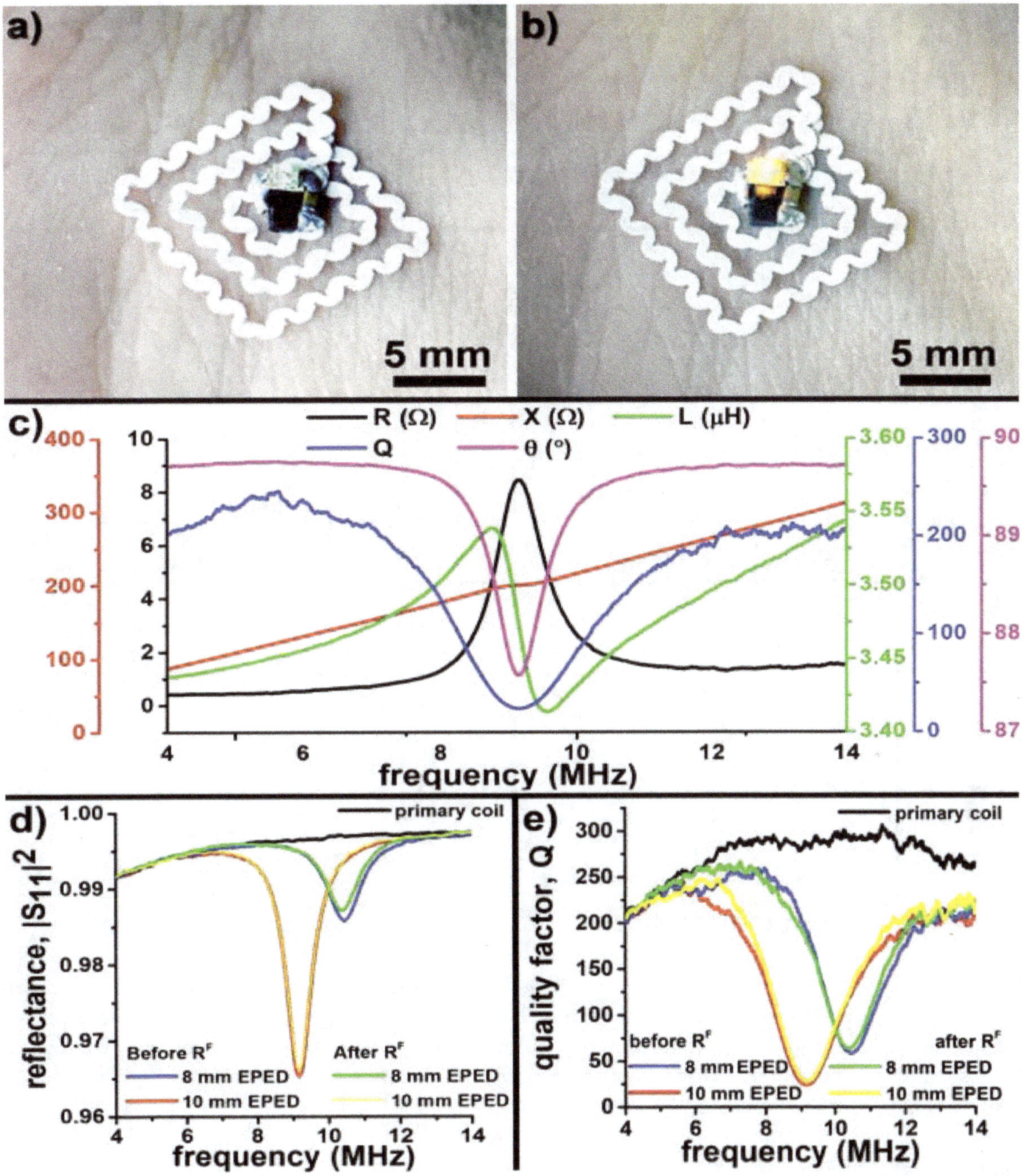

Figure 6. Electrical characteristics of wirelessly powered copper-based EPEDs: (**a**) Square EPED antenna (10 mm side) coupled with an SMD rectifier-LED circuit and attached to the skin of the wrist; (**b**) The LED is wirelessly powered using a primary coil 15 cm away (not shown in picture) running alternating currents at a resonant frequency of 9.0 MHz; (**c**) Frequency-dependent passive characteristics (Resistance, R, Reactance, X, Phase, θ, Inductance, L, and Quality factor, Q) of the EPED shown in (**a**,**b**). The frequency at which the phase θ is minimum is the resonant frequency of the EPED; (**d**,**e**) Effect of EPED size and silanization on the wireless power transfer efficiency ($\eta = 1 - |S_{11}|^2$) and the quality factor (Q). The silanization process has a negligible effect on η and Q. η and Q decrease when the size of the EPED is reduced.

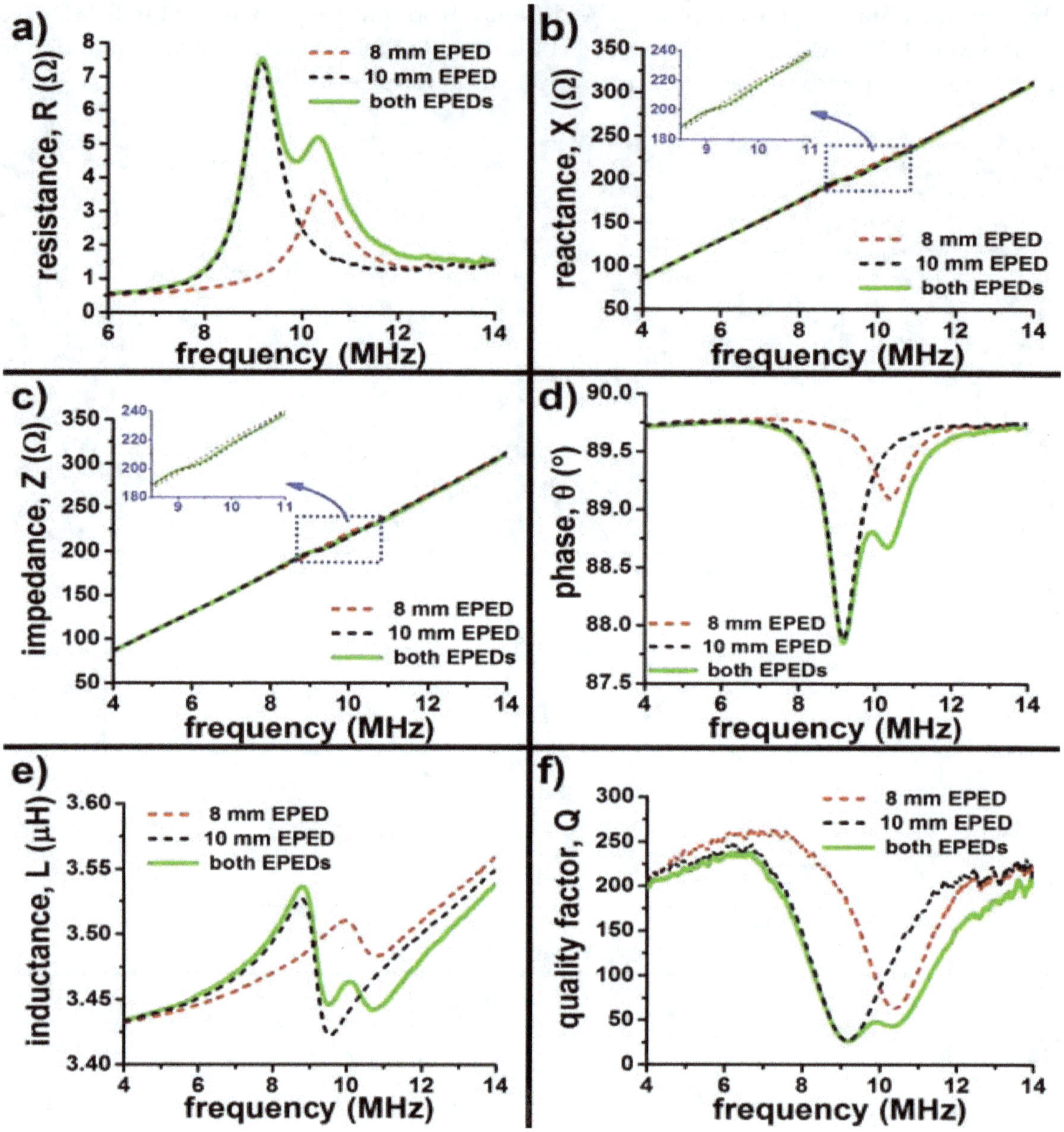

Figure 7. Selective powering of multiple EPEDs by varying excitation frequency, using the same primary coil. The green solid curves in all panels correspond to the frequency dependent passive characteristics when both EPEDs (8 mm and 10 mm side) are 15 cm away from the primary coil. The dashed lines represent individual characteristics: (**a**) Resistance, R; (**b**) Reactance, X; (**c**) Impedance, Z; (**d**) Phase, θ; (**e**) Inductance, L; (**f**) Quality factor, Q.

3.6. Localized Heat Therapy

Epidermal heat therapy is commonly employed in cancer treatments [43] and in orthopedics for alleviating joint pain [44]. Figure 8a shows a copper-based EPED fabricated to apply localized heat therapy on the skin. Heat is produced in accordance with Joule's law of heating by running D.C. power through the EPED. The serpentine layout of this EPED was designed according to previously reported epidermal electronic devices [10,11] and razor printed with a minimum linewidth of 200 μm. We monitored the temperature distribution produced by the EPED using an IR camera, limiting the maximum temperature applied to the skin to 42 °C (Figure 8b). Each of the quadrants of the EPED has two independent contact pads that enable their individual activation to provide localized doses of heat (Figure 8b inset). Figure 8c shows the time-dependence of the heating process for different

D.C. powers. The low specific heat of copper ensures that the EPED temperature rapidly stabilizes as power is applied, and restores quickly to room temperature once the power supply is turned off. The omniphobic properties of the EPED remain unaffected after the heating cycles.

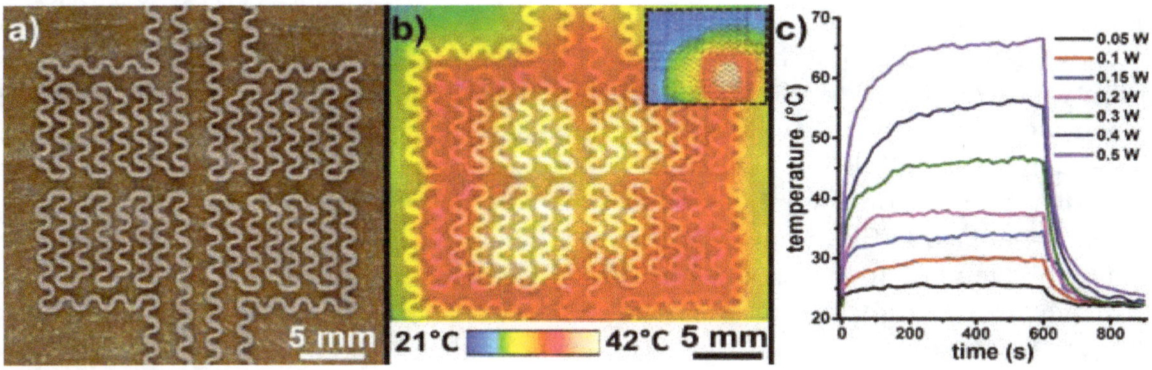

Figure 8. Application of localized heat therapy using razor printed EPEDs: (**a**) Copper-based thermotherapy EPED mounted on skin; (**b**) IR image of the EPED shown in (a) during thermotherapy (inset shows the application of localized heat by the selective activation of only one quadrant of the EPED); (**c**) Temperature-time response of the thermotherapy EPED for different D.C. powers.

3.7. Thermal Sensing

The temperature dependence of the resistivity of the conductive layer of the EPEDs enables their use as wearable thermometers (Figure 9). The linear relationship between the increment in resistance of the EPED and its temperature allowed us to calculate the sensitivity of Ag/AgCl- and copper-based EPED thermometers (Figure 9a). The sensitivity of Ag/AgCl- and copper-based EPEDs are 0.01 Ω/°C and 0.001 Ω/°C, respectively. We characterized the time response of the Ag/AgCl EPEDs by placing them over a surface at room temperature (t = 0 s) and placing an aluminum cylinder preheated to different reference temperatures (t = 10 s). We observed that Ag/AgCl EPEDs required less than 1 s to reach reference temperatures in the clinically relevant range (Figure 9b).

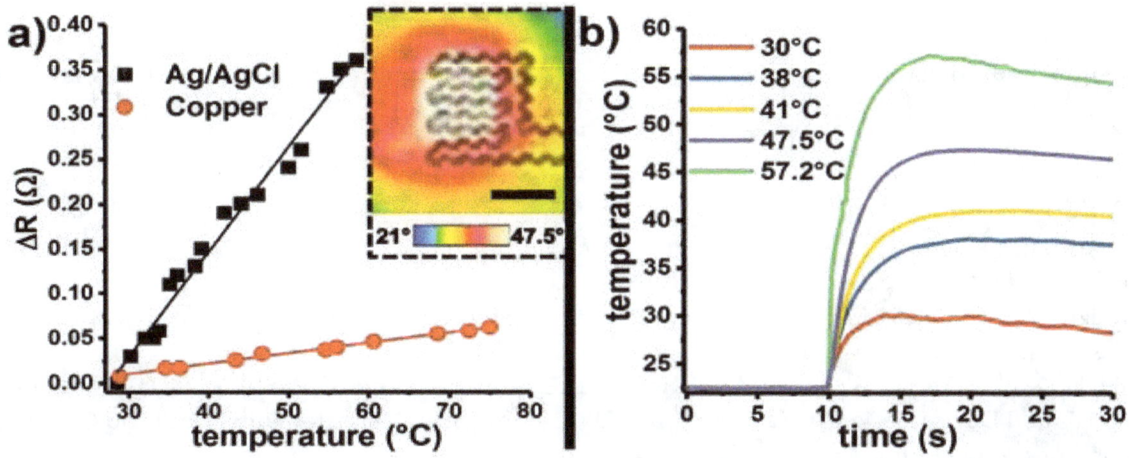

Figure 9. Sensing temperature using EPEDs: (**a**) Change of the EPED resistance as a function of temperature for EPEDs with a thin copper film (solid red dots) and Ag/AgCl ink (solid black squares) as conductive layers. The inset shows an IR image of an Ag/AgCl-based EPED when a small aluminum cylinder at 47.5 °C is placed on its surface for 10 s and then removed. Scale bar is 5 mm; (**b**) Response of the Ag/AgCl-based EPED thermometers when an aluminum cylinder at different temperatures is placed in contact with the EPED at t = 10 s.

4. Conclusions

This work reports the simple, inexpensive, and scalable, fabrication of epidermal paper-based electronic devices (EPEDs) using a bench-top razor printer. EPEDs fabricated using silanized paper can be used as moisture-insensitive epidermal electrodes, with a cost so low that it makes them compatible with single-use applications (see Table S1). Razor printed EPEDs fabricated using copper film or Ag/AgCl ink are easy to mount on skin, conforming to its natural moves, and exhibit good mechanical contact with the user and a stable electrical performance upon stretching. Copper-based EPEDs exhibit low resistivity values (~20 nΩ m), enabling their use as efficient electrophysiological monitors, thermotherapeutic devices, and wirelessly powered systems. The low resistance of copper-based EPEDs, however, makes it difficult to detect changes in the resistance caused by environmental temperature. Ag/AgCl-based EPEDs have higher resistivity values (~110 nΩ m), facilitating their use as temperature sensors since small changes in the environmental temperature induce larger changes in the resistance of the devices. The fibrous structure of the paper substrates of the EPEDs makes them breathable when their conductive layer is porous, such as Ag/AgCl-based EPEDs. The adhesion of a continuous copper film to the paper, however, compromises the passage of gases across the EPED. We demonstrated the omniphobic character of razor printed EPEDs by efficiently recording ECGs, EMGs, and EOGs in air and under water without any significant decrease in performance. The use of razor printing to fabricate EPEDs, at its present level of development, also has two limitations: (i) The minimum line width of the serpentine traces is 200 μm; (ii) The shear forces applied during high-speed cutting processes can lead to the delamination of the conductive layer from the omniphobic paper support if the adhesive used to secure both layers is not properly chosen. The wide range of adhesive materials and films compatible with razor printing, however, can ameliorate this limitation. We expect that the proposed method to fabricate inexpensive wearable electrodes will facilitate the adoption of epidermal electronics in personalized medicine, especially in resource-limited and home environments.

Author Contributions: R.V.M and B.S. conceived the research and designed the experiments. B.S. fabricated the EPEDs and performed the experiments related to physiological signal monitoring, thermal stimulation, and mechanical characterization. D.G. performed electrical and mechanical characterization experiments as well as the structural characterization of the devices using electron microscopy. B.S., D.G. and R.V.M. co-wrote the paper.

Acknowledgments: The authors gratefully acknowledge start-up funding from Purdue University. B.S. also acknowledges the funding from Procter & Gamble (grant no. 209621). D.G. thanks the Ross Fellowship program at Purdue University for providing partial support of his work.

References

1. Hayward, J.; Chansin, G.; Zervos, H. *Wearable Technology 2017–2027: Markets, Players, Forecasts*; IDTechEx: Santa Clara, CA, USA, 2017.
2. Lobodzinski, S.S.; Laks, M.M. New devices for very long-term ECG monitoring. *Cardiol. J.* **2012**, *19*, 210–214. [CrossRef] [PubMed]
3. Araki, H.; Kim, J.; Zhang, S.; Banks, A.; Crawford, K.E.; Sheng, X.; Gutruf, P.; Shi, Y.; Pielak, R.M.; Rogers, J.A. Materials and Device Designs for an Epidermal UV Colorimetric Dosimeter with Near Field Communication Capabilities. *Adv. Funct. Mater.* **2017**, *27*, 1604465. [CrossRef]
4. Choi, H.; Luzio, S.; Beutler, J.; Porch, A. Microwave noninvasive blood glucose monitoring sensor: Human clinical trial results. In Proceedings of the 2017 IEEE MTT-S International Microwave Symposium (IMS), Honololu, HI, USA, 4–9 June 2017; pp. 876–879.
5. Lee, H.; Song, C.; Hong, Y.S.; Kim, M.S.; Cho, H.R.; Kang, T.; Shin, K.; Choi, S.H.; Hyeon, T.; Kim, D.-H. Wearable/disposable sweat-based glucose monitoring device with multistage transdermal drug delivery module. *Sci. Adv.* **2017**, *3*, e1601314. [CrossRef] [PubMed]

6. Glavan, A.C.; Martinez, R.V.; Maxwell, E.J.; Subramaniam, A.B.; Nunes, R.M.D.; Soh, S.; Whitesides, G.M. Rapid fabrication of pressure-driven open-channel microfluidic devices in omniphobic R^F paper. *Lab Chip* **2013**, *13*, 2922–2930. [CrossRef] [PubMed]
7. Jeong, J.-W.; Kim, M.K.; Cheng, H.; Yeo, W.-H.; Huang, X.; Liu, Y.; Zhang, Y.; Huang, Y.; Rogers, J.A. Capacitive Epidermal Electronics for Electrically Safe, Long-Term Electrophysiological Measurements. *Adv. Healthc. Mater.* **2014**, *3*, 642–648. [CrossRef] [PubMed]
8. Park, S.I.; Brenner, D.S.; Shin, G.; Morgan, C.D.; Copits, B.A.; Chung, H.U.; Pullen, M.Y.; Noh, K.N.; Davidson, S.; Oh, S.J.; et al. Soft, stretchable, fully implantable miniaturized optoelectronic systems for wireless optogenetics. *Nat. Biotechnol.* **2015**, *33*, 1280–1286. [CrossRef] [PubMed]
9. Yeo, W.-H.; Kim, Y.-S.; Lee, J.; Ameen, A.; Shi, L.; Li, M.; Wang, S.; Ma, R.; Jin, S.H.; Kang, Z.; et al. Multifunctional Epidermal Electronics Printed Directly onto the Skin. *Adv. Mater.* **2013**, *25*, 2773–2778. [CrossRef] [PubMed]
10. Zhang, Y.; Wang, S.; Li, X.; Fan, J.A.; Xu, S.; Song, Y.M.; Choi, K.-J.; Yeo, W.-H.; Lee, W.; Nazaar, S.N.; et al. Experimental and Theoretical Studies of Serpentine Microstructures Bonded To Prestrained Elastomers for Stretchable Electronics. *Adv. Funct. Mater.* **2014**, *24*, 2028–2037. [CrossRef]
11. Fan, J.A.; Yeo, W.-H.; Su, Y.; Hattori, Y.; Lee, W.; Jung, S.-Y.; Zhang, Y.; Liu, Z.; Cheng, H.; Falgout, L.; et al. Fractal design concepts for stretchable electronics. *Nat. Commun.* **2014**, *5*, 3266. [CrossRef] [PubMed]
12. Efimenko, K.; Rackaitis, M.; Manias, E.; Vaziri, A.; Mahadevan, L.; Genzer, J. Nested self-similar wrinkling patterns in skins. *Nat. Mater.* **2005**, *4*, 293–297. [CrossRef] [PubMed]
13. Webb, R.C.; Bonifas, A.P.; Behnaz, A.; Zhang, Y.; Yu, K.J.; Cheng, H.; Shi, M.; Bian, Z.; Liu, Z.; Kim, Y.-S.; et al. Ultrathin conformal devices for precise and continuous thermal characterization of human skin. *Nat. Mater.* **2013**, *12*, 938–944. [CrossRef] [PubMed]
14. Kim, J.; Banks, A.; Cheng, H.; Xie, Z.; Xu, S.; Jang, K.-I.; Lee, J.W.; Liu, Z.; Gutruf, P.; Huang, X.; et al. Epidermal Electronics with Advanced Capabilities in Near-Field Communication. *Small* **2015**, *11*, 906–912. [CrossRef] [PubMed]
15. Liu, Y.; Pharr, M.; Salvatore, G.A. Lab-on-Skin: A Review of Flexible and Stretchable Electronics for Wearable Health Monitoring. *ACS Nano* **2017**, *11*, 9614–9635. [CrossRef] [PubMed]
16. Tobjörk, D.; Österbacka, R. Paper Electronics. *Adv. Mater.* **2011**, *23*, 1935–1961. [CrossRef] [PubMed]
17. Russo, A.; Ahn, B.Y.; Adams, J.J.; Duoss, E.B.; Bernhard, J.T.; Lewis, J.A. Pen-on-Paper Flexible Electronics. *Adv. Mater.* **2011**, *23*, 3426–3430. [CrossRef] [PubMed]
18. Lessing, J.; Glavan, A.C.; Walker, S.B.; Keplinger, C.; Lewis, J.A.; Whitesides, G.M. Inkjet Printing of Conductive Inks with High Lateral Resolution on Omniphobic "R^F Paper" for Paper-Based Electronics and MEMS. *Adv. Mater.* **2014**, *26*, 4677–4682. [CrossRef] [PubMed]
19. Bollström, R.; Määttänen, A.; Tobjörk, D.; Ihalainen, P.; Kaihovirta, N.; Österbacka, R.; Peltonen, J.; Toivakka, M. A multilayer coated fiber-based substrate suitable for printed functionality. *Org. Electron.* **2009**, *10*, 1020–1023. [CrossRef]
20. Cui, L.-F.; Ruffo, R.; Chan, C.K.; Peng, H.; Cui, Y. Crystalline-Amorphous Core–Shell Silicon Nanowires for High Capacity and High Current Battery Electrodes. *Nano Lett.* **2009**, *9*, 491–495. [CrossRef] [PubMed]
21. Hamedi, M.M.; Ainla, A.; Güder, F.; Christodouleas, D.C.; Fernández-Abedul, M.T.; Whitesides, G.M. Integrating Electronics and Microfluidics on Paper. *Adv. Mater.* **2016**, *28*, 5054–5063. [CrossRef] [PubMed]
22. Calvert, P. Inkjet Printing for Materials and Devices. *Chem. Mater.* **2001**, *13*, 3299–3305. [CrossRef]
23. Kim, D.-H.; Kim, Y.-S.; Wu, J.; Liu, Z.; Song, J.; Kim, H.-S.; Huang, Y.Y.; Hwang, K.-C.; Rogers, J.A. Ultrathin Silicon Circuits With Strain-Isolation Layers and Mesh Layouts for High-Performance Electronics on Fabric, Vinyl, Leather, and Paper. *Adv. Mater.* **2009**, *21*, 3703–3707. [CrossRef]
24. Hyun, W.J.; Secor, E.B.; Hersam, M.C.; Frisbie, C.D.; Francis, L.F. High-Resolution Patterning of Graphene by Screen Printing with a Silicon Stencil for Highly Flexible Printed Electronics. *Adv. Mater.* **2015**, *27*, 109–115. [CrossRef] [PubMed]
25. Huang, J.; Zhu, H.; Chen, Y.; Preston, C.; Rohrbach, K.; Cumings, J.; Hu, L. Highly Transparent and Flexible Nanopaper Transistors. *ACS Nano* **2013**, *7*, 2106–2113. [CrossRef] [PubMed]
26. Liu, H.; Crooks, R.M. Paper-Based Electrochemical Sensing Platform with Integral Battery and Electrochromic Read-Out. *Anal. Chem.* **2012**, *84*, 2528–2532. [CrossRef] [PubMed]

27. Hu, L.; Zheng, G.; Yao, J.; Liu, N.; Weil, B.; Eskilsson, M.; Karabulut, E.; Ruan, Z.; Fan, S.; Bloking, J.T.; et al. Transparent and conductive paper from nanocellulose fibers. *Energy Environ. Sci.* **2013**, *6*, 513–518. [CrossRef]
28. Zhao, R.; Zhang, X.; Xu, J.; Yang, Y.; He, G. Flexible paper-based solid state ionic diodes. *RSC Adv.* **2013**, *3*, 23178–23183. [CrossRef]
29. Pal, A.; Cuellar, H.E.; Kuang, R.; Caurin, H.F.N.; Goswami, D.; Martinez, R.V. Self-Powered, Paper-Based Electrochemical Devices for Sensitive Point-of-Care Testing. *Adv. Mater. Technol.* **2017**, *2*, 1700130. [CrossRef]
30. Anagnostou, D. *Organic Paper-Based Antennas*; WIT Press: Billerica, MA, USA, 2014.
31. Teisala, H.; Tuominen, M.; Kuusipalo, J. Superhydrophobic Coatings on Cellulose-Based Materials: Fabrication, Properties, and Applications. *Adv. Mater. Interfaces* **2014**, *1*, 1300026. [CrossRef]
32. Li, X.; Ballerini, D.R.; Shen, W. A perspective on paper-based microfluidics: Current status and future trends. *Biomicrofluidics* **2012**, *6*, 011301. [CrossRef] [PubMed]
33. Carrilho, E.; Martinez, A.W.; Whitesides, G.M. Understanding Wax Printing: A Simple Micropatterning Process for Paper-Based Microfluidics. *Anal. Chem.* **2009**, *81*, 7091–7095. [CrossRef] [PubMed]
34. Glavan, A.C.; Martinez, R.V.; Subramaniam, A.B.; Yoon, H.J.; Nunes, R.M.D.; Lange, H.; Thuo, M.M.; Whitesides, G.M. Omniphobic "R^F Paper" Produced by Silanization of Paper with Fluoroalkyltrichlorosilanes. *Adv. Funct. Mater.* **2014**, *24*, 60–70. [CrossRef]
35. Liu, X.; Mwangi, M.; Li, X.; O'Brien, M.; Whitesides, G.M. Paper-based piezoresistive MEMS sensors. *Lab Chip* **2011**, *11*, 2189–2196. [CrossRef] [PubMed]
36. Chitnis, G.; Ding, Z.; Chang, C.-L.; Savran, C.A.; Ziaie, B. Laser-treated hydrophobic paper: An inexpensive microfluidic platform. *Lab Chip* **2011**, *11*, 1161–1165. [CrossRef] [PubMed]
37. Glavan, A.C.; Christodouleas, D.C.; Mosadegh, B.; Yu, H.D.; Smith, B.S.; Lessing, J.; Fernández-Abedul, M.T.; Whitesides, G.M. Folding Analytical Devices for Electrochemical ELISA in Hydrophobic R^H Paper. *Anal. Chem.* **2014**, *86*, 11999–12007. [CrossRef] [PubMed]
38. Giokas, D.L.; Tsogas, G.Z.; Vlessidis, A.G. Programming Fluid Transport in Paper-Based Microfluidic Devices Using Razor-Crafted Open Channels. *Anal. Chem.* **2014**, *86*, 6202–6207. [CrossRef] [PubMed]
39. Huang, X.; Liu, Y.; Cheng, H.; Shin, W.-J.; Fan, J.A.; Liu, Z.; Lu, C.-J.; Kong, G.-W.; Chen, K.; Patnaik, D.; et al. Materials and Designs for Wireless Epidermal Sensors of Hydration and Strain. *Adv. Funct. Mater.* **2014**, *24*, 3846–3854. [CrossRef]
40. Maiti, R.; Gerhardt, L.-C.; Lee, Z.S.; Byers, R.A.; Woods, D.; Sanz-Herrera, J.A.; Franklin, S.E.; Lewis, R.; Matcher, S.J.; Carré, M.J. In vivo measurement of skin surface strain and sub-surface layer deformation induced by natural tissue stretching. *J. Mech. Behav. Biomed. Mater.* **2016**, *62*, 556–569. [CrossRef] [PubMed]
41. Sardini, E.; Serpelloni, M. Passive and Self-Powered Autonomous Sensors for Remote Measurements. *Sensors* **2009**, *9*, 943–960. [CrossRef] [PubMed]
42. Thackston, K.A.; Mei, H.; Irazoqui, P.P. Coupling Matrix Synthesis and Impedance-Matching Optimization Method for Magnetic Resonance Coupling Systems. *IEEE Trans. Microw. Theory Tech.* **2018**, *66*, 1536–1542. [CrossRef]
43. Schroeder, A.; Heller, D.A.; Winslow, M.M.; Dahlman, J.E.; Pratt, G.W.; Langer, R.; Jacks, T.; Anderson, D.G. Treating metastatic cancer with nanotechnology. *Nat. Rev. Cancer* **2012**, *12*, 39–50. [CrossRef] [PubMed]
44. Brosseau, L.; Yonge, K.; Welch, V.; Marchand, S.; Judd, M.; Wells, G.A.; Tugwell, P. Thermotherapy for treatment of osteoarthritis. *Cochrane Database Syst. Rev.* **2003**. [CrossRef] [PubMed]

6

Recent Advancements in Liquid Metal Flexible Printed Electronics: Properties, Technologies and Applications

Xuelin Wang [1] and Jing Liu [1,2,*]

1 Department of Biomedical Engineering, School of Medicine, Tsinghua University, Beijing 100084, China; wang-xl15@mails.tsinghua.edu.cn

2 Beijing Key Lab of CryoBiomedical Engineering and Key Lab of Cryogenics, Technical Institute of Physics and Chemistry, Chinese Academy of Sciences, Beijing 100190, China

* Correspondence: jliubme@tsinghua.edu.cn

Academic Editors: Seung Hwan Ko, Daeho Lee and Zhigang Wu

Abstract: This article presents an overview on typical properties, technologies, and applications of liquid metal based flexible printed electronics. The core manufacturing material—room-temperature liquid metal, currently mainly represented by gallium and its alloys with the properties of excellent resistivity, enormous bendability, low adhesion, and large surface tension, was focused on in particular. In addition, a series of recently developed printing technologies spanning from personal electronic circuit printing (direct painting or writing, mechanical system printing, mask layer based printing, high-resolution nanoimprinting, etc.) to 3D room temperature liquid metal printing is comprehensively reviewed. Applications of these planar or three-dimensional printing technologies and the related liquid metal alloy inks in making flexible electronics, such as electronical components, health care sensors, and other functional devices were discussed. The significantly different adhesions of liquid metal inks on various substrates under different oxidation degrees, weakness of circuits, difficulty of fabricating high-accuracy devices, and low rate of good product—all of which are challenges faced by current liquid metal flexible printed electronics—are discussed. Prospects for liquid metal flexible printed electronics to develop ending user electronics and more extensive applications in the future are given.

Keywords: liquid metal; flexible printed electronics; 3D printing; functional device; additive manufacture; consumer electronics

---

1. Introduction

Printed electronics is the technology to fabricate electronic devices based on the principle of printing [1]. Unlike traditional printing technology, the inks used in the printing machine are electronic materials with the properties of being conductive, dielectric, semi-conductive, or magnetic. Printed electronics is closely related to diverse fields such as organic electronics, plastic electronics, flexible electronics, and paper electronics [2], which indicates that one can not only print circuits on silicon and glass, but also on plastic, paper, and more flexible substrates. One of the methods to produce flexible electronics is to directly print conductive materials onto flexible substrates.

Classical flexible electronics refers to the technology by which organic/inorganic electronic devices are deposited on a flexible substrate to form a circuit [3]. Although the rigid circuit board can protect the electronic components from being damaged, it restricts the ductility and flexibility of the electronics. With intriguing properties such as softness, ductility, and low-cost fabrication, flexible electronics has broad application prospects in the information, energy, medicine, and defense technology fields through providing smart sensors, actuators, flexible displays, organic light-emitting diodes (OLEDs), and so on. The most obvious characteristic of flexible electronics lies in their flexibility compared with traditional rigid microelectronics, which makes them stretchable, conformal, portable, wearable, and easy to print quickly [4–8]. Because of their unique merits in terms of electrical, printable, biomedical, and sensing properties, flexible electronics can find diverse applications in electronic components [9], printing technology, implantable devices [10], and health monitors [11], with specific uses such as antennas [12], eyeball cameras [13] and pressure sensors [14], etc. Nowadays, two common strategies have been adopted to improve the softness of the electronics [15]. One is embedding conductive materials, which are stiff and rigid, onto a stretchable substrate, such as poly(dimethylsiloxane) (PDMS). Rogers and coworkers introduced a complex wavy structure to keep the circuit stretchable, which can absorb the major tension when stress acts on the soft substrate [16,17]. The other is using inherently stretchable conductors to form the circuit [18]. Russo and coworkers demonstrated a way to connect the circuit using conductive silver ink, which can directly write conductive text to interconnect light-emitting diode (LED) arrays and three-dimensional (3D) antennas on paper [12].

Clearly, flexible printed electronics (FPE) combines features of flexible electronics and printed electronics. In this way, one can quickly manufacture functional flexible electronic devices. Along with the maturation of printing technology, flexible printing has become a hot topic in research. Several typical printing technologies have been emerging, such as tapping mode printing [18], aerosol jet printing, roll-to-roll technology [19], inkjet printing [20], and micropen and brush printing [21,22]. Among them, micropen is perhaps the simplest method: pump the electrical ink into a cartridge to directly write out conductive texts. This makes it possible to draw a circuit diagram on an A4 sheet of paper. A micropen evidently reduces the complexity of circuit production and improves the efficiency of electronic circuit manufacturing. FPE have contributed to significant achievements in different fields, such as flexible display devices [23], thin-film solar cells [24,25], large area sensors and drivers [26,27], electronic skin [28], wearable electronics and biological prosthetic devices [29], self-charging system [30], self-powered wireless monitoring [31], etc. If a circuit benefits from the properties of softness, conformability, stretchability and portability, one can call it a flectional electronic device. Meanwhile, many laboratories have developed electronic devices with flexibility, such as warped display [32], folding battery [9], soft field effect transistor [33], stretchable wire [34], and so on. FPE can also be used in health monitoring, medical examination, vital signs detection, and other daily life needs. Flexible biomedical equipment has been widely utilized in the field of implantable devices [10], nerve connection [35,36], health monitoring [11], and wearable medical devices. FPE has also been used for biological sensing, the most important application of which is electronic skin (E-skin), such as pressure mapping [37], self-healing sensor [38], prosthesis [39], pressure sensor [14], etc. There are many research organizations working on different materials applied on flexible electronic skin. In Bao's laboratory, they invented a skin-inspired artificial mechanoreceptor system and the receptor can transform the pressure stimulation signal into electrophysiological signal that human neurons can perceive [40]. In Rogers's laboratory, they initiated a wavy structure to realize stretchable epidermal electronics that can maintain the original shape and function after compressing or stretching [16,41,42], In Chiolerio's laboratory, the researchers invented a printable spin-coated silver nanocomposite ink to manufacture a resistive switching devices for neuromorphic applications and a nanocomposite flexible liquid state device in a synthetic colloidal suspension [43,44].

Generally, traditional nanoparticle-based conductive ink does not have intrinsic conductivity, and needs special post-processing (e.g., sintering, annealing) to remove solvent from the conductive ink to achieve conductive capacity, such as silver nanoparticle, PEDOT:PSS, polyaniline-based ink, nickel and copper conductive ink [45–49]. The newly emerging liquid metal ink has intrinsically high conductivity, which enables it to be a kind of ideal conductive ink (Table 1). (In this paper, EGaIn refers to Ga–In, which contains the metal elements gallium and indium, commonly in a mixture of 75.5% gallium and 24.5% indium by weight; Galinstan refers to Ga–In–Sn, which contains the metal elements gallium, indium, and tin, commonly in a mixture of 62.5% gallium, 21.5% indium and 16% tin by weight.)

Liquid metal, just as its name suggests, is a kind of metal that remains in liquid phase from room temperature up to 2000 °C. (However, 2000 °C is a generic number; it depends upon the composition.) It is superior to many other liquid materials in terms of thermal conductivity and resistivity at low temperatures or around room temperature [50]. One traditional liquid metal—mercury—is not accepted on account of its high toxicity to the human body. However, because of its non-toxicity and benign biocompatibility, a liquid metal eutectic alloy of gallium indium has made important progresses in the biomedical arena such as being used as the material of bone cement [51], as a vascular contrast agent [52], and in drug delivery nanomedicine [53]. All of these indicate that liquid metal can be widely used in the fields of electronics, materials, and biology, which significantly broadens its impact. Figure 1 shows the typical applications of personalized flexible printed electronics based on liquid metal, such as implantable devices [35,36,54], electrical skin [55–57] and wearable bioelectronics [58–60] etc. Along with the development of material science and technology, liquid metal flexible printed electronics are quickly shaping the field of flexible electronic circuits and allied machines, serving as a basic way of quickly making functional devices. In this review, we are dedicated to integrating the liquid metal alloy inks with flexible printed electronics and focus our attention on interpreting the recent advancements in liquid metal flexible printed electronics.

Table 1. A comparison of the electrical conductivities of several typical conductive inks [61].

Ink Type	Ink Composition	Conductivity
Carbon conductive ink	Carbon CNT	1.8×10^3 S/m $(5.03 \pm 0.05) \times 10^3$ S/m
Polymer conductive ink	PEDOT:PSS	8.25×10^3 S/m
Nano-silver ink	Ag-DDA Ag-PVP	3.45×10^7 S/m 6.25×10^6 S/m
Liquid metal ink	EGaIn $Bi_{35}In_{48.6}Sn_{16}Zn_{0.4}$	3.4×10^6 S/m 7.3×10^6 S/m

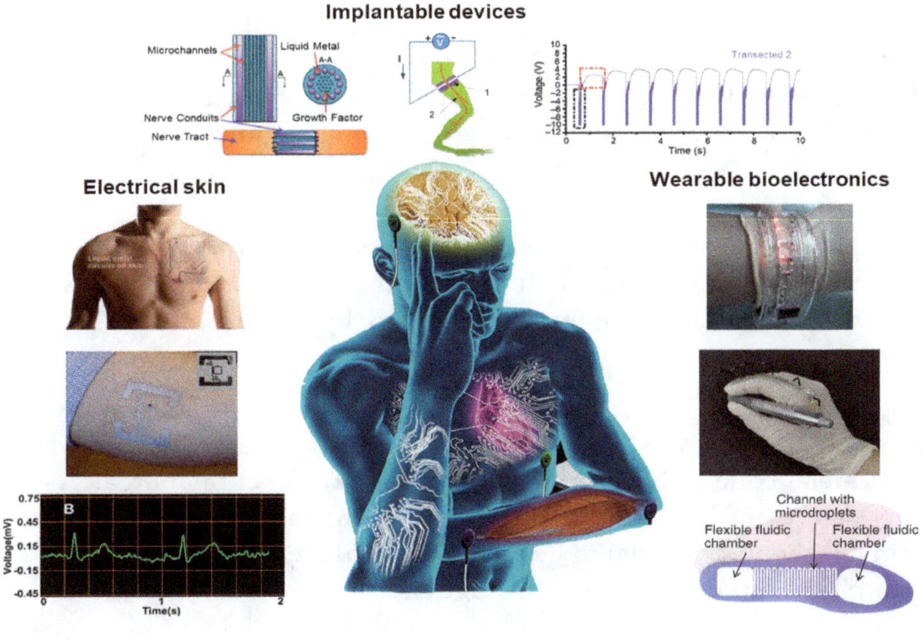

Figure 1. Personalized flexible printed electronics based on liquid metal. "Implantable devices": Liquid metal nerve connection [35] (https://arxiv.org/abs/1404.5931); Injectable 3D bioelectrode [54] (Reproduced with permission from Jin, C.; Liu, J. et al., Injectable 3-D fabrication of medical electronics at the target biological tissues; published by Scientific Reports, 2013.); Electrophysiological measurement of liquid metal reconnected nerve [36]; "Electrical skin": Liquid metal circuits on human body [55] (Reproduced with permission from Guo, C.; Liu, J. et al., Rapidly patterning conductive components

on skin substrates as physiological testing devices via liquid metal spraying and pre-designed mask; published by Journal of Mterials Chemistry B, 2014.); Liquid metal tag on human skin [56] (Reproduced with permission from Jeong, S.H.; Wu, Z. et al., Liquid alloy printing of microfluidic stretchable electronics; published by Lab on a Chip, 2012.); ECG test by liquid metal electrode [57] (Reproduced with permission from Yu, Y.; Zhang, J.; Liu, J., Biomedical implementation of liquid metal ink as drawable ECG electrode and skin circuit; published by PLoS ONE, 2013.); "Wearable bioelectronics": Liquid metal wristband [58] (Reproduced with permission from Wang, Q.; Yu, Y.; Yang, J.; Liu, J., Fast fabrication of flexible functional circuits based on liquid metal dual-trans printing; published by Advanced Material, 2015.); A wearable data glove [59] (Reproduced with permission from Matsuzaki, R.; Tabayashi, K., Highly stretchable, global, and distributed local strain sensing line using gainsn electrodes for wearable electronics; published by Advanced Functional Material, 2015.); Footwear-embedded microfluidic energy harvest [60] (Reproduced with permission from Krupenkin, T.; Taylor, J.A., Reverse electrowetting as a new approach to high-power energy harvesting; published by Nature Communication, 2011.).

2. Basic Properties of Liquid Metal

Generally speaking, alloy elements of low melting point include gallium, bismuth, lead, tin, cadmium, and indium. Among them, $GaIn_{24.5}$ is commonly used, which is a mixture of 75.5% gallium and 24.5% indium by weight. For the same composition, a different proportion may lead to quite varied properties in the alloys. Sometimes, even a small change of mass ratio would cause strong variation in the material behaviors. If loaded with certain microelements, the alloy may also display additional properties accordingly. Therefore, one can change the ratio of chemical materials of alloys or add some microelements to adjust the melting point and other properties to fulfill various specific needs. Room-temperature liquid metal material, generally represented by gallium and its alloys, offers a unique for to manufacturing flexible circuits, due to the combined nature of metallicity and fluidity [62,63], which makes it especially suitable for printing on soft substrates. (In this paper, the liquid metal alloy refers to gallium and its alloys.) From [62], one can get information on the main physical properties of frequently used liquid metal alloys, such as combination, melting point, density, resistivity, and thermal conductivity. In this review, we mainly discuss the resistivity, flexibility, adhesion, and wettability of liquid metal in flexible printed electronics.

2.1. Resistivity

Resistivity is the ability of an object to conduct an electric current, and generally the conductivity of metal is better than that of a non-metal. If the liquid metal is applied to print a flexible circuit, we can calculate the resistance value of a liquid metal wire by the formula:

$$R = \frac{\rho \cdot L}{A}, \tag{1}$$

where R is the resistance; ρ is the resistivity and for the same material ρ is a fixed value; L is the length of measured liquid metal conductive wire; and A is the cross-sectional area. As can be seen from Equation (1), the resistance of the material is directly proportional to the length of the material and the resistivity, and is inversely proportional to its cross-sectional area.

The relationship between the volume and length is:

$$A = \frac{V}{L}. \tag{2}$$

Therefore, one can get the factors associated with resistance change:

$$\frac{R}{R_0} = \frac{\frac{\rho \cdot L}{A}}{\frac{\rho_0 \cdot L_0}{A_0}} = \frac{\rho V_0}{\rho_0 V}\left(\frac{L}{L_0}\right)^2. \tag{3}$$

The volume of the liquid metal conductor is constant when drawing $V = V_0$; the resistivity is also considered constant, $\rho = \rho_0$.

Consequently, one can simplify Equation (3) to:

$$\frac{R}{R_0} = \left(\frac{L}{L_0}\right)^2. \tag{4}$$

The resistance variation of a liquid metal conductor has a linear relationship with the square of the length change of the conductor before and after stretching, confirming the relationship between resistance and tensile length, which provides theoretical support for the development of a liquid metal flexible sensor [55].

However, the resistivity c of the liquid metal has a significant relationship with the thermal conductivity [64,65]:

$$\frac{\Lambda}{\rho T} = \frac{\pi^2 \kappa_B^2}{3e^2} = 2.45 \times 10^{-8} (\mathrm{W} \cdot \Omega / \mathrm{K}^2). \tag{5}$$

Here, Λ is the thermal conductivity; ρ is the resistivity; and T is the thermodynamic temperature. Figure 2B shows the $\rho - T$ and the $\Lambda - T$ curves for $Bi_{35}In_{48.6}Sn_{16}Zn_{0.4}$. With the increase in temperature from -150 °C to 50 °C, Λ shows a linear increase while ρ shows a linear decrease. ρ_B is the Boltzmann constant. e is the electron charge, $e = -1.6 \times 10^{-19}$C. Equation (5) is the famous Wiedemann–Franz–Lorenz equation, from which one can deduce that the thermal conductivity of the liquid metal is proportional to the product of the resistivity and the thermodynamic temperature. One can get the resistance value of some number of wires connected in parallel by Ohm's Law [66]:

$$R = \left(\sum_{i=1}^{n} \frac{1}{R_i}\right)^{-1}. \tag{6}$$

Here, n is the number of wires connected in parallel. R_i is the individual resistance values, i.e.,

$$R_i = \rho \int_0^L \frac{dx}{A_i(x)}, \tag{7}$$

where ρ is the resistivity and the resistivity of EGaIn is $29.4 \times 10^{-6}\,\Omega \cdot \mathrm{cm}$. Gozen et al. demonstrates that the measured and predicted resistance values show a high degree of agreement with the number of wires changing [66]. However, there are some non-Ohm's Law phenomena that the current does not linearly vary with the voltage (Figure 3A) [67,68]. In Reference [67], the liquid metal marble is coated with WO_3 and the substrate is gold. It forms an electrical interface across the n-type semiconducting coating (WO_3) without the native oxide layer on the surface of the liquid metal, and the flow of current is owing to the combination of electron and hole carriers, which forms a n-type semiconductor layer. Therefore the measured current–voltage (I–V) shows the non-Ohm's Law phenomena. They have similar I–V curves between Reference [67] and Figure 3A, however different in theory. In Figure 3A, the electronic device is composed of one gallium droplet and a layer of gold nanoparticles, and the gallium droplet is the electrode. The nanoparticle and its ligands form a capacitor. This is the process of capacitor charging and discharging. Under the threshold voltage, the circuit is not on and the current is zero. Once up the threshold voltage, the circuit is switched on. Meanwhile, some data demonstrate the relationships between the resistance, the reactance, and the frequency [57]. Figure 2A shows that when the frequency changes from 1 Hz to 10 kHz, Ga resistance remains at 0.225 Ω. However, Ga reactance closes to 0 Ω from 1 Hz to 100 Hz and has analogously exponential growth from 100 Hz to 10 kHz. Liquid metal resistance invariance under different frequencies can be applied to measure some circuit frequency changes.

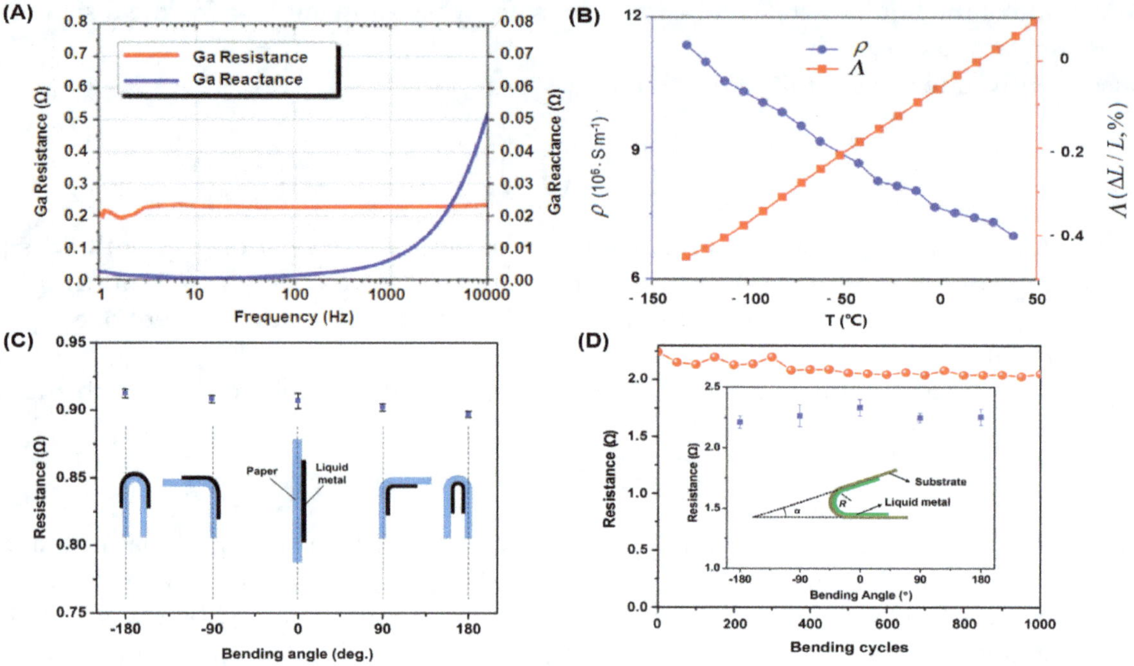

Figure 2. (**A**) The resistance and reactance curves from 1 Hz to 10 kHz of the liquid metal Ga [57] (Reproduced with permission from Yu, Y.; Zhang, J.; Liu, J., Biomedical implementation of liquid metal ink as drawable ECG electrode and skin circuit; (**B**) ρ (the resistivity)-*T* (temperature) and Λ (the thermal conductivity)-*T* (temperature) curves of liquid metal $Bi_{35}In_{48.6}Sn_{16}Zn_{0.4}$ [64] (Reproduced with permission from Wang, L.; Liu, J., Compatible hybrid 3D printing of metal and nonmetal inks for direct manufacture of end functional devices; published by Science China Technological Sciences, 2014.); (**C**) resistance values of liquid metal line at five different bending angles [69] (Reproduced with permission from Zheng, Y.; He, Z.; Gao, Y.; Liu, J., Direct desktop printed-circuits-on-paper flexible electronics; published by Scientific Reports, 2013.); (**D**) resistance values of liquid metal line at different bending angles and different bending cycles [70] (Reproduced with permission from Wang, L.; Liu, J., Pressured liquid metal screen printing for rapid manufacture of high resolution electronic patterns; published by RSC Advances, 2015.).

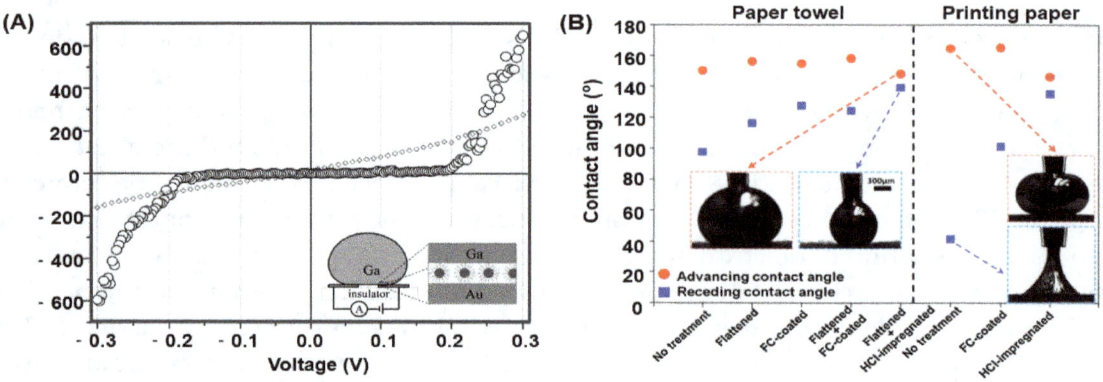

Figure 3. (**A**) Plot of the measured current and voltage of the electronic device composed of one gallium droplet and a layer of gold nanoparticles [68] (Reproduced with permission from Du, K.; Glogowski, E.; Tuominen, M.T.; Emrick, T.; Russell, T.P.; Dinsmore, A.D., Self-assembly of gold nanoparticles on gallium droplets: Controlling charge transport through microscopic devices; published by Langmuir, 2013.); (**B**) the advancing and receding contact angle of the liquid metal droplets on the paper towel and printing paper [71] (Reproduced with permission from Kim, D.; Lee, Y.; Lee, D.-W.; Choi, W.; Yoo, K.; Lee, J.-B., Hydrochloric acid-impregnated paper for gallium-based liquid metal microfluidics; published by Sensors and Actuators B: Chemical, 2015.).

2.2. Superior Flexibility

In order to illustrate the feasibility of liquid metal in manufacturing flexible circuits, Zheng et al. carried out a test to study the influence of bending on the stability of the circuit through measuring the resistance value when bending the printed liquid metal wire at $-180°$, $-90°$, $0°$, $90°$, and $180°$. Figure 2C evidently indicates that the resistance variation is very little and the resistance stability of the liquid metal when bending manifests that liquid metal printed wire can be well applied for making flexible electronics [69]. Furthermore, Wang et al. bent the PVC plastic film printed by liquid metal at $-180°$, $-90°$, $0°$, $90°$, and $180°$ for one cycle, and the resistance value was measured every 50 bending cycles (Figure 2D). The resistance values fluctuated slightly from 1 to 1000 cycles, which demonstrated the stability of the liquid metal resistor line [70]. References [72,73] demonstrate that when straining the liquid metal circuits, the resonance frequency and the resistance value can be changed. All the bending and stretching tests suggest that liquid metal printed wire has reliable flexibility when employed in flexible printed electronics.

2.3. Tunable Adhesion

Adhesion is the attraction among different molecules, such as the adhesion between the particle and the substrate [64–66]. When printing a circuit, the liquid metal is ejected from the spray gun in the form of droplets [65]. When put in air, the surface of liquid metal droplets may generate a layer of oxide whose composition is Ga_2O_3/Ga_2O [74]. As is disclosed on the proportional relationship between the surface tension of liquid metal droplets and oxide content [75], the smaller the metal droplets the higher the oxide content, thus the larger the surface tension and the better the substrate adhesion. Starting from this point, Zhang et al. established a generalized methodology for ubiquitous printed electronics [65] whereby, through atomized spraying of liquid metal droplets, a circuit can be quickly fabricated on almost any desired solid substrate surface, whether smooth or rough [65]. Additionally, the greater the oxide content of the metal droplets, the smaller the contact angle between the droplet and substrate [76].

Based on the Young–Dupre equation [77], one has

$$W_{SL} = \gamma_L (1 + \cos\theta), \quad (8)$$

where W_{SL} is the adhesion work of the liquid metal drop, γ_L is its surface tension, and θ is the contact angle between the droplet and substrate. The smaller the droplet, the larger the γ_L and the smaller the θ, hence the greater the W_{SL}, representing better adhesion. In other words, one can regulate the adhesion of liquid metal ink on a substrate by controlling its oxide layer [78]. Another possible strategy would be realizing a controlled de-wetting after the metal layer has solidified by a slight temperature increase [79]. However, the liquid metal alloy ink has strong adhesion to a majority of substrates, which makes it a challenge to manufacture micro-precision circuits [80].

2.4. Prominent Wettability

Wettability is the ability of liquid to spread on a solid surface. The contact angle usually expresses the wettability: the smaller the contact angle the better the wettability. The dynamic contact angle is the contact angle of a liquid moving over a surface. The dynamic contact angle (sliding angle and advancing-receding angle) plays an important role in quantifying the wetting property of the oxidized Galinstan droplets compared with the static contact angle [81,82]:

$$\sin\alpha = \gamma_{LG} \frac{Rk}{mg} \left[\cos\theta_{rec} - \cos\theta_{adv}\right]. \quad (9)$$

Here, α is the sliding angle; R is the liquid metal droplet radius; m is the mass of the droplet; θ_{rec} is the receding contact angle; and θ_{adv} is the advancing contact angle. The liquid metal droplet on the

HCl-impregnated flattened paper has the lowest contact angle, while for the non-treated printing paper it has the highest contact angle (Figure 3B) [71]. Gao et al. have studied the excellent wettability of liquid metal $GaIn_{10}$ with different substrate materials including smooth polyvinylchloride (Figure 4A), porous rubber (Figure 4B), rough polyvinylchloride (Figure 4C), tree leaf (Figure 4D), epoxy resin board (Figure 4E), typing paper (Figure 4F), glass (Figure 4G), cotton paper (Figure 4H), plastic (Figure 4I), cotton cloth (Figure 4J), silica gel plate (Figure 4K), glass fiber cloth (Figure 4L), and other substrates with different surface roughness and material properties [63,75]. Kramer et al. investigated the relationship between the microtextured surface topography of the liquid metal marbles and the wetting behaviors [83]. Doudrick et al. demonstrated that the oxide surface of the liquid metal alloy can improve the wettability between the metal material and substrate [80]. When increasing the oxygen content, the resistivity of the liquid metal alloy decreases while the viscosity and the wettability increase. For this reason, we can control the thickness of the oxide layer and then the wettability of the liquid metal wire on the soft substrate to obtain a superior performance of the flexible printed circuit.

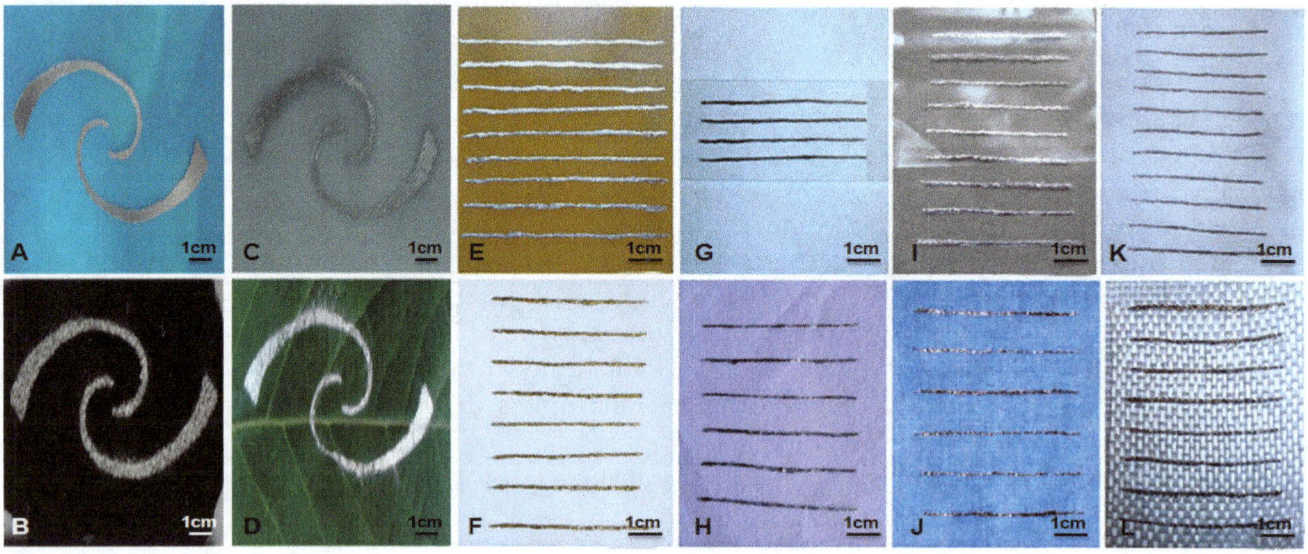

Figure 4. Demonstrated wettability of liquid metal printed on different substrate materials [63,75]; (**A**) Smooth polyvinylchloride; (**B**) porous rubber; (**C**) rough polyvinylchloride; (**D**) tree leaf; (**E**) epoxy resin board; (**F**) typing paper; (**G**) glass; (**H**) cotton paper; (**I**) plastic; (**J**) cotton cloth; (**K**) silica gel plate; (**L**) glass fiber cloth (Reproduced with permission from Gao, Y.; Li, H.; Liu, J., Direct writing of flexible electronics through room temperature liquid metal ink.) (Reproduced with permission from Zhang, Q.; Gao, Y.; Liu, J., Atomized spraying of liquid metal droplets on desired substrate surfaces as a generalized way for ubiquitous printed electronics; published by Applied Physics A: Materials Science & Processing, 2013.).

3. Printing Technologies and Apparatuses

In ancient China, Sheng Bi, a brilliant inventor who lived around 1041 AD, invented a new type of printing method: movable-type printing, the convenience of which led to the spread of knowledge and culture; the basic principle and method of modern letterpress stamping is the same as movable-type printing in spite of its different equipment and technical conditions. Thus, the invention of an excellent printing method and the selection of appropriate printing ink can have a huge impact on human life. Table 1 shows a conductivity comparison of several typical electroconductive inks [61]. The corrosivity of the liquid metal ink for different substrates (Table 2) prevents liquid gallium from being printed on specific flexible substrates.

Table 2. The corrosion between liquid gallium and other metals [84].

Corrosive Behavior	Easy to React with Gallium	Better Resistance for Corrosion	Best Resistance for Corrosion
Metal	Iron, nickel, cadmium, cerium, copper, aluminum, gold, manganese, platinum, silver, tin and vanadium, zirconium	Titanium, tantalum, niobium, molybdenum, beryllium	Tungsten, rhenium, sintering BeO, Al_2O_3, and fused quartz, graphite
Temperature range	Appearing corrosion at 200 °C or even lower	Appearing corrosion above 400 °C	No corrosion even at 800 °C

3.1. *Printing Electronic Circuit*

Nowadays, due to the high surface tension and good conductive ability of liquid metal, printing circuit using liquid metal ink has attracted more and more attention. The concept of DREAM ink (direct writing or printing of electronics based on alloy and metal ink) [62], as defined by Liu et al., can lead to alternative electronic devices in different fields [63,85,86]. So far, these are the different methods of printing a personal electronic circuit: screen printing [70], atomized spraying [75], microcontact printing (μCP) [87], masked deposition [88], and inkjet printing [89].

3.1.1. Direct Painting or Writing

Because of its good wettability with multifarious substrates, room-temperature liquid metal (RTLM) can be directly painted or written on paper, glass, or cloth as liquid metal ink [63]. Sheng et al. directly painted the liquid metal ink on the surface of VHB 4905 acrylic films to manufacture a capacitor sensor (Figure 5A) [90]. Because of its gravity and adhesion to substrates, the liquid metal alloy can be directly installed into the refill as a kind of ink to write conductive texts or lines [91,92]. Zheng et al. developed the liquid metal roller-ball pen (LMRP), which can write conductive lines or electronic devices on soft plates with the tip diameters ranging from 200 μm to 1000 μm (Figure 5B) [93]. Boley et al. illustrated a direct writing system, using the liquid metal alloy to fabricate small-scale stretchable electronics [73].

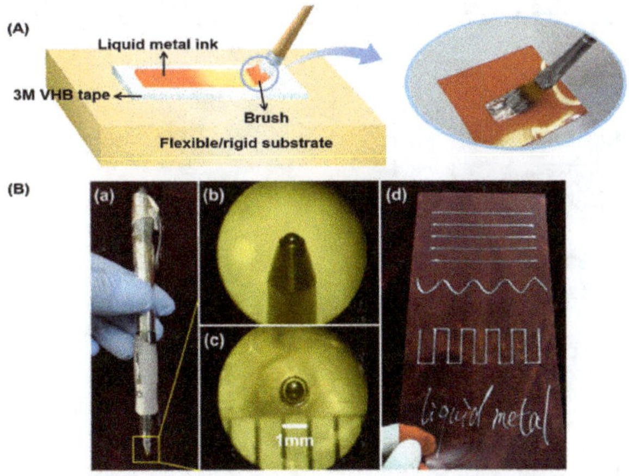

Figure 5. Printing methods of direct painting or writing without machine. (**A**) Direct painting of the liquid metal ink on the surface of VHB 4905 acrylic films to manufacture capacitor sensor [90]; (**B**) the liquid metal ink roller-ball pen and conductive tracks written by it [93] (Reproduced with permission from Zheng, Y.; Zhang, Q.; Liu, J., Pervasive liquid metal based direct writing electronics with roller-ball pen; published by AIP Advances, 2013.).

3.1.2. Mechanical Printing Methods

Compared to traditional direct painting or writing, mechanical printing methods can realize digital control, which makes the liquid metal flexible electronical circuits more accurate (Figure 6).

Zheng et al. have manufactured and demonstrated a versatile desktop liquid metal printer that can print either a plane circuit or a 3D metal object. The machine is made up of a syringe, a nitrogen gas tube, a pressure controller, a teach pendant, a stage, the *X* axis, the *Y* axis, and the *Z* axis (Figure 6B) [69]. In order to achieve a match between different material substrate and printing ink, they have designed a brush-like porous needle to print files of different types. A desktop printer can print a variety of electrical patterns via computer software. In particular, paper is low-cost and recyclable, which makes it a common substrate material and allows the concept of printed-circuits-on-paper (PCP) [69]. Along with that, the same group created another fully automatic liquid metal printer for pervasive electronic circuit printing [18]. Its theory of tapping mode printing is fixing an ink storage tube on the printing driver, clamping the driver on the guide rail using a sliding wheel, which can make the printer head move along the guide rail in the *X* direction, and driving the base plate along in the *Y* direction using a motor (Figure 6A). Consequently, one can control the position of the printer head in the printing area of the base plate. The tapping mode means that the nib can move a small distance in the *Z* direction. When printing, the printer head falls to contact the base, uplifts a certain height over the printing substrate, so as to avoid the printed line. In the process of printing, the printing header bead rotates due to the basal friction, which makes the liquid metal ink in the pen container run downwards. One can use the tapping mode method to manufacture various printed-circuit-on-board (PCB) circuits, antennae, and so on. In addition, other mechanical system printing methods to print personal electronic circuits using liquid metal include inkjet printing and coelectrospinning (Figure 6C,D) [89,94,95].

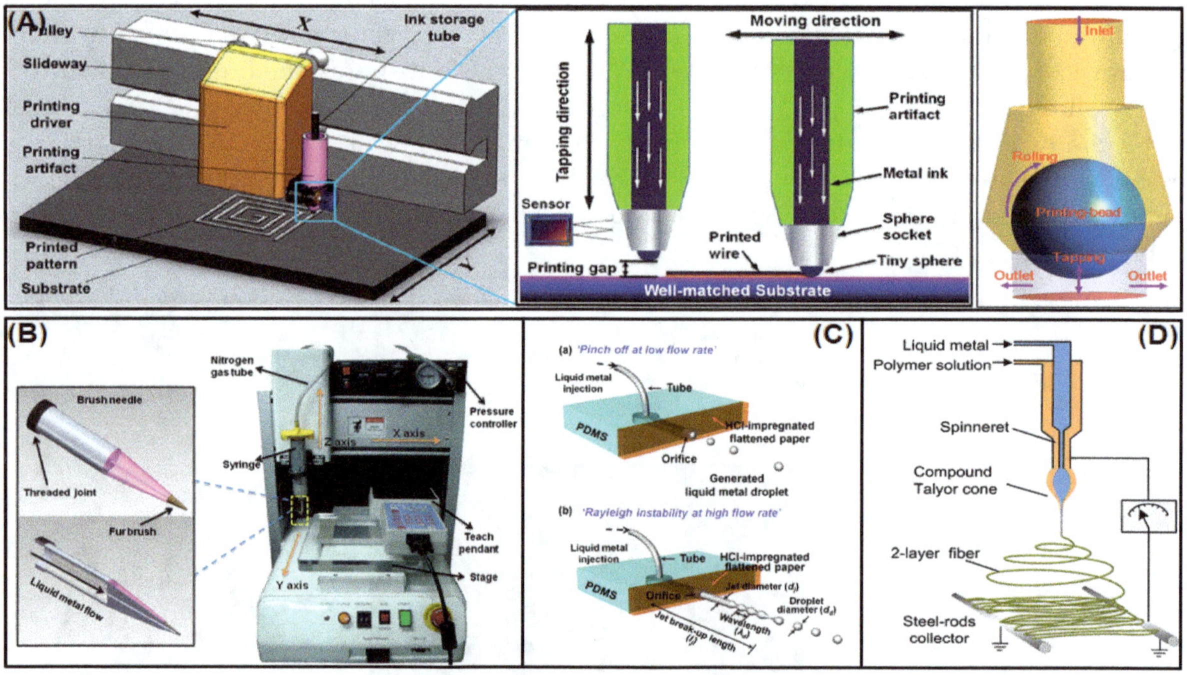

Figure 6. Mechanical printing methods. (**A**) The tapping mode printer and its printing process [18] (Reproduced with permission from Zheng, Y.; He, Z.-Z.; Yang, J.; Liu, J., Personal electronics printing via tapping mode composite liquid metal ink delivery and adhesion mechanism; published by Scientific Reports, 2014.); (**B**) The desktop printer to directly print circuits on paper [69] (Reproduced with permission from Zheng, Y.; He, Z.; Gao, Y.; Liu, J., Direct desktop printed-circuits-on-paper flexible electronics; published by Scientific Reports, 2013.); (**C**) inkjet printing of the liquid metal droplet based on pinch-off and Rayleigh instability [94] (Reproduced with permission from Kim, D.; Yoo, J.H.; Lee, Y.; Choi, W.; Yoo, K.; Lee, J.B.J., Gallium-Based Liquid Metal Inkjet Printing; published by IEEE Proceedings, 2014.); (**D**) coelectrospinning to fabricate a light-emitting coaxial nanofiber with a liquid metal core and polymer shell [95] (Reproduced with permission from Yang, H.; Lightner, C.R.; Dong, L., Light-emitting coaxial nanofibers; published by I ACS NANO, 2012.).

3.1.3. Mask-Based Printing Methods

Printing methods based on masks are the most commonly used method to manufacture liquid metal flexible electronics in the laboratory. Kramer et al. provided a masked deposition method to fabricate liquid metal hyperelastic electronic circuits [88] (Figure 7A). At the same time, Gozen et al. used the masked deposition method to prepare micro-scale high-density soft-matter electronics [66]. Wang et al. proposed a rapid fabrication of flexible functional circuits based on liquid metal dual-trans printing, which can be applied to quickly fabricate a flexible functional electronic device fitting to any complex surface shape (Figure 7B) [58]. First one can print out a liquid metal circuit on PVC membrane, then use polydimethylsiloxane (PDMS) solution to cover the circuit. After a period of time, PDMS solution can be cured and when it is liquid, arbitrary shape objects can be immersed into PDMS solution. Finally, cool the whole object to make liquid metal solid and one can completely transfer the initial liquid metal circuit to a PDMS flexible substrate. In the process, after PDMS curing, one can peel off the PVC membrane and target object to obtain PDMS device embedded liquid metal flexible circuits. Because the shape of the PDMS substrate completely fits with objects, one can achieve highly conformal electronic devices. This technology has significant implications for sensing, monitoring in health care, and households as a corresponding device to conduct specific functions. Common methods mentioned in the literature also include microcontact printing [87], atomized spraying [75], screen printing [70], and CO_2 laser ablation [96] (Figure 8D). The theory of the atomized spraying and the screen printing methods is that the gun sprays the liquid metal ink onto the substrate through a specific mask to form a circuit which has the same shape as the mask (Figure 8B,C).

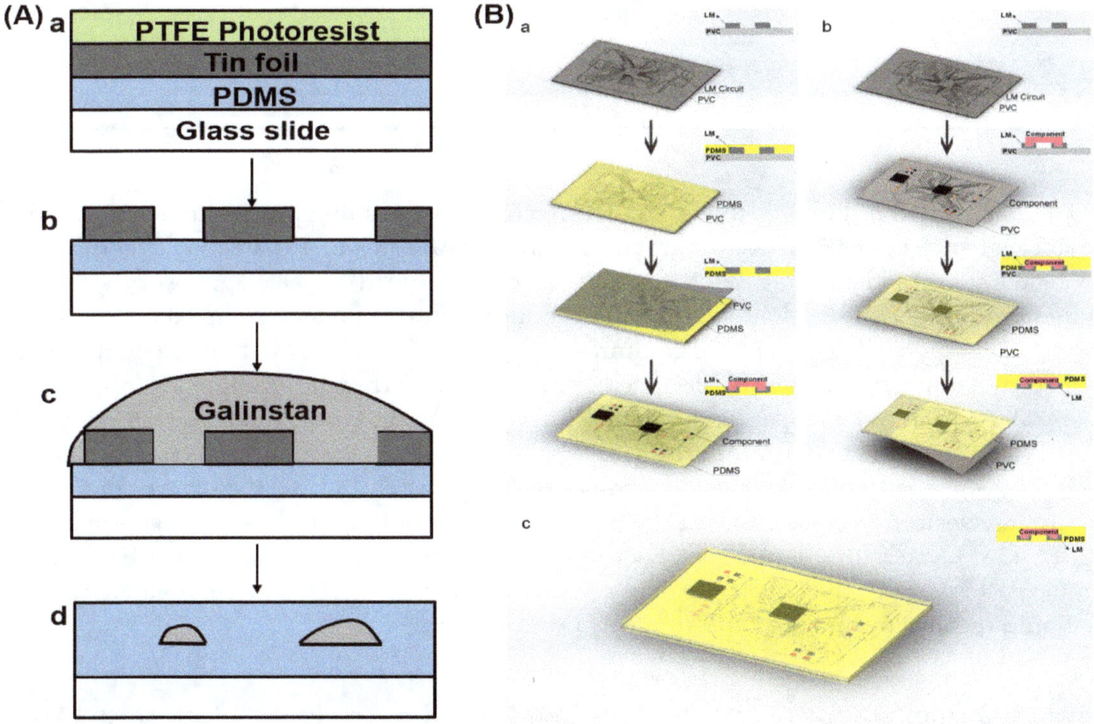

Figure 7. Mask-based printing methods. (**A**) The masked deposition process of gallium–indium alloys for liquid-embedded elastomer conductors, (**a**) painting photoresist onto tin foil; (**b**) micro-channel on the elastomer surface; (**c**) flooding the surface with galinstan; (**d**) removing excessgalinstan with a thin film, cooling the liquid metal in a freezer and encapsulating it by coating (Modified from [88]); (**B**) the process of dual-trans printing fabrication [58] (Reproduced with permission from Wang, Q.; Yu, Y.; Yang, J.; Liu, J., Fast fabrication of flexible functional circuits based on liquid metal dual-trans printing; published by Advanced Material, 2015.).

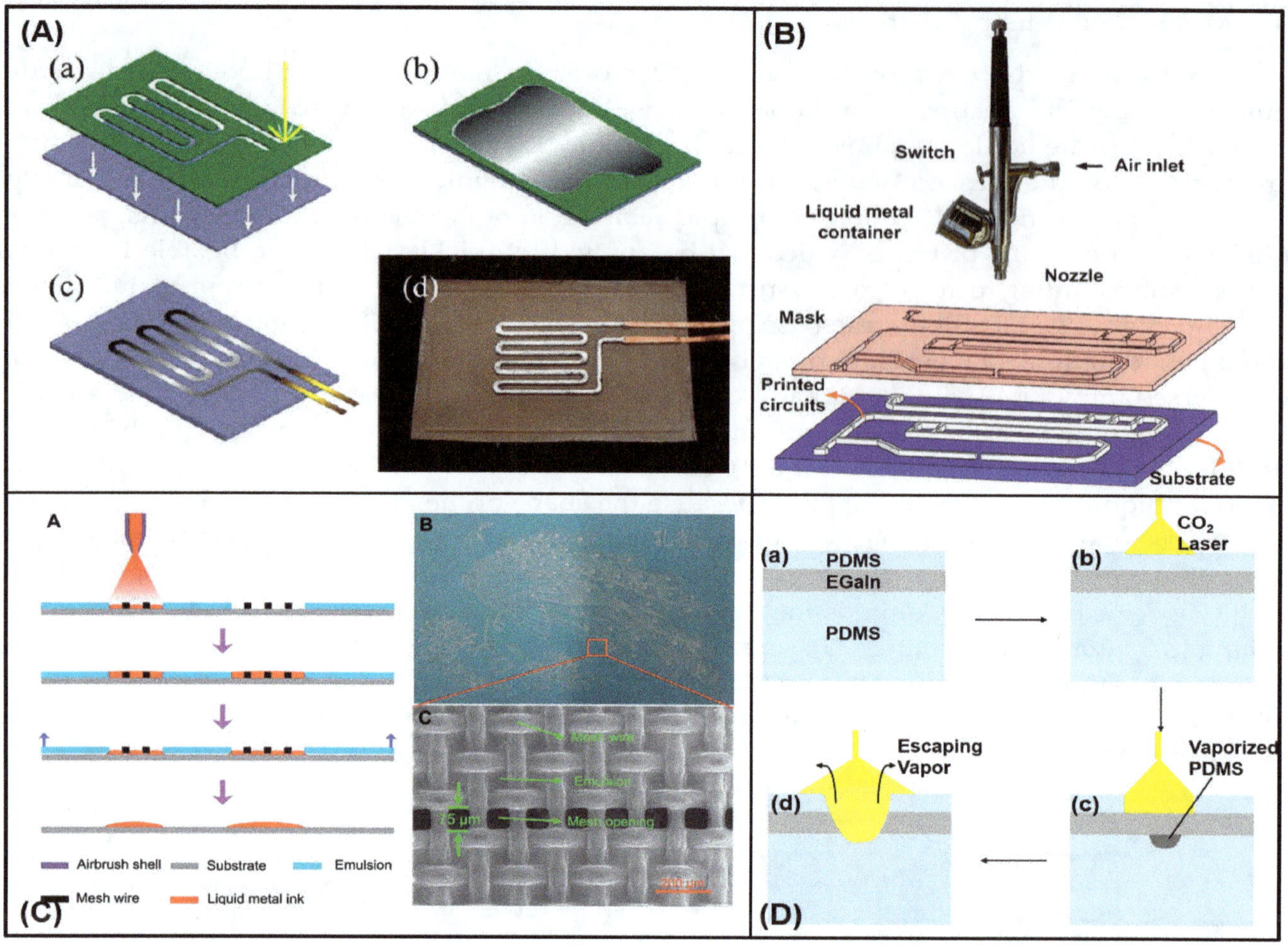

Figure 8. Mask-based printing methods. (**A**) The microcontact printing process to fabricate the liquid metal soft-matter circuit [87], (**a**) laser engraving a thin film; (**b**) spreading GaIn alloy over the stencil; (**c**) removing the stencil and inserting the copper wires before sealing; (**d**) unsealed galinstan heater on VHB elastomer produced with stencil lithography (Reproduced with permission from Tabatabai, A.; Fassler, A.; Usiak, C.; Majidi, C., Liquid-phase gallium–indium alloy electronics with microcontact printing; published by Langmuir, 2013.); (**B**) atomized spraying of liquid metal droplets to rapid prototype circuits [75] (Reproduced with permission from Zhang, Q.; Gao, Y.; Liu, J., Atomized spraying of liquid metal droplets on desired substrate surfaces as a generalized way for ubiquitous printed electronics; published by Applied Physics A: Materials Science & Processing, 2013.); (**C**) the screen printing process and scanning electron microscopy image of the screen mesh [70] (Reproduced with permission from Wang, L.; Liu, J., Pressured liquid metal screen printing for rapid manufacture of high resolution electronic patterns; published by RSC Advances, 2015.); (**D**) proposed CO_2 laser ablation mechanism of the liquid metal alloy on the PDMS substrate (Modified from [96]).

The common method for preparation of liquid metal microfluidic electronics is pouring the liquid metal alloy into a specific shape of channel to form the functional circuit. Therefore, from the strict definition, the preparation method of microfluidic electronics cannot be classified as fabrication of liquid metal flexible printed electronics. However, Figure 9 demonstrates the processing steps of manufacturing liquid alloy microfluidic wireless power transfer, in which the mask layer and atomization printing technology are applied and the power transfer displays large flexibility in its rolled state and can attach to the human arm [97].

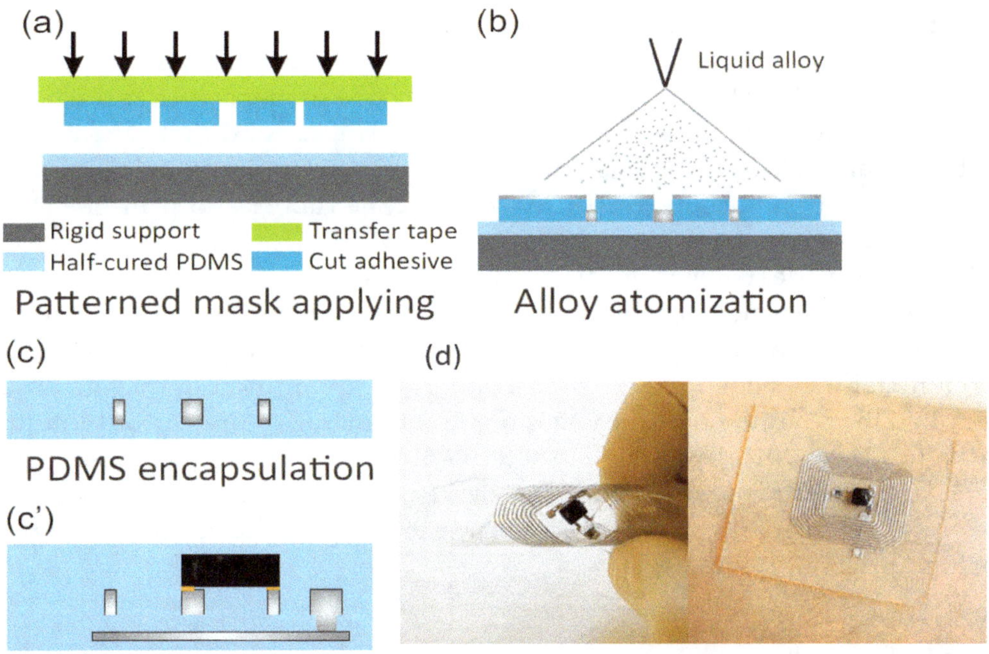

Figure 9. Processing steps of manufacturing liquid alloy microfluidic wireless power transfer (**a**–**c**), an active component mounted in the circuit followed by encapsulation (**c'**) and in rolled state and attached to human arm (**d**) [97] (Reproduced with permission from Jeong, S.H.; Hjort, K.; Wu, Z., Tape transfer atomization patterning of liquid alloys for microfluidic stretchable wireless power transfer; published by Scientific Reports, 2015.).

3.1.4. High-Resolution Nanoimprint Lithography

Unlike the relatively large liquid metal droplets, one can also prepare liquid metal nanoparticles to print micro-scale or nano-scale wires. John et al. produced small features through mechanically sintering gallium–indium nanoparticles to realize high-resolution nanoimprint lithography [98]. As shown in Reference [98], 1 μm coalesced EGaIn line arrays were formed by reducing the size of the sintering tool. Via inkjet printing EGaIn nanoparticles dispersions on the surface of nitrile glove, the functionalized elastomer glove with liquid metal micro-wire arrays represents excellent stretchability when holding a tennis ball.

3.2. 3D Printing

3D printing is a technology to construct an object by means of printing layer by layer; in other words, it is a kind of typical additive manufacturing technology. Different from traditional subtractive technology, additive manufacturing is a top-down, layer-by-layer manufacturing process. It can not only avoid corrosion and contribute to environmental protection, but also saves a lot of raw materials. This significantly reduces environmental pollution and the cost of materials. Figure 10A shows the difference between subtractive and additive manufacturing technologies [99]. Next we will introduce several 3D printing technologies using low melting point metal ink based on the additive mode.

Wang et al. first put forward the concept of liquid phase 3D printing to employ gallium-, bismuth-, and indium-based alloys whose melting points are higher than room temperature and lower than 300 °C to accomplish the printing process [100]. The liquid phase 3D printing is achieved in a liquid environment, and the liquid can be water, ethanol, kerosene, electrolyte solution, etc. However, the melting point of liquid metal ink must be lower than that of the liquid environment to ensure the printed item run. In the experiments, Wang et al. introduced $Bi_{35}In_{48.6}Sn_{16}Zn_{0.4}$ as printing ink because it does not absorb/release as much heat as normal metal in the process of phase change (Figure 11B) [100]. Shortly after, they originated a late-model 3D printing technology, hybrid 3D printing, which means interactive printing of various inks or combination of multiple printing methods. In Figure 10B, $Bi_{35}In_{48.6}Sn_{16}Zn_{0.4}$ and 705 silicone rubber (nonmetal) inks were adopted to accomplish hybrid 3D printing [64]. The 705 silicone rubber serves as a kind of waterproof, anticorrosive, transparent, and insulating adhesive, and can be solidified when absorbing water vapor at room temperature. Therefore, it is often used as an electrical packaging material. First, one can print the bottom layer on the substrate using 705 silicone rubber and print the middle layer on the 705 silicone rubber layer using $Bi_{35}In_{48.6}Sn_{16}Zn_{0.4}$. After that, one can print the top layer using 705 silicone rubber. When adequately solidified, a sandwich structure can be obtained once taking off the printed item from the basement. Compatible hybrid 3D printing of metal and nonmetal inks fully embodies the good mechanical strength and electrical/thermal conductive of metal ink, and excellent insulation of nonmetal ink, which makes the printing circuit applicable in a harsh environment. Fassler et al. applied the common freeze-casting method to manufacture various liquid metal based 3D structures (Figure 11A) [101]. Ladd et al. tested the direct-writing microcomponents of the liquid metal alloy for 3D free-standing liquid metal microstructures at room temperature [102]. Liquid metal can also be employed as an injectable 3D bio-electrode to measure the electrocardiograph (ECG) or electroencephalogram (EEG) [54,57,103].

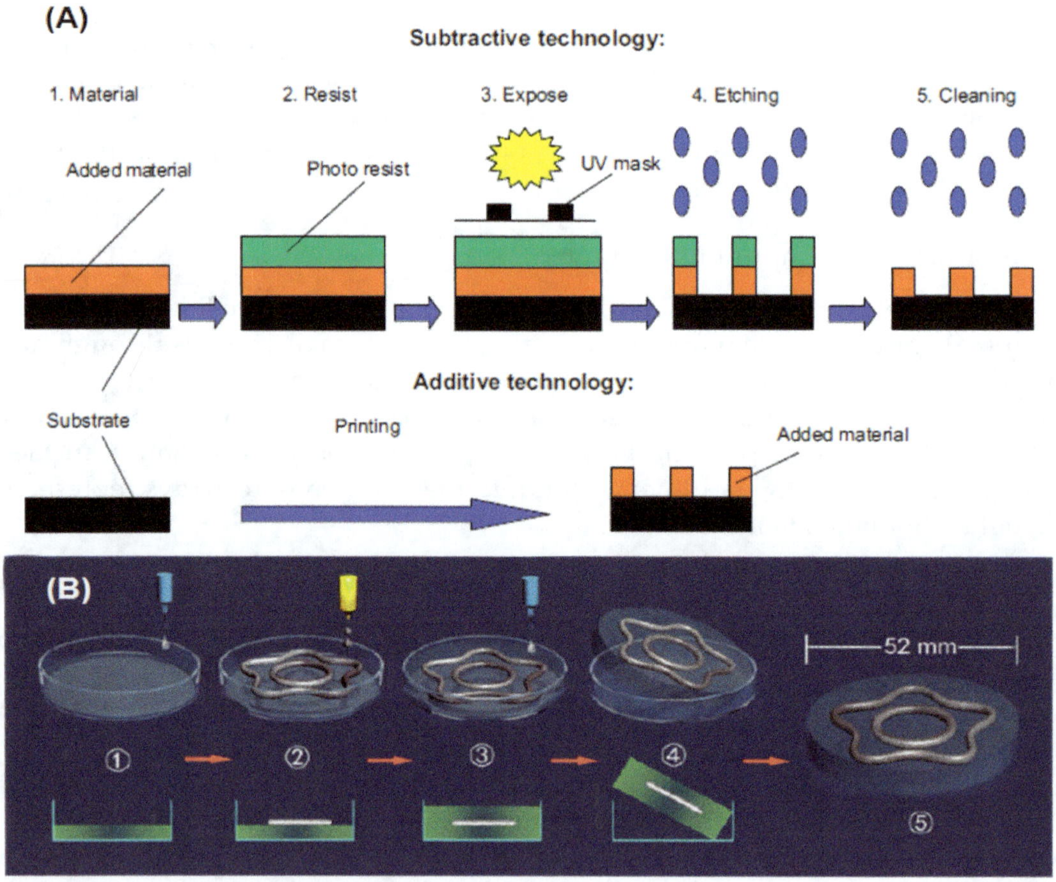

Figure 10. (**A**) Difference between subtractive and additive manufacturing technologies [99] (Reproduced with permission from Kunnari, E.; Valkama, J.; Keskinen, M.; Mansikkamäki, P., Environmental evaluation of new technology: Printed electronics case study.; published by Journal of Cleaner Production, 2009.); (**B**) hybrid 3D printing process with $Bi_{35}In_{48.6}Sn_{16}Zn_{0.4}$ and 705 silicone rubber inks [64] (Reproduced with permission from Wang, L.; Liu, J., Compatible hybrid 3D printing of metal and nonmetal inks for direct manufacture of end functional devices; published by Science China Technological Sciences, 2014.).

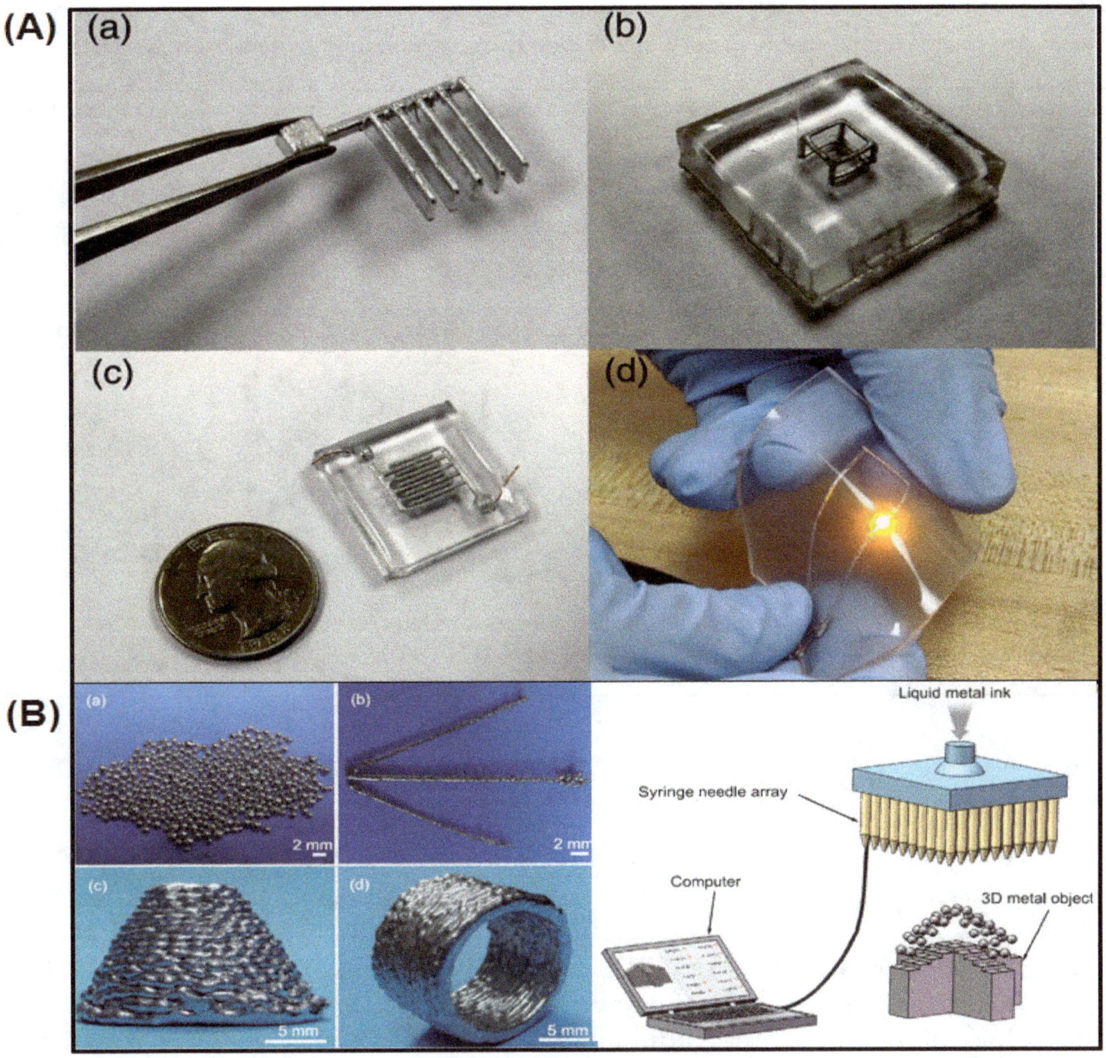

Figure 11. (**A**) The electronical components based on the liquid metal alloy [101] (Reproduced with permission from Fassler, A.; Majidi, C., 3D structures of liquid-phase gain alloy embedded in pdms with freeze casting; published by Lab on a Chip, 2013.); (**B**) liquid phase 3D printing method [100], (**a**) liquid metal balls; (**b**) liquid metal rods; (**c**) cone structure; (**d**) cylinder structure (Reproduced with permission from Wang, L.; Liu, J., Liquid phase 3D printing for quickly manufacturing conductive metal objects with low melting point alloy ink; published by Science China Technological Sciences, 2014.)

3.3. Printing Devices

Liu et al. have delivered printers that have now been put into practical use (Table 3) [18,100]. Liquid metal circuit printers have completely changed the traditional mode, breaking the technical bottleneck on personal electronics manufacturing, which makes it a reality to quickly fabricate electronic circuits with extremely low costs. Through computer control, even people with no electronics experience can download a circuit diagram from the Internet and print it directly. Inkjet printing equipment is based on the liquid metal screen printing process. Through a USB cable, the printer can connect with a computer to realize online printing, and can also print offline via an SD card. Before printing one needs to prepare .gcode files to control the nozzle movement. A 3D Metal Printer is a desktop fused deposition modeling (FDM) device; it substitutes traditional metal parts processing to prepare metal parts with a complex spatial structure.

Table 3. Three kinds of printers [69] (Reproduced with permission from Zheng, Y.; He, Z.; Gao, Y.; Liu, J., Direct desktop printed-circuits-on-paper flexible electronics; published by Scientific Reports, 2013.).

Liquid Metal Printer	Personal Circuit Printer	Ubiquitous Inkjet Printer	3D Metal Printer
Device type	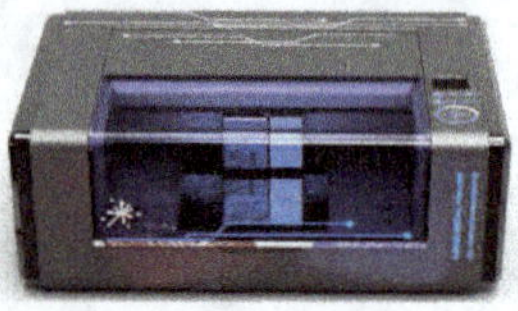	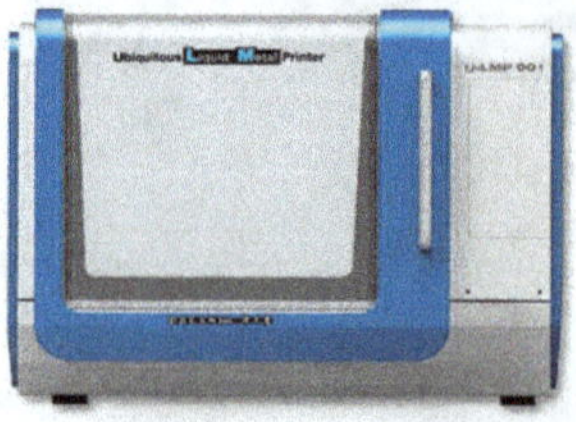	
Printed out items		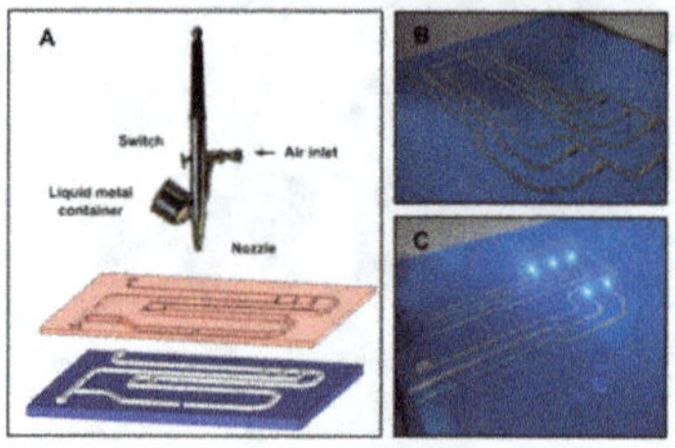	
Application Fields	circuit; art; devices; rapid prototyping.	electronic circuit; art; radio frequency identification label; antenna.	circuit; radar; antenna; molds.

4. Typical Applications

As a novel electronic ink material, the flexibility and shape-preserving of the liquid metal alloy allows it to play an important role in fabricating components and soft circuit. Its non-toxicity and benign biocompatibility make it an implant material in the body, with uses such as the selected low melting point metal skeleton material [51], high-resolution angiography contrast [52], biological micro-droplets [95], nerve connection [35,36], human exoskeleton [104], and electronic skin [55,56,105].

The technology of directly printing liquid metal electronic circuit transforms conventional rigid circuits process to flexible desktop manufacturing [106]. With a way to rapidly manufacture various flexible household electronics with different functions, one can envisage that daily life will undergo big changes in the near future. Imagine a quilt with flexible functional circuits containing temperature or pressure sensors that can monitor human sleep quality and adjust the quilt temperature according to body temperature to achieve intelligent heating or cooling. In fact, all of these needs can be implemented by liquid metal flexible circuits. Liquid metal flexible printed electronics can be directly printed on many soft substrates, and could be used to produce lots of complex artwork. Zhang et al. demonstrated that even leaves can become a circuit board, which provides a solid theoretical basis for the universality of liquid metal atomized spraying manufacturing flexible electronics [75].

Through the direct-writing method, Li et al. drew a liquid metal thermocouple on paper [85]. The liquid metal thermocouple has many special advantages, such as small contact resistance and thermal resistance, no need for welding, and a wide working temperature zone. The thermoelectric effect is the theoretical basis of thermocouple temperature measurement. The thermoelectric power produced by the thermocouple has nothing to do with its shape or size, but relates to the composition of the thermoelectric materials and the temperature difference between the two ends. The thermocouple composed of Ga and EGaIn not only has temperature sensing ability but also bending capacity.

Fassler et al. prepared two kinds of capacitors—the square wave capacitor and the spiral capacitor—and a square planar spiral inductor [107]. They calculated the values of the capacitors or inductor of various shapes via formula derivation. A soft-matter diode, a new type of flexible functional equipment component with liquid metal electrodes, was developed by So et al. [29]. Because the liquid metal EGaIn electrodes' sandwich layers are conductive forward from Ga_xO_y to Ga while nonconductive backward from Ga to Ga_xO_y, the soft-matter diode has the function of limiting current and can pass the current one-way through.

At the same time, because of its flexibility and fluidity, liquid alloy is widely used in sensors and functional devices; Wu's group made important contributions to the field [108]. They revealed a liquid alloy microfluidic stretchable large-area wireless strain antenna, consisting of two layers of liquid metal alloy filled microfluidic channels in a silicone elastomer [109]. A stretchable ultra-high frequency radio frequency identity tag printed on human skin demonstrated that liquid alloy microfluidic electronics can be applied to electronic skin [56]. Liquid alloy microfluidic electronics can also be fabricated as antennae with strong stretchability, rolled space [110,111], and liquid alloy microfluidic wireless power transfer [97]. Matsuzaki et al. designed a wearable data glove with local strain sensing using GaInSn elestrodes (Figure 12A) [59]; the data glove can judge the degree of finger bending according to the change in resistance at different monitoring points. Krupenkin et al. modeled a footwear-embedded microfluidic energy harvester, which is a wearable mechanical-to-electrical energy conversion based on liquid metal microdroplets [60]. Meanwhile, the liquid metal Galinstan can fabricate a stretchable loudspeaker to apply audio frequency electric signal (Figure 13A) [112]. When the liquid metal circuit is cut off, once reconnected, it can work well without additional repairmen (Figure 13B) [113].

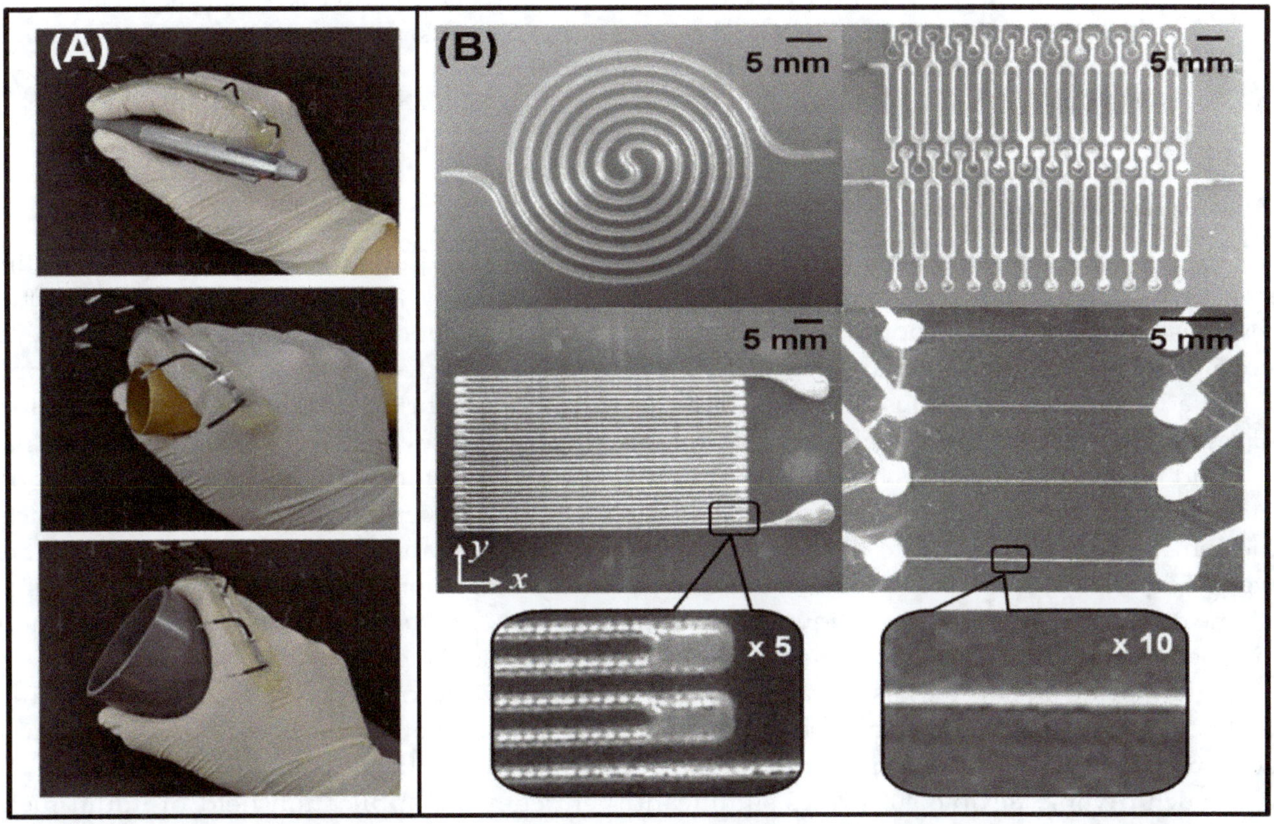

Figure 12. Sensors based on liquid metal deformation. (**A**) A wearable data glove with local strain sensing using GaInSn elestrodes [59] (Reproduced with permission from Matsuzaki, R.; Tabayashi, K., Highly stretchable, global, and distributed local strain sensing line using gainsn electrodes for wearable electronics; published by Advanced Functional Material, 2015.); (**B**) hyperelastic pressure sensors with microchannels of conductive liquid metal alloy [114] (Reproduced with permission from Park, Y.-L.; Majidi, C.; Kramer, R.; Berard, P.; Wood, R.J., Hyperelastic pressure sensing with a liquid-embedded elastomer; published by Journal of Micromechanics and Microengineering, 2010.).

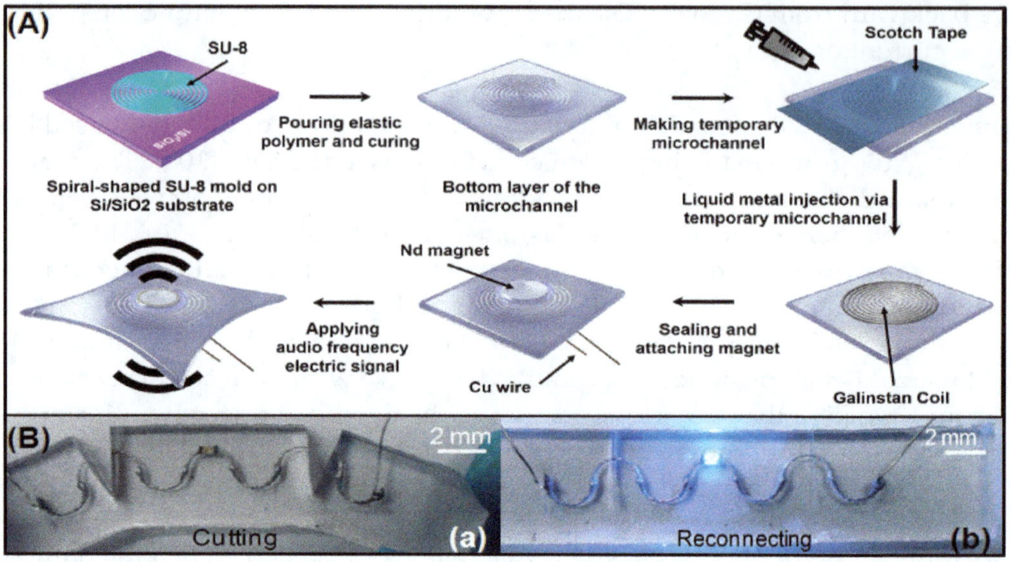

Figure 13. The liquid metal flexible printed electronics functional devices. (**A**) The fabrication process of the liquid metal stretchable loudspeaker [112] (Reproduced with permission from Jin, S.W.; Park, J.; Hong, S.Y.; Park, H.; Jeong, Y.R.; Park, J.; Lee, S.-S. ; Ha, J.S., Stretchable loudspeaker using liquid metal microchannel; published by Scientific Reports, 2015.); (**B**) Physical separation (**a**) and reconnection (**b**) of the liquid metal circuit with an LED [113] (Reproduced with permission from Li, G.; Wu, X.; Lee, D.-W., A galinstan-based inkjet printing system for highly stretchable electronics with self-healing capability; published by Lab on a Chip, 2016.).

5. Discussion and Conclusions

As new-generation inks for flexible printed electronics, liquid metals possess outstanding versatility and specific merits of fluidity, conductivity, metallicity, and electromagnetic properties. They thus offer tremendous opportunities for making future flectional electronic devices. However, there are still challenges in preparing, processing, and utilizing high-performance inks. For example, the low adhesion to many substrates, easy oxidation, and high surface tension of liquid alloy inks still restrict the realization of high-precision, stable circuits. To solve such key issues, we can mix the liquid metal with materials that are metal or non-metal. Therefore, new hybrid functional liquid metal materials can be made with desired properties of semiconductivity, semi-flexibility, enhanced adhesion, antioxidant ability, etc. This would enable the inks to adapt well to practical situations and have more functions. For example, the liquid metal nanocrystallization can improve printing precision, and one can modify the surface of liquid metal nanoparticles, e.g., by combining with the luminous substance through metal bond to achieve the self-luminous colorful printing.

So far, liquid metal ink has already been printed on a wide range of soft substrates, such as plate, paper, cloth, skin, leaves, and so on, which opens up many opportunities for making flexible electronical components like sensors, actuators, wearable bioelectronics, electrical skin, etc. The use of such diverse flexible substrates significantly enriches the application field of liquid metal flexible printed electronics, and one can thus expand related practices from human beings to other organisms such as animals and plants. For example, one can even print flexible circuits with features of energy harvesting and light emission on leaves to capture the energy of swinging leaves to supply light at night.

Various kinds of liquid metal printing technologies are enabling efficient circuit fabrication. At this stage, a series of technological problems still need to be tackled, such as precise and reliable connection between the printed liquid metal circuits with the traditional integrated circuits. High precision printing is critical to achieve the liquid metal integrated circuits. Regarding the specific fabrication, hybrid flexible printing with planar printing and 3D printing are worth trying. Furthermore, cooling of complex liquid metal circuits should be carefully monitored to avoid damage from the "thermal barrier." Meanwhile, encapsulation is often a necessity for making ending user devices. For instance, long-lasting liquid metal skin electronics mainly depend on packaging technology. PDMS is often applied to package liquid metal flexible circuits. However, for various substrates in different environments, one needs to consider diverse encapsulation technologies to make the circuit more secure in order to achieve the desired functionality. For applications in organisms, the non-toxicity of packaging materials especially needs to be considered. Sometimes, liquid metal's movement and shape-shifting under the electric field can produce more self-configurational operations of the circuits [106,107]. These unconventional, dynamically changing or self-destructive circuits are also worth pursuing.

In summary, liquid metal flexible printed electronics break through the limitations of traditional flexible electronics. The process of circuits manufacturing is straightforward, fast, stable, and relatively compatible with existing integrated circuit technology. Liquid metal flexible printed electronics would allow us to make high-quality products such as wearable devices, electronic skin, medical implants, flexible display, and solar panels that will meet the coming demands. The existing liquid metal flexible printed technology is a combination of a liquid metal soft circuit and rigid components or chips, so not completely flexible electronics in the larger sense. At the same time, this kind of semi-flexible circuit may partially limit its application. A basic significance of liquid metal flexible printed electronics lies in the popularity of liquid metal printing technology: people can discretionarily print their personalized flexible functional devices on any surface, which can significantly expand traditional electronics engineering. By overcoming the problems in obtaining a high rate of good product, self-repair of damaged circuits, long service life, low environmental pollution, and waste recycling, liquid metal flexible printed electronics can contribute to a bright future.

Acknowledgments: This work was partially supported by Beijing Municipal Science and Technology Funding (Under Grant No. Z151100003715002) and Key Project Funding of the Chinese Academy of Sciences.

Author Contributions: Xuelin Wang wrote the whole manuscript. Jing Liu supervised the work and wrote part of the manuscript.

References

1. Cui, Z. *Printed Electronics: Materials, Technologies and Applications*; Higher Education Press: Beijing, China, 2012.
2. Lin, Y.; Gritsenko, D.; Liu, Q.; Lu, X.; Xu, J. Recent advancements in functionalized paper-based electronics. *ACS Appl. Mater. Interfaces* **2016**, *8*, 20501–20515. [CrossRef] [PubMed]
3. Yin, Z.; Huang, Y.; Bu, N.; Wang, X.; Xiong, Y. Inkjet printing for flexible electronics: Materials, processes and equipments. *Chin. Sci. Bull.* **2010**, *55*, 3383–3407. [CrossRef]
4. Wong, W.S.; Salleo, A. *Flexible Electronics: Materials and Applications*; Springer: New York, NY, USA, 2009.
5. Reuss, R.H.; Chalamala, B.R.; Moussessian, A.; Kane, M.G.; Kumar, A.; Zhang, D.C.; Rogers, J.A.; Hatalis, M.; Temple, D.; Moddel, G.; et al. Macroelectronics: Perspectives on technology and applications. *Proc. IEEE* **2005**, *93*, 1239–1256. [CrossRef]
6. Kim, D.-H.; Ahn, J.-H.; Choi, W.M.; Kim, H.-S.; Kim, T.-H.; Song, J.; Huang, Y.Y.; Liu, Z.; Lu, C.; Rogers, J.A. Stretchable and foldable silicon integrated circuits. *Science* **2008**, *320*, 507–511. [CrossRef] [PubMed]
7. Crawford, G.P. *Flexible Flat Panel Displays*; John Wiley & Sons, Ltd.: New York, NY, USA, 2005.
8. Stoppa, M.; Chiolerio, A. Wearable electronics and smart textiles: A critical review. *Sensors* **2014**, *14*, 11957–11992. [CrossRef] [PubMed]
9. Koo, M.; Park, K.-I.; Lee, S.H.; Suh, M.; Jeon, D.Y.; Cho, J.W.; Kang, K.; Lee, K.J. Bendable inorganic thin-film battery for fully flexible electronic systems. *NANO Lett.* **2012**, *12*, 4810–4816. [CrossRef] [PubMed]
10. Reeder, J.; Kaltenbrunner, M.; Ware, T.; Arreaga-Salas, D.; Avendano-Bolivar, A.; Yokota, T.; Inoue, Y.; Sekino, M.; Voit, W.; Sekitani, T.; et al. Mechanically adaptive organic transistors for implantable electronics. *Adv. Mater.* **2014**, *26*, 4967–4973. [CrossRef] [PubMed]
11. Kang, D.; Pikhitsa, P.V.; Choi, Y.W.; Lee, C.; Shin, S.S.; Piao, L.; Park, B.; Suh, K.-Y.; Kim, T.-I.; Choi, M. Ultrasensitive mechanical crack-based sensor inspired by the spider sensory system. *Nature* **2014**, *516*, 222–226. [CrossRef] [PubMed]

12. Russo, A.; Ahn, B.Y.; Adams, J.J.; Duoss, E.B.; Bernhard, J.T.; Lewis, J.A. Pen-on-paper flexible electronics. *Adv. Mater.* **2011**, *23*, 3426–3430. [CrossRef] [PubMed]
13. Jung, I.; Shin, G.; Malyarchuk, V.; Ha, J.S.; Rogers, J.A. Paraboloid electronic eye cameras using deformable arrays of photodetectors in hexagonal mesh layouts. *Appl. Phys. Lett.* **2010**, *96*, 021110. [CrossRef]
14. Schwartz, G.; Tee, B.C.K.; Mei, J.; Appleton, A.L.; Kim, D.H.; Wang, H.; Bao, Z. Flexible polymer transistors with high pressure sensitivity for application in electronic skin and health monitoring. *Nat. Commun.* **2013**, *4*, 54–56. [CrossRef] [PubMed]
15. Wang, X.; Dong, L.; Zhang, H.; Yu, R.; Pan, C.; Wang, Z.L. Recent progress in electronic skin. *Adv. Sci.* **2015**, *2*, 10. [CrossRef]
16. Kim, D.-H.; Lu, N.; Ma, R.; Kim, Y.-S.; Kim, R.-H.; Wang, S.; Wu, J.; Won, S.M.; Tao, H.; Islam, A.; et al. Epidermal electronics. *Science* **2011**, *333*, 838–843. [CrossRef] [PubMed]
17. Choi, W.M.; Song, J.; Khang, D.-Y.; Jiang, H.; Huang, Y.Y.; Rogers, J.A. Biaxially stretchable "wavy" silicon nanomembranes. *NANO Lett.* **2007**, *7*, 1655–1663. [CrossRef] [PubMed]
18. Zheng, Y.; He, Z.-Z.; Yang, J.; Liu, J. Personal electronics printing via tapping mode composite liquid metal ink delivery and adhesion mechanism. *Sci. Rep.* **2014**, *4*, 101–104. [CrossRef] [PubMed]
19. Ghosh, D.; Maki, S.P.; Lyons, C.; Theiss, S.D.; Owings, R.R. Hybrid dielectric thin films on flexible substrates for embedded capacitor applications. *IEEE Trans. Compon. Packag. Manuf. Technol.* **2016**, *6*, 941–945. [CrossRef]
20. Singh, M.; Haverinen, H.M.; Dhagat, P.; Jabbour, G.E. Inkjet printing-process and its applications. *Adv. Mater.* **2010**, *22*, 673–685. [CrossRef] [PubMed]
21. MicroPen Technologies. Available online: http://www.Micropen.com/ (accessed on 1 September 2016).
22. Kim, S.S.; Na, S.I.; Jo, J.; Tae, G.; Kim, D.Y. Efficient polymer solar cells fabricated by simple brush painting. *Adv. Mater.* **2007**, *19*, 4410–4415. [CrossRef]
23. Jang, J. Displays develop a new flexibility. *Mater. Today* **2006**, *9*, 46–52. [CrossRef]
24. Krebs, F.C. Fabrication and processing of polymer solar cells: A review of printing and coating techniques. *Sol. Energy Mater. Sol. Cells* **2009**, *93*, 394–412. [CrossRef]
25. Mayer, A.C.; Scully, S.R.; Hardin, B.E.; Rowell, M.W.; McGehee, M.D. Polymer-based solar cells. *Mater. Today* **2007**, *10*, 28–33. [CrossRef]
26. Someya, T.; Kato, Y.; Sekitani, T.; Iba, S.; Noguchi, Y.; Murase, Y.; Kawaguchi, H.; Sakurai, T. Conformable, flexible, large-area networks of pressure and thermal sensors with organic transistor active matrixes. *Proc. Natl. Acad. Sci. USA* **2005**, *102*, 12321–12325. [CrossRef] [PubMed]
27. Madden, P.G.A. *Development and Modeling of Conducting Polymer Actuators and the Fabrication of a Conducting Polymer Based Feedback Loop*; Massachusetts Institute of Technology: Cambridge, MA, USA, 2003.
28. Hammock, M.L.; Chortos, A.; Tee, B.C.K.; Tok, J.B.H.; Bao, Z. 25th anniversary article: The evolution of electronic skin (e-skin): A brief history, design considerations, and recent progress. *Adv. Mater.* **2013**, *25*, 5997–6038. [CrossRef] [PubMed]
29. So, J.-H.; Koo, H.-J.; Dickey, M.D.; Velev, O.D. Ionic current rectification in soft-matter diodes with liquid-metal electrodes. *Adv. Funct. Mater.* **2012**, *22*, 625–631. [CrossRef]
30. Niu, S.; Wang, X.; Yi, F.; Zhou, Y.S.; Wang, Z.L. A universal self-charging system driven by random biomechanical energy for sustainable operation of mobile electronics. *Nat. Commun.* **2015**, *6*, 8975. [CrossRef] [PubMed]
31. Zheng, Q.; Zhang, H.; Shi, B.; Xue, X.; Liu, Z.; Jin, Y.; Ma, Y.; Zou, Y.; Wang, X.; An, Z.; et al. In vivo self-powered wireless cardiac monitoring via implantable triboelectric nanogenerator. *ACS Nano* **2016**, *10*, 6510–6518. [CrossRef] [PubMed]
32. Rogers, J.A.; Bao, Z.; Baldwin, K.; Dodabalapur, A.; Crone, B.; Raju, V.R.; Kuck, V.; Katz, H.; Amundson, K.; Ewing, J.; et al. Paper-like electronic displays: Large-area rubber-stamped plastic sheets of electronics and microencapsulated electrophoretic inks. *Proc. Natl. Acad. Sci. USA* **2001**, *98*, 4835–4840. [CrossRef] [PubMed]
33. Wang, X.; Tian, H.; Mohammad, M.A.; Li, C.; Wu, C.; Yang, Y.; Ren, T.-L. A spectrally tunable all-graphene-based flexible field-effect light-emitting device. *Nat. Commun.* **2015**, *6*, 37–42. [CrossRef] [PubMed]
34. Lessing, J.; Morin, S.A.; Keplinger, C.; Tayi, A.S.; Whitesides, G.M. Stretchable conductive composites based on metal wools for use as electrical vias in soft devices. *Adv. Funct. Mater.* **2015**, *25*, 1418–1425. [CrossRef]

35. Zhang, J.; Sheng, L.; Jin, C.; Liu, J. Liquid Metal as Connecting or Functional Recovery Channel for the Transected Sciatic Nerve. Available online: https://arxiv.org/abs/1404.5931 (accessed on 28 September 2016).
36. Liu, F.; Yu, Y.; Yi, L.; Liu, J. Liquid metal as reconnection agent for peripheral nerve injury. *Sci. Bull.* **2016**, *61*, 939–947. [CrossRef]
37. Wang, X.; Zhang, H.; Yu, R.; Dong, L.; Peng, D.; Zhang, A.; Zhang, Y.; Liu, H.; Pan, C.; Wang, Z.L. Dynamic pressure mapping of personalized handwriting by a flexible sensor matrix based on the mechanoluminescence process. *Adv. Mater.* **2015**, *27*, 2324–2331. [CrossRef] [PubMed]
38. Tee, B.C.K.; Wang, C.; Allen, R.; Bao, Z. An electrically and mechanically self-healing composite with pressure- and flexion-sensitive properties for electronic skin applications. *Nat. Nano* **2012**, *7*, 825–832. [CrossRef] [PubMed]
39. Kim, J.; Lee, M.; Shim, H.J.; Ghaffari, R.; Cho, H.R.; Son, D.; Jung, Y.H.; Soh, M.; Choi, C.; Jung, S.; et al. Stretchable silicon nanoribbon electronics for skin prosthesis. *Nat. Commun.* **2014**, *5*, 5747. [CrossRef] [PubMed]
40. Tee, B.C.-K.; Chortos, A.; Berndt, A.; Nguyen, A.K.; Tom, A.; McGuire, A.; Lin, Z.C.; Tien, K.; Bae, W.-G.; Wang, H.; et al. A skin-inspired organic digital mechanoreceptor. *Science* **2015**, *350*, 313–316. [CrossRef] [PubMed]
41. Wang, S.; Li, M.; Wu, J.; Kim, D.-H.; Lu, N.; Su, Y.; Kang, Z.; Huang, Y.; Rogers, J.A. Mechanics of epidermal electronics. *J. Appl. Mech.* **2012**, *79*, 031022. [CrossRef]
42. Webb, R.C.; Ma, Y.; Krishnan, S.; Li, Y.; Yoon, S.; Guo, X.; Feng, X.; Shi, Y.; Seidel, M.; Cho, N.H.; et al. Epidermal devices for noninvasive, precise, and continuous mapping of macrovascular and microvascular blood flow. *Sci. Adv.* **2015**, *1*, e1500701. [CrossRef] [PubMed]
43. Rajan, K.; Bocchini, S.; Chiappone, A.; Roppolo, I.; Perrone, D.; Bejtka, K.; Ricciardi, C.; Pirri, C.F.; Chiolerio, A. Spin-coated silver nanocomposite resistive switching devices. *Microelectron. Eng.* **2017**, *168*, 27–31. [CrossRef]
44. Chiolerio, A.; Roppolo, I.; Bejtka, K.; Asvarov, A.; Pirri, C.F. Resistive hysteresis in flexible nanocomposites and colloidal suspensions: Interfacial coupling mechanism unveiled. *RSC Adv.* **2016**, *6*, 56661–56667. [CrossRef]
45. Tai, Y.L. Preparation and Application of Conductive Ink for Printed Electronics. Ph.D. Thesis, Fudan University, Shanghai, China, 2012.
46. Rajan, K.; Roppolo, I.; Chiappone, A.; Bocchini, S.; Perrone, D.; Chiolerio, A. Silver nanoparticle ink technology: State of the art. *Nanotechnol. Sci. Appl.* **2016**, *9*, 1–13. [PubMed]
47. Xiong, Z.; Liu, C. Optimization of inkjet printed PEDOT:PSS thin films through annealing processes. *Org. Electron.* **2012**, *13*, 1532–1540. [CrossRef]
48. Chiolerio, A.; Bocchini, S.; Scaravaggi, F.; Porro, S.; Perrone, D.; Beretta, D.; Caironi, M.; Pirri, C.F. Synthesis of polyaniline-based inks for inkjet printed devices: Electrical characterization highlighting the effect of primary and secondary doping. *Semicond. Sci. Technol.* **2015**, *30*, 104001. [CrossRef]
49. Petukhov, D.I.; Kirikova, M.N.; Bessonov, A.A.; Bailey, M.J.A. Nickel and copper conductive patterns fabricated by reactive inkjet printing combined with electroless plating. *Mater. Lett.* **2014**, *132*, 302–306. [CrossRef]
50. Plevachuk, Y.; Sklyarchuk, V.; Yakymovych, A.; Svec, P.; Janickovic, D.; Illekova, E. Electrical conductivity and viscosity of liquid Sn–Sb–Cu alloys. *J. Mater. Sci. Mater. Electron.* **2011**, *22*, 631–638. [CrossRef]
51. Yi, L.; Jin, C.; Wang, L.; Liu, J. Liquid-solid phase transition alloy as reversible and rapid molding bone cement. *Biomaterials* **2014**, *35*, 9789–9801. [CrossRef] [PubMed]
52. Wang, Q.; Yu, Y.; Pan, K.; Liu, J. Liquid metal angiography for mega contrast X-ray visualization of vascular network in reconstructing in vitro organ anatomy. *IEEE Trans. Biomed. Eng.* **2014**, *61*, 2161–2166. [CrossRef] [PubMed]
53. Lu, Y.; Hu, Q.; Lin, Y.; Pacardo, D.B.; Wang, C.; Sun, W.; Ligler, F.S.; Dickey, M.D.; Gu, Z. Transformable liquid-metal nanomedicine. *Nat. Commun.* **2015**, *6*, 10066. [CrossRef] [PubMed]
54. Jin, C.; Zhang, J.; Li, X.; Yang, X.; Li, J.; Liu, J. Injectable 3-D fabrication of medical electronics at the target biological tissues. *Sci. Rep.* **2013**, *3*, 3442. [CrossRef] [PubMed]
55. Guo, C.; Yu, Y.; Liu, J. Rapidly patterning conductive components on skin substrates as physiological testing devices via liquid metal spraying and pre-designed mask. *J. Mater. Chem. B* **2014**, *2*, 5739–5745. [CrossRef]
56. Jeong, S.H.; Hagman, A.; Hjort, K.; Jobs, M.; Sundqvist, J.; Wu, Z. Liquid alloy printing of microfluidic stretchable electronics. *Lab Chip* **2012**, *12*, 4657–4664. [CrossRef] [PubMed]

57. Yu, Y.; Zhang, J.; Liu, J. Biomedical implementation of liquid metal ink as drawable ecg electrode and skin circuit. *PLoS ONE* **2013**, *8*, e58771. [CrossRef] [PubMed]
58. Wang, Q.; Yu, Y.; Yang, J.; Liu, J. Fast fabrication of flexible functional circuits based on liquid metal dual-trans printing. *Adv. Mater.* **2015**, *27*, 7109–7116. [CrossRef] [PubMed]
59. Matsuzaki, R.; Tabayashi, K. Highly stretchable, global, and distributed local strain sensing line using gainsn electrodes for wearable electronics. *Adv. Funct. Mater.* **2015**, *25*, 3806–3813. [CrossRef]
60. Krupenkin, T.; Taylor, J.A. Reverse electrowetting as a new approach to high-power energy harvesting. *Nat. Commun.* **2011**, *2*, 73–86. [CrossRef] [PubMed]
61. Wang, L.; Liu, J. Research advancement of liquid metal printed electronics ink (in Chinese). *Imaging Sci. Photochem.* **2014**, *32*, 382–392.
62. Zhang, Q.; Zheng, Y.; Liu, J. Direct writing of electronics based on alloy and metal (dream) ink: A newly emerging area and its impact on energy, environment and health sciences. *Front. Energy* **2012**, *6*, 311–340. [CrossRef]
63. Gao, Y.; Li, H.; Liu, J. Direct writing of flexible electronics through room temperature liquid metal ink. *PLoS ONE* **2012**, *7*, e45485. [CrossRef] [PubMed]
64. Wang, L.; Liu, J. Compatible hybrid 3D printing of metal and nonmetal inks for direct manufacture of end functional devices. *Sci. China Technol. Sci.* **2014**, *57*, 2089–2095. [CrossRef]
65. Tripathi, V.; Loh, Y.L. Thermal conductivity of a granular metal. *Phys. Rev. Lett.* **2006**, *96*, 046805. [CrossRef] [PubMed]
66. Gozen, B.A.; Tabatabai, A.; Ozdoganlar, O.B.; Majidi, C. High-density soft-matter electronics with micron-scale line width. *Adv. Mater.* **2014**, *26*, 5211–5216. [CrossRef] [PubMed]
67. Sivan, V.; Tang, S.-Y.; O'Mullane, A.P.; Petersen, P.; Eshtiaghi, N.; Kalantar-zadeh, K.; Mitchell, A. Liquid metal marbles. *Adv. Funct. Mater.* **2013**, *23*, 144–152. [CrossRef]
68. Du, K.; Glogowski, E.; Tuominen, M.T.; Emrick, T.; Russell, T.P.; Dinsmore, A.D. Self-assembly of gold nanoparticles on gallium droplets: Controlling charge transport through microscopic devices. *Langmuir* **2013**, *29*, 13640–13646. [CrossRef] [PubMed]
69. Zheng, Y.; He, Z.; Gao, Y.; Liu, J. Direct desktop printed-circuits-on-paper flexible electronics. *Sci. Rep.* **2013**, *3*, 1786. [CrossRef]
70. Wang, L.; Liu, J. Pressured liquid metal screen printing for rapid manufacture of high resolution electronic patterns. *RSC Adv.* **2015**, *5*, 57686–57691. [CrossRef]
71. Kim, D.; Lee, Y.; Lee, D.-W.; Choi, W.; Yoo, K.; Lee, J.-B. Hydrochloric acid-impregnated paper for gallium-based liquid metal microfluidics. *Sens. Actuators B Chem.* **2015**, *207*, 199–205. [CrossRef]
72. Kubo, M.; Li, X.; Kim, C.; Hashimoto, M.; Wiley, B.J.; Ham, D.; Whitesides, G.M. Stretchable microfluidic radiofrequency antennas. *Adv. Mater.* **2010**, *22*, 2749–2752. [CrossRef] [PubMed]
73. Boley, J.W.; White, E.L.; Chiu, G.T.C.; Kramer, R.K. Direct writing of gallium-indium alloy for stretchable electronics. *Adv. Funct. Mater.* **2014**, *24*, 3501–3507. [CrossRef]
74. Kim, D.; Thissen, P.; Viner, G.; Lee, D.-W.; Choi, W.; Chabal, Y.J.; Lee, J.-B. Recovery of nonwetting characteristics by surface modification of gallium-based liquid metal droplets using hydrochloric acid vapor. *ACS Appl. Mater. Interfaces* **2013**, *5*, 179–185. [CrossRef] [PubMed]
75. Zhang, Q.; Gao, Y.; Liu, J. Atomized spraying of liquid metal droplets on desired substrate surfaces as a generalized way for ubiquitous printed electronics. *Appl. Phys. A* **2014**, *116*, 1091–1097. [CrossRef]
76. Xu, Q.; Oudalov, N.; Guo, Q.; Jaeger, H.M.; Brown, E. Effect of oxidation on the mechanical properties of liquid gallium and eutectic gallium-indium. *Phys. Fluids* **2012**, *24*, 063101. [CrossRef]
77. Schrader, M.E. Young-dupre revisited. *Langmuir* **1995**, *11*, 3585–3589. [CrossRef]
78. Liu, Y.; Gao, M.; Mei, S.; Han, Y.; Liu, J. Ultra-compliant liquid metal electrodes with in-plane self-healing capability for dielectric elastomer actuators. *Appl. Phys. Lett.* **2013**, *103*, 064101. [CrossRef]
79. Tiberto, P.; Gupta, S.; Bianco, S.; Celegato, F.; Martino, P.; Chiolerio, A.; Tagliaferro, A.; Allia, P. Morphology and magnetic properties of island-like Co and Ni films obtained by de-wetting. *J. Nanopart. Res.* **2011**, *13*, 245–255. [CrossRef]
80. Doudrick, K.; Liu, S.; Mutunga, E.M.; Klein, K.L.; Damle, V.; Varanasi, K.K.; Rykaczewski, K. Different shades of oxide: From nanoscale wetting mechanisms to contact printing of gallium-based liquid metals. *Langmuir* **2014**, *30*, 6867–6877. [CrossRef] [PubMed]

81. Li, G.; Parmar, M.; Kim, D.; Lee, J.-B.; Lee, D.-W. Pdms based coplanar microfluidic channels for the surface reduction of oxidized galinstan. *Lab Chip* **2014**, *14*, 200–209. [CrossRef] [PubMed]
82. Furmidge, C.G.L. Studies at phase interfaces. I. The sliding of liquid drops on solid surfaces and a theory for spray retention. *J. Colloid Sci.* **1962**, *17*, 309–324. [CrossRef]
83. Kramer, R.K.; Boley, J.W.; Stone, H.A.; Weaver, J.C.; Wood, R.J. Effect of microtextured surface topography on the wetting behavior of eutectic gallium–indium alloys. *Langmuir* **2014**, *30*, 533–539. [CrossRef] [PubMed]
84. Yi, L.T. Injectable Bone Cement Based on Low-Melting-Point Alloy with Liquid-Solid Phase Switchable Capability. Ph.D. Thesis, Tsinghua University, Beijing, China, 2015.
85. Li, H.; Yang, Y.; Liu, J. Printable tiny thermocouple by liquid metal gallium and its matching metal. *Appl. Phys. Lett.* **2012**, *101*, 073511. [CrossRef]
86. Liu, J.; Li, H. A Liquid Metal Based Printed Circuit Board and Its Fabrication Method. China Patent 201110140156, 14 October 2011.
87. Tabatabai, A.; Fassler, A.; Usiak, C.; Majidi, C. Liquid-phase gallium–indium alloy electronics with microcontact printing. *Langmuir* **2013**, *29*, 6194–6200. [CrossRef] [PubMed]
88. Kramer, R.K.; Majidi, C.; Wood, R.J. Masked deposition of gallium-indium alloys for liquid-embedded elastomer conductors. *Adv. Funct. Mater.* **2013**, *23*, 5292–5296. [CrossRef]
89. Jeong, J.-A.; Kim, H.-K. Characteristics of inkjet-printed nano indium tin oxide particles for transparent conducting electrodes. *Curr. Appl. Phys.* **2010**, *10*, e105–e108. [CrossRef]
90. Sheng, L.; Teo, S.; Liu, J. Liquid-metal-painted stretchable capacitor sensors for wearable healthcare electronics. *J. Med. Biol. Eng.* **2016**, *36*, 265–272. [CrossRef]
91. Wang, L.; Liu, J. Printing low-melting-point alloy ink to directly make a solidified circuit or functional device with a heating pen. *Proc. R. Soc. A Math. Phys. Eng. Sci.* **2014**, *470*, 0609. [CrossRef] [PubMed]
92. Yang, J.; Liu, J. Direct printing and assembly of fm radio at the user end via liquid metal printer. *Circ. World* **2014**, *40*, 134–140. [CrossRef]
93. Zheng, Y.; Zhang, Q.; Liu, J. Pervasive liquid metal based direct writing electronics with roller-ball pen. *AIP Adv.* **2013**, *3*, 6459–6463. [CrossRef]
94. Kim, D.; Yoo, J.H.; Lee, Y.; Choi, W.; Yoo, K.; Lee, J.B.J. Gallium-Based Liquid Metal Inkjet Printing. In Proceedings of the 2014 IEEE 27th International Conference on Micro Electro Mechanical Systems (MEMS), San Francisco, CA, USA, 26–30 January 2014; pp. 967–970.
95. Yang, H.; Lightner, C.R.; Dong, L. Light-emitting coaxial nanofibers. *ACS Nano* **2012**, *6*, 622–628. [CrossRef] [PubMed]
96. Lu, T.; Finkenauer, L.; Wissman, J.; Majidi, C. Rapid prototyping for soft-matter electronics. *Adv. Funct. Mater.* **2014**, *24*, 3351–3356. [CrossRef]
97. Jeong, S.H.; Hjort, K.; Wu, Z. Tape transfer atomization patterning of liquid alloys for microfluidic stretchable wireless power transfer. *Sci. Rep.* **2015**, *5*, 8419. [CrossRef] [PubMed]
98. Boley, J.W.; White, E.L.; Kramer, R.K. Mechanically sintered gallium–indium nanoparticles. *Adv. Mater.* **2015**, *27*, 2355–2360. [CrossRef] [PubMed]
99. Kunnari, E.; Valkama, J.; Keskinen, M.; Mansikkamäki, P. Environmental evaluation of new technology: Printed electronics case study. *J. Clean. Prod.* **2009**, *17*, 791–799. [CrossRef]
100. Wang, L.; Liu, J. Liquid phase 3D printing for quickly manufacturing conductive metal objects with low melting point alloy ink. *Sci. China Technol. Sci.* **2014**, *57*, 1721–1728. [CrossRef]
101. Fassler, A.; Majidi, C. 3D structures of liquid-phase gain alloy embedded in pdms with freeze casting. *Lab Chip* **2013**, *13*, 4442–4450. [CrossRef] [PubMed]
102. Ladd, C.; So, J.-H.; Muth, J.; Dickey, M.D. 3D printing of free standing liquid metal microstructures. *Adv. Mater.* **2013**, *25*, 5081–5085. [CrossRef] [PubMed]
103. Liu, R.; Yang, X.; Jin, C.; Fu, J.; Chen, W.; Liu, J. Development of three-dimension microelectrode array for bioelectric measurement using the liquidmetal-micromolding technique. *Appl. Phys. Lett.* **2013**, *103*, 193701. [CrossRef]
104. Deng, Y.; Liu, J. Flexible mechanical joint as human exoskeleton using low-melting-point alloy. *J. Med. Devices* **2014**, *8*, 044506. [CrossRef]
105. Li, J.; Guo, C.; Wang, Z.; Gao, K.; Shi, X.; Liu, J. Electrical stimulation towards melanoma therapy via liquid metal printed electronics on skin. *Clin. Transl. Med.* **2016**, *5*, 1–7. [CrossRef] [PubMed]

106. Wang, L.; Liu, J. Ink spraying based liquid metal printed electronics for directly making smart home appliances. *ECS J. Solid State Sci. Technol.* **2015**, *4*, 3057–3062. [CrossRef]
107. Fassler, A.; Majidi, C. Soft-matter capacitors and inductors for hyperelastic strain sensing and stretchable electronics. *Smart Mater. Struct.* **2013**, *22*, 955–966. [CrossRef]
108. Cheng, S.; Wu, Z. Microfluidic electronics. *Lab Chip* **2012**, *12*, 2782–2791. [CrossRef] [PubMed]
109. Cheng, S.; Wu, Z. A microfluidic, reversibly stretchable, large-area wireless strain sensor. *Adv. Funct. Mater.* **2011**, *21*, 2282–2290. [CrossRef]
110. Wu, Z.; Hjort, K.; Jeong, S.H. Microfluidic stretchable radio-frequency devices. *Proc. IEEE* **2015**, *103*, 1211–1225. [CrossRef]
111. Jeong, S.; Hjort, K.; Wu, Z. Tape transfer printing of a liquid metal alloy for stretchable RF electronics. *Sensors* **2014**, *14*, 16311–16321. [CrossRef] [PubMed]
112. Jin, S.W.; Park, J.; Hong, S.Y.; Park, H.; Jeong, Y.R.; Park, J.; Lee, S.-S.; Ha, J.S. Stretchable loudspeaker using liquid metal microchannel. *Sci. Rep.* **2015**, *5*, 11695. [CrossRef] [PubMed]
113. Li, G.; Wu, X.; Lee, D.-W. A galinstan-based inkjet printing system for highly stretchable electronics with self-healing capability. *Lab Chip* **2016**, *16*, 1366–1373. [CrossRef] [PubMed]
114. Park, Y.-L.; Majidi, C.; Kramer, R.; Berard, P.; Wood, R.J. Hyperelastic pressure sensing with a liquid-embedded elastomer. *J. Micromech. Microeng.* **2010**, *20*, 125029–125034. [CrossRef]

7

Large-Area Compatible Laser Sintering Schemes with a Spatially Extended Focused Beam

Habeom Lee [1,†], Jinhyeong Kwon [1,†], Woo Seop Shin [2], Hyeon Rack Kim [2], Jaeho Shin [1], Hyunmin Cho [1], Seungyong Han [3], Junyeob Yeo [4,*] and Sukjoon Hong [2,*]

1 Applied Nano and Thermal Science Lab, Department of Mechanical Engineering, Seoul National University, 1 Gwanak-ro, Gwanak-gu, Seoul 151-742, Korea; habeom.lee@snu.ac.kr (H.L.); jhs0909k@snu.ac.kr (J.K.); jayz.shin84@gmail.com (J.S.); augustinus310@snu.ac.kr (H.C.)

2 Department of Mechanical Engineering, Hanyang University, 55 Hanyangdaehak-ro, Sangnok-gu, Ansan Gyeonggi-do 15588, Korea; caribou11@hanyang.ac.kr (W.S.S.); kihll1004@hanyang.ac.kr (H.R.K.)

3 Department of Mechanical Engineering, Ajou University, San 5, Woncheon-Dong, Yeongtong-Gu, Suwon 16499, Korea; sy84han@ajou.ac.kr

4 Novel Applied Nano Optics (NANO) Lab, Department of Physics, Kyungpook National University, 80 Daehak-ro, Bukgu, Daegu 41566, Korea

* Correspondence: junyeob@knu.ac.kr (J.Y.); sukjoonhong@hanyang.ac.kr (S.H.)

† These authors contributed equally to this work.

Academic Editor: Zhigang Wu

Abstract: Selective laser sintering enables the facile production of metal nanoparticle-based conductive layers on flexible substrates, but its application towards large-area electronics has remained questionable due to the limited throughput of the laser process that originates from the direct writing nature. In this study, modified optical schemes are introduced for the fabrication of (1) a densely patterned conductive layer and (2) a thin-film conductive layer without any patterns. In detail, a focusing lens is substituted by a micro lens array or a cylindrical lens to generate multiple beamlets or an extended focal line. The modified optical settings are found to be advantageous for the creation of repetitive conducting patterns or areal sintering of the silver nanoparticle ink layer. It is further confirmed that these optical schemes are equally compatible with plastic substrates for its application towards large-area flexible electronics.

Keywords: laser sintering; metal nanoparticle; micro lens array; cylindrical lens; flexible electrode

---

1. Introduction

A metal electrode layer, either patterned or in a thin-film form, is an inevitable component of an electronic device. Fabrication of a metal electrode layer based on conventional photolithographic means with vacuum evaporation techniques has achieved tremendous success with silicon-based electronics thus far, however, it has often been found to be improper for flexible and stretchable electronics. New types of conductive layers based on metal nanoparticle (NP) ink have provided a possible solution to this problem. As the melting temperature of metal NPs decreases according to its diameter due to the size effect, [1,2] a metal electrode layer is easily created on an arbitrary substrate at a mild temperature in ambient conditions through the direct coating and sintering of metal NP ink. However, despite its advantages, the minimum feature size of the resultant electrode prepared by metal NPs was relatively coarse due to the limited resolution of the conventional printing techniques which were employed for the selective deposition of metal NP ink [3]. At the same time, the sintering scheme remained questionable for a number of heat-sensitive substrates, as it was experimentally verified that the conventional sintering step with substantial sintering time can damage the underlying substrate [4].

As an alternative to the conventional sintering scheme, selective laser sintering of metal NPs was introduced to conduct patterning and sintering of metal NP ink simultaneously using a focused laser beam [5,6]. The focused laser beam is employed as a localized heat source based on the photothermal reaction, [7] and the area subject to heating is mainly determined by the spatial intensity distribution of the focused laser beam. Yeo et al. [8] demonstrated that the feature size of the resultant electrode can be easily reduced down to several micrometers using a laser beam focused by a telecentric lens module, which is difficult to achieve with other printing techniques. In addition, it was confirmed that the thermal damage on the underlying substrate can be minimized owing to the reduced heat-affected zone generated by a focused laser. These results suggest that laser sintering can be a convenient technique for the creation of metal patterns on flexible substrates. At the early stage, noble metal NPs such as gold (Au) [5,6,9–15] and silver (Ag) NP ink [1,2,4,8,16–23] were employed as the target materials for laser sintering, but recent studies demonstrate that the application of laser sintering can be extended to other oxidation-sensitive metal NPs such as copper (Cu) [24–26], even in ambient conditions, by reducing the local heating time through rapid scanning of the focused laser beam.

Laser sintering, however, is often considered to be inappropriate for mass production since it is a direct writing method in principle. As the electrode pattern becomes denser, the time required for the scanning increases linearly. At the same time, a thin-film electrode is another form of electrode that is difficult to be manufactured by the laser sintering method. Laser sintered thin-film electrodes not only require raster scanning of the entire area, but also often show imperfect electrical conductivity due to the discontinuities in electrical path originated from separate scanning steps. In this study, we introduce extended laser sintering schemes with a spatially modified focused beam to assist in the facile production of the electrodes with denser patterns or thin-film metal layers. In detail, a focusing lens is substituted by a micro lens array (MLA) and a cylindrical lens, designated for the fabrication of densely patterned electrodes and thin-film metal layers, respectively. It is also confirmed that the proposed optical schemes are still compatible with plastic substrates for their application in large-area flexible electronics.

2. Materials and Methods

Ag NP ink is firstly synthesized with the two-phase method, as reported in previous studies [8,17]. The resultant Ag NP is ~5 nm in diameter and encapsulated with self-assembled monolayer (SAM) to prevent agglomeration between NPs. It is confirmed that the Ag NP ink experiences melting when the temperature reaches ~150 °C, which is significantly lower than its bulk counterpart, due to the melting point depression from the size effect [2]. The synthesized Ag NP ink is coated on arbitrary substrates by spin-coating at 1000 rpm, and dried in a convection oven at 70 °C for 3 min to evaporate the excessive solvent. The target substrate can be either a rigid substrate such as a silicon wafer or a slide glass, or a flexible substrate. Polyimide (PI) thin film with a thickness of 150 μm is selected as the flexible substrate throughout the study.

A thin-film composed of Ag NPs is prepared on the substrate after the coating and drying steps, as shown in Figure 1a. Although Ag NPs are closely packed together, the as-prepared NP ink layer does not exhibit good electrical conductivity since the Ag NPs exist as separate entities. These Ag NPs can be transformed into a continuous conductive layer once a focused laser beam is scanned along the designated path, as shown in Figure 1b. The local temperature of the area subject to the laser irradiation increases rapidly and the Ag NPs experience melting and solidification steps as a consequence. The Ag NPs after the laser scanning subsequently show different physical properties from the as-deposited area, such as reflective color and high electrical conductivity.

An optical system is required to focus the laser beam on a designated spot, and an objective lens is frequently employed for the focusing, as shown in Figure 2a. The scanning path is controlled by a programmable motorized stage. Instead of moving the stage, a galvano-mirror together with a telecentric lens can be employed to achieve fast scanning of the laser beam over a large area [8,17]. Together with single focusing scheme, an MLA and cylindrical lens have been exploited in this study as new optical schemes for large-area compatible laser sintering, as depicted in Figure 2b,c. For single focusing, a green wavelength continuous wave laser (Millenia V, Spectra-Physics, Santa Clara, CA, USA) is scanned by a 2D galvano-mirror scanning system (hurrySCAN II, Scanlab GmbH, Puchheim, Germany) while the laser is focused by an f-theta telecentric lens with f = 100 mm. The laser scanner system is controlled by a computer with CAD software (SAMLight, SCAPS GmbH, Oberhaching, Germany) to draw arbitrary patterns. For the generation of multiple beamlets, an MLA with 400–900 nm anti-reflective coating is employed together with a 5× objective lens. More detailed information on the optical system is included in the Results and Discussions section. The MLA has a pitch of 300 μm and a focal length of 18.6 mm. For a focal line, the MLA is replaced by a plano-convex cylindrical lens with f = 50 mm. The laser scanning speed is fixed at 5 mm/s in every case.

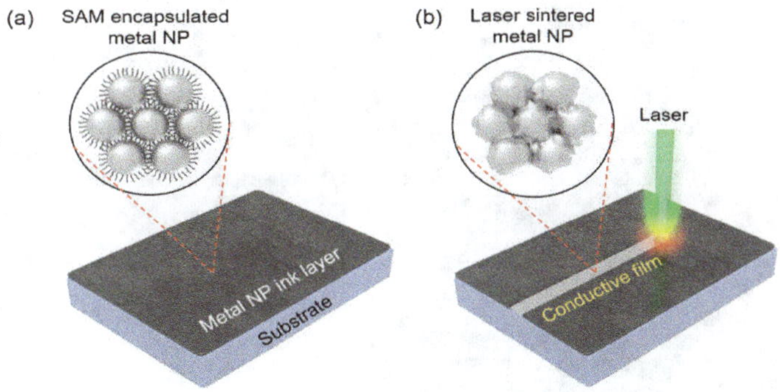

Figure 1. Schematic illustration of metal nanoparticle (NP) ink layer on a substrate (**a**) before and (**b**) after laser sintering.

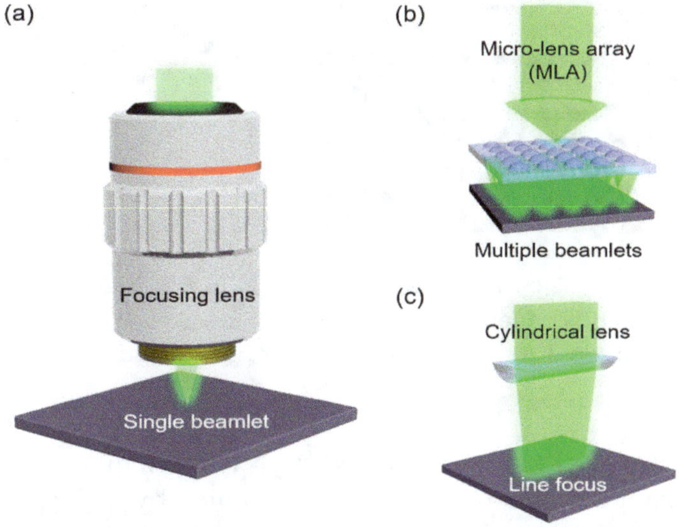

Figure 2. Optical schemes for selective laser sintering with (**a**) a single beamlet, (**b**) multiple beamlets, and (**c**) a focal line.

3. Results and Discussion

Figure 3 shows the sintering results created by the 2D galvano-mirror scanning system with the single focusing beamlet on a glass substrate in terms of their optical, scanning electron microscope (SEM) and atomic force microscopy (AFM) images. More information on the optical setting can be found in the previous studies [8,17]. Laser sintering is firstly conducted on an Ag NP ink layer according to the irradiated laser power. Figure 3a shows the optical microscope image of the Ag NP ink layer after the scanning, whereas the left and the right columns are the bright and dark field images of the same region, respectively. From the bright field image, it is noticeable that the optical transmission starts to change at the laser power of ~20 mW, while the dark field image shows no difference until the power reaches ~60 mW. Electrical measurement reveals that the resultant conductor line does not exhibit substantial electrical conductivity until it becomes reflective in the dark field image. We therefore anticipate that the change in optical transmission in the bright field image is due to the evaporation of the trapped solvent, or from the incomplete sintering between a small fraction of Ag NPs. The width of the area affected by the laser sintering is slightly bigger in the bright field image, and it is highly dependent to the drying condition.

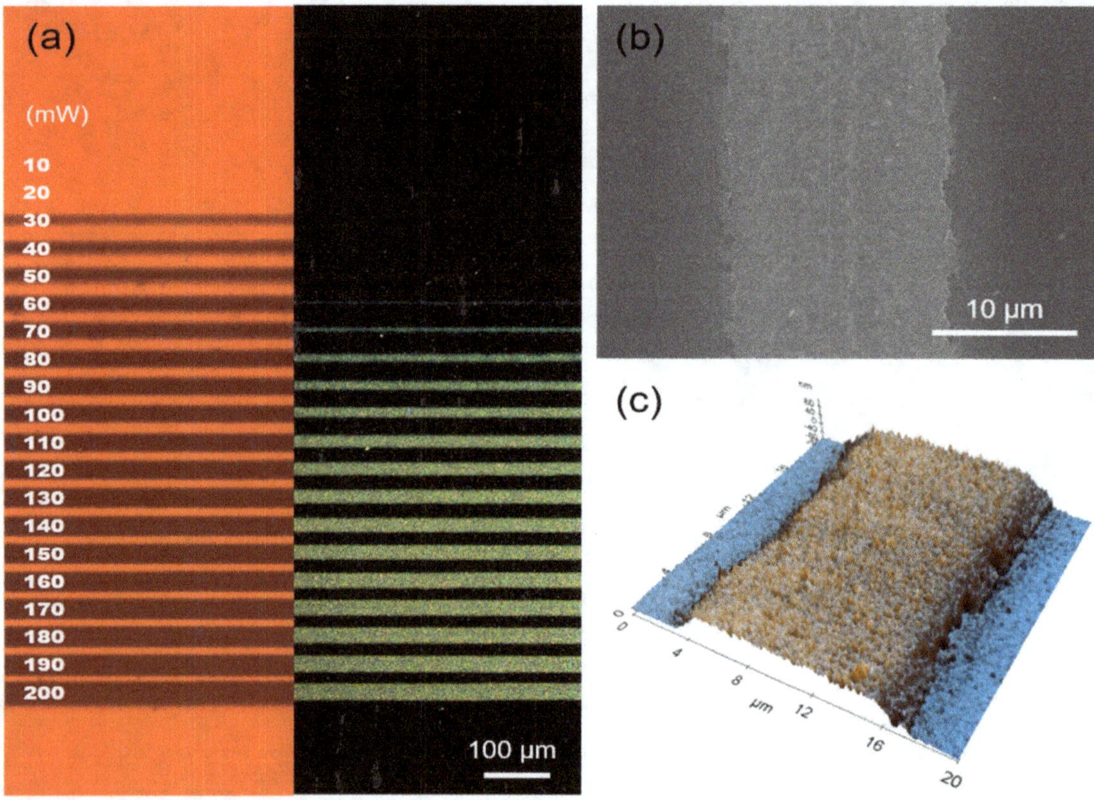

Figure 3. Laser sintering results by a single beamlet. (**a**) Optical images of Ag NP ink sintered at different laser power. (Left) Bright field; (Right) Dark field images. (**b**) Scanning electron microscope (SEM) image of the laser-sintered Ag NP at 100 mW laser power, and (**c**) its atomic force microscopy (AFM) image.

Figure 3b is the SEM image of the laser-sintered Ag NP at 100 mW laser power after the removal of remaining Ag NP ink from the substrate. The area irradiated by the laser remains on the substrate as a thin electrode while the other Ag NPs are washed away. More detailed information about its 3D morphology can be confirmed from its AFM image in Figure 3c. The height of the resultant conductor is ~ 120 nm, and this value can be further controlled by changing the wet processing conditions including spin-coating speed and drying conditions [17].

In order to generate a spatially extended focused beam, we first modified the optical setting, as shown in Figure 4a. Since our objective is to split the incident beam into multiple beamlets or to extend the focus into a focal line, incident laser intensity becomes an important issue. Two achromatic lenses with different focal lengths have been added in a Keplerian telescope configuration in order to reduce the beam size for sufficient laser intensity. At the same time, two adjustable slits are installed in the x- and y- directions to cut the incident laser beam, since a flat-top intensity profile is demanded in large-area applications instead of a TEM_{00} (Fundamental Transverse Mode) Gaussian beam to ensure spatial uniformity in processing. The truncated laser beam shows a relatively flat-top intensity profile, but it is not quite perfect and other optical components such as a homogenizer will be required for further advancement of the proposed technique [18].

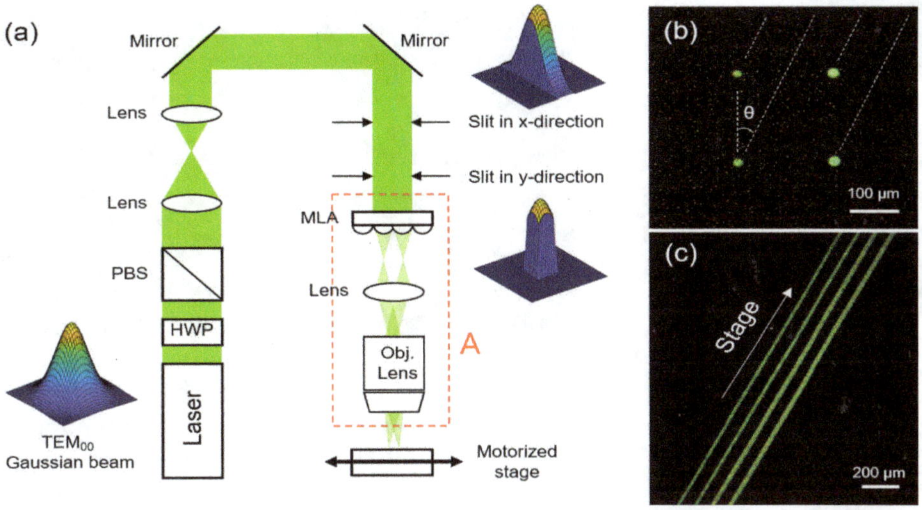

Figure 4. Laser sintering results by multiple beamlets. (**a**) Optical setting for multiple beamlets and a focal line (HWP: Half-wave plate, PBS: polarized beam splitter, MLA: Micro lens array). (**b**) Optical image of Ag NP ink by sintering with the multiple beamlets at a single exposure and (**c**) after moving the stage at a slanted angle.

Figure 4b is the optical image of the Ag NP ink layer after a single exposure to the multiple beamlets. The multiple beamlets are generated by opening each slit at a width of 600 μm, so that the incident beam covers $2 \times 2 = 4$ cells in the 300-μm pitch MLA. Different from the single beamlet case, four distinct spots are transformed into the conductive electrode with a single exposure as confirmed from the reflective color on the dark field optical image. Their relative positions can be further controlled by changing the parameters in the optical setting. Using the multiple beamlets, repetitive patterns can be easily created. As a representative example, parallel conductive lines, which are basic components for various applications including wire-grid polarizer and grid-type transparent conductors, [17] are created by moving the stage at a slanted angle. Figure 4c shows the resultant parallel lines produced by a single translational movement of the motorized stage.

For the sintering of metal NP ink over the entire area as a thin-film conductor without any patterns, a focal line is more suitable compared to the multiple beamlets. The optical components in Figure 4a (denoted as "A") are replaced by a cylindrical lens to create a focal line. In the current configuration, slit width in the y- direction (L_y) determines the laser intensity at the focus, while the slit width in the x-direction (L_x) controls the length of the focal line, as shown in Figure 5a. In this experiment, L_y has been fixed at ~300 μm throughout the study, while L_x is altered from 50 μm to 250 μm. Although the intensity of the incoming laser remains the same, it is found that the sintering characteristics of the resultant Ag NP ink are different in each case, as shown in Figure 5b. We estimate that the inconsistency comes from the different heat transfer conditions. For an extended focal line, an equal amount of heat is generated along the line and hence the major heat dissipation should occur towards a perpendicular direction to the focal line. However, as the length of the focal line becomes smaller, heat starts to dissipate in every lateral direction so that the resultant temperature reached by a shorter focal line should be lower at $L_x = 50$ μm. Figure 5b shows that the Ag NP ink at a width of >150 μm is successfully sintered by a single scanning with the cylindrical lens. It is worth mentioning that the resultant conductive layer does not have apparent boundaries which can be found in the thin-film conductors produced by raster scanning of a single beamlet, as shown in the inset.

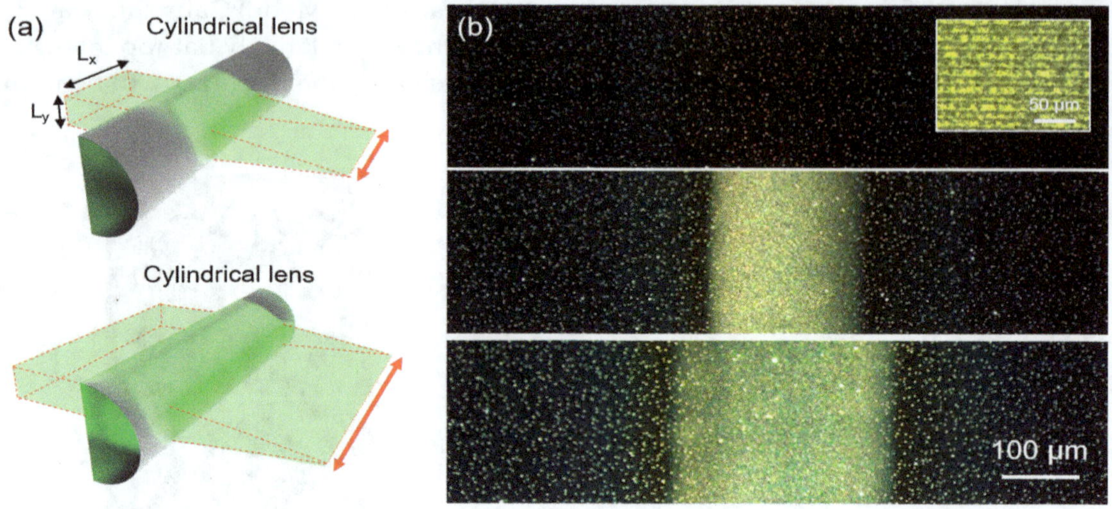

Figure 5. (**a**) Schematic illustration of the laser sintering by a focal line. (**b**) The optical image of the resultant Ag NP ink after scanning at (Top) $L_x = 50$ μm, (Middle) $L_x = 150$ μm and (Bottom) $L_x = 250$ μm (Inset: Ag NP thin-film conductor created by raster scanning of a single beamlet).

In order to verify that the proposed optical schemes are equally compatible with flexible substrates, parallel Ag NP lines are created on a flexible PI substrate with the modified optical setting. It is apparent from Figure 6a,b that consistent lines are created on the PI substrate without any apparent thermal damage. The current-voltage (IV) curve of the resultant conductor is measured to ensure that their electrical characteristics are suitable for the application of flexible electronics. The IV curve in Figure 6c shows that the sintered lines exhibit ohmic behavior and the calculated resistivity is estimated to be 7~8 times higher than bulk Ag, while the as-prepared NP film displays insignificant electrical conductivity.

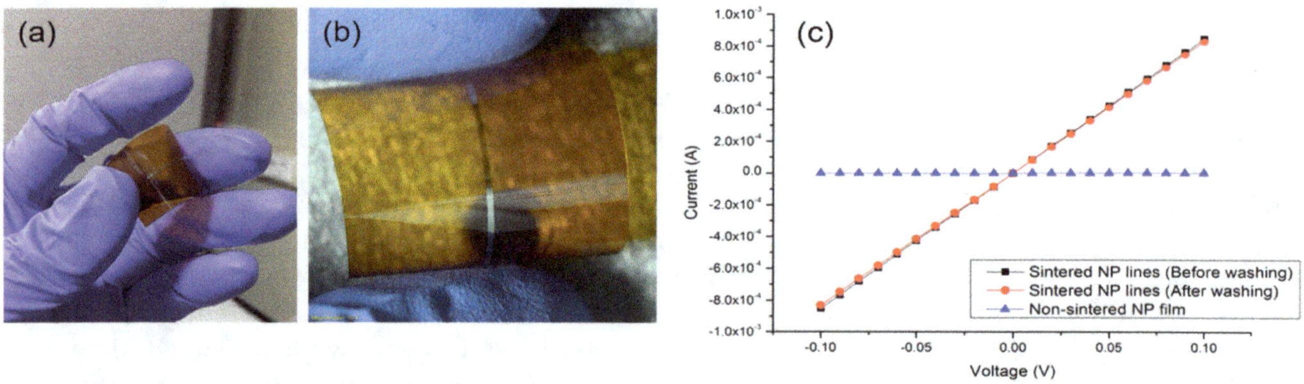

Figure 6. Application of the proposed optical schemes for laser sintering of Ag NP ink on a flexible substrate. Optical image at (**a**) low magnification and (**b**) high magnification; (**c**) current-voltage (IV) curve of the resultant Ag electrode on the flexible substrate.

In summary, we propose two forms of a spatially extended focused beam intended for the fabrication of Ag NP conducting layers with dense patterns or in thin-film configuration. It is demonstrated that the modified optical setting can produce multiple beamlets for repetitive patterns or focal lines for thin-film conductors. Since these optical schemes are equally compatible with flexible substrates, it is expected that the proposed techniques will supplement fabrication of large-area flexible electronics.

Acknowledgments: This work is supported by the National Research Foundation of Korea (NRF) Grant funded through the Basic Science Research Program (NRF-2016R1C1B1014729) and the research fund of Hanyang University (HY-2016-00000002097).

Author Contributions: S.H. and J.Y. designed the experiments; J.K. and H.L. synthesized the material; J.K., H.L. and S.H. conducted the experiments; W.S.S. and H.R.K. assisted in analyzing the experimental result; J.S. and H.C. built the optical setting; S.H., J.Y., J.H. and H.L. wrote the manuscript.

References

1. Son, Y.; Yeo, J.; Ha, C.W.; Lee, J.; Hong, S.; Nam, K.H.; Yang, D.-Y.; Ko, S.H. Application of the specific thermal properties of Ag nanoparticles to high-resolution metal patterning. *Thermochim. Acta* **2012**, *542*, 52–56. [CrossRef]
2. Son, Y.; Yeo, J.; Moon, H.; Lim, T.W.; Hong, S.; Nam, K.H.; Yoo, S.; Grigoropoulos, C.P.; Yang, D.-Y.; Ko, S.H. Nanoscale electronics: Digital fabrication by direct femtosecond laser processing of metal nanoparticles. *Adv. Mater.* **2011**, *23*, 3176–3181. [CrossRef] [PubMed]
3. Cummins, G.; Desmulliez, M.P.Y. Inkjet printing of conductive materials: A review. *Circuit World* **2012**, *38*, 193–213. [CrossRef]
4. Yeo, J.; Kim, G.; Hong, S.; Kim, M.S.; Kim, D.; Lee, J.; Lee, H.B.; Kwon, J.; Suh, Y.D.; Kang, H.W.; et al. Flexible supercapacitor fabrication by room temperature rapid laser processing of roll-to-roll printed metal nanoparticle ink for wearable electronics application. *J. Power Sources* **2014**, *246*, 562–568. [CrossRef]
5. Chung, J.; Bieri, N.R.; Ko, S.; Grigoropoulos, C.P.; Poulikakos, D. In-tandem deposition and sintering of printed gold nanoparticle inks induced by continuous Gaussian laser irradiation. *Appl. Phys. A* **2004**, *79*, 1259–1261. [CrossRef]
6. Bieri, N.R.; Chung, J.; Haferl, S.E.; Poulikakos, D.; Grigoropoulos, C.P. Microstructuring by printing and laser curing of nanoparticle solutions. *Appl. Phys. Lett.* **2003**, *82*, 3529–3531. [CrossRef]
7. Bäuerle, D.W. *Laser Processing and Chemistry*; Springer Science & Business Media: Berlin, Germany, 2013.
8. Yeo, J.; Hong, S.; Lee, D.; Hotz, N.; Lee, M.-T.; Grigoropoulos, C.P.; Ko, S.H. Next generation non-vacuum, maskless, low temperature nanoparticle ink laser digital direct metal patterning for a large area flexible electronics. *PLoS ONE* **2012**, *7*, e42315. [CrossRef] [PubMed]
9. Choi, T.Y.; Poulikakos, D.; Grigoropoulos, C.P. Fountain-pen-based laser microstructuring with gold nanoparticle inks. *Appl. Phys. Lett.* **2004**, *85*, 13–15. [CrossRef]
10. Chung, J.; Ko, S.; Bieri, N.R.; Grigoropoulos, C.P.; Poulikakos, D. Conductor microstructures by laser curing of printed gold nanoparticle ink. *Appl. Phys. Lett.* **2004**, *84*, 801–803. [CrossRef]
11. Ko, S.H.; Pan, H.; Grigoropoulos, C.P.; Luscombe, C.K.; Fréchet, J.M.J.; Poulikakos, D. Air stable high resolution organic transistors by selective laser sintering of ink-jet printed metal nanoparticles. *Appl. Phys. Lett.* **2007**, *90*, 141103. [CrossRef]
12. Chung, J.; Ko, S.; Grigoropoulos, C.P.; Bieri, N.R.; Dockendorf, C.; Poulikakos, D. Damage-free low temperature pulsed laser printing of gold nanoinks on polymers. *J. Heat Transf.* **2005**, *127*, 724–732. [CrossRef]
13. Ko, S.H.; Pan, H.; Grigoropoulos, C.P.; Luscombe, C.K.; Fréchet, J.M.; Poulikakos, D. All-inkjet-printed flexible electronics fabrication on a polymer substrate by low-temperature high-resolution selective laser sintering of metal nanoparticles. *Nanotechnology* **2007**, *18*, 345202. [CrossRef]
14. Ko, S.H.; Chung, J.; Pan, H.; Grigoropoulos, C.P.; Poulikakos, D. Fabrication of multilayer passive and active electric components on polymer using inkjet printing and low temperature laser processing. *Sens. Actuators A Phys.* **2007**, *134*, 161–168. [CrossRef]
15. Pan, H.; Hwang, D.J.; Ko, S.H.; Clem, T.A.; Fréchet, J.M.J.; Bäuerle, D.; Grigoropoulos, C.P. High-throughput near-field optical nanoprocessing of solution-deposited nanoparticles. *Small* **2010**, *6*, 1812–1821. [CrossRef] [PubMed]
16. An, K.; Hong, S.; Han, S.; Lee, H.; Yeo, J.; Ko, S.H. Selective sintering of metal nanoparticle ink for maskless fabrication of an electrode micropattern using a spatially modulated laser beam by a digital micromirror device. *ACS Appl. Mater. Interfaces* **2014**, *6*, 2786–2790. [CrossRef] [PubMed]
17. Hong, S.; Yeo, J.; Kim, G.; Kim, D.; Lee, H.; Kwon, J.; Lee, H.; Lee, P.; Ko, S.H. Nonvacuum, maskless fabrication of a flexible metal grid transparent conductor by low-temperature selective laser sintering of nanoparticle ink. *ACS Nano* **2013**, *7*, 5024–5031. [CrossRef] [PubMed]

18. Ko, S.H.; Pan, H.; Ryu, S.G.; Misra, N.; Grigoropoulos, C.P.; Park, H.K. Nanomaterial enabled laser transfer for organic light emitting material direct writing. *Appl. Phys. Lett.* **2008**, *93*, 151110. [CrossRef]
19. Son, Y.; Yeo, J.; Ha, C.W.; Hong, S.; Ko, S.H.; Yang, D.-Y. Fabrication of submicron-sized metal patterns on a flexible polymer substrate by femtosecond laser sintering of metal nanoparticles. *Int. J. Nanomanuf.* **2013**, *9*, 468–476. [CrossRef]
20. Lee, H.; Hong, S.; Kwon, J.; Suh, Y.D.; Lee, J.; Moon, H.; Yeo, J.; Ko, S.H. All-solid-state flexible supercapacitors by fast laser annealing of printed metal nanoparticle layers. *J. Mater. Chem. A* **2015**, *3*, 8339–8345. [CrossRef]
21. Suh, Y.D.; Jung, J.; Lee, H.; Yeo, J.; Hong, S.; Lee, P.; Lee, D.; Ko, S.H. Nanowire reinforced nanoparticle nanocomposite for highly flexible transparent electrodes: Borrowing ideas from macrocomposites in steel-wire reinforced concrete. *J. Mater. Chem. C* **2017**, *5*, 791–798. [CrossRef]
22. Lee, M.-T.; Lee, D.; Sherry, A.; Grigoropoulos, C.P. Rapid selective metal patterning on polydimethylsiloxane (PDMS) fabricated by capillarity-assisted laser direct write. *J. Micromech. Microeng.* **2011**, *21*, 095018. [CrossRef]
23. Yang, M.; Chon, M.-W.; Kim, J.-H.; Lee, S.-H.; Jo, J.; Yeo, J.; Ko, S.H.; Choa, S.-H. Mechanical and environmental durability of roll-to-roll printed silver nanoparticle film using a rapid laser annealing process for flexible electronics. *Microelectron. Reliab.* **2014**, *54*, 2871–2880. [CrossRef]
24. Kwon, J.; Cho, H.; Eom, H.; Lee, H.; Suh, Y.D.; Moon, H.; Shin, J.; Hong, S.; Ko, S.H. Low-temperature oxidation-free selective laser sintering of cu nanoparticle paste on a polymer substrate for the flexible touch panel applications. *ACS Appl. Mater. Interfaces* **2016**, *8*, 11575–11582. [CrossRef] [PubMed]
25. Suh, Y.D.; Kwon, J.; Lee, J.; Lee, H.; Jeong, S.; Kim, D.; Cho, H.; Yeo, J.; Ko, S.H. Maskless fabrication of highly robust, flexible transparent Cu conductor by random crack network assisted Cu nanoparticle patterning and laser sintering. *Adv. Electron. Mater.* **2016**, *2*. [CrossRef]
26. Michael, Z.; Oleg, E.; Amir, S.; Zvi, K. Laser sintering of copper nanoparticles. *J. Phys. D Appl. Phys.* **2014**, *47*, 025501.

8

Tunnel Encapsulation Technology for Durability Improvement in Stretchable Electronics Fabrication

Kangmin Leng [1], Chuanfei Guo [2], Kang Wu [1,*] and Zhigang Wu [1,*]

1 State Key Laboratory of Digital Manufacturing Equipment and Technology, Huazhong University of Science and Technology, Wuhan 430074, China; lengkangmin@hust.edu.cn

2 Department of Materials Science and Engineering, Southern University of Science and Technology, Shenzhen 518055, China; guocf@sustc.edu.cn

* Correspondence: wuk16@hust.edu.cn (K.W.); zgwu@hust.edu.cn (Z.W.)

Abstract: Great diversity of process technologies and materials have been developed around stretchable electronics. A subset of them, which are made up of zigzag metal foil and soft silicon polymers, show advantages of being easy to manufacture and low cost. However, most of the circuits lack durability due to stress concentration of interconnects entirely embedded in elastic polymer silicone such as polydimethylsiloxane (PDMS). In our demonstration, tunnel encapsulation technology was introduced to relieve stress of these conductors when they were stretched to deform in and out of plane. It was realized by dissolving the medium of Polyvinyl Alcohol (PVA), previous cured together with circuits in polymer, to form the micro-tunnel which not only guarantee the stretchability of interconnect, but also help to improve the durability. With the protection of tunnel, the serpentine could stably maintain the designed shape and electrical performance after 50% strain cycling over 20,000 times. Finally, different materials for encapsulation were employed to provide promising options for applications in portable biomedical devices which demand duplicate distortion.

Keywords: stretchable electronics; tunnel encapsulation; Polyvinyl Alcohol; durability

---

1. Introduction

With a significant development of fabrication processes and patterning technologies, stretchable electronic devices, based on inorganic materials, have achieved high performance in bioinspired and biointegrated systems in the past few years [1–3]. To achieve conformal contact with human skin and maintain stable electrical performance, these devices need to endure strain and stress from many repetitive deformation movements in or out of plane. Special designed structures such as zigzag and serpentine interconnects [4–6] provide the allowance for deformation. This brings the fabrication feasibility for stretchable electronics based on thin inorganic materials [7–9]. Moreover, the prestrain strategy [10] has been reported extensively. It significantly increased the stretchability of interconnects of different shapes for biomedical and healthcare applications has been reported in recent years [11]. Among those studies, very thin metal wires were patterned on the soft substrates through evaporation or transfer printing [12,13]. The finished wires and substrates were encapsulated by pouring a thin soft polymer directly to form a protection layer [5,14,15]. During these processes, encapsulation technologies are closely related to the lifetime and durability when these devices are up against mechanical damage in practical application. The differences of modulus and morphology between metal wires and polymer matrix bring problems when deformation happens. For example, the out-of-plane wavy and 3D buckling deformation for metal wires are constrained by those closely embedded polymers as there are no spaces for these movements [6,14]. They come out of the reducing of desirable stretchability and durability of the systems. Recently, an option involving microfluidic suspensions in thin elastomeric enclosures at system level was proposed [16]. However, this technology was faced with fluid leakage and shape distortion under large deformation. However, research focusing on the stretchability and durability of interconnects with encapsulation needs further supplementary research and perfection [17–19].

Here, we present a new encapsulation technology for the life time improvement of stretchable interconnects. A three-dimensional tunnel space was made to enable slipping and buckling for deformed interconnects. We describe the newly invented encapsulation technology of embedding the serpentine interconnect in three-dimensional tunnel, called tunnel encapsulation. In this study, the micro-tunnel space was realized by dissolving PVA, previous cured together with interconnects in polymer. Nevertheless, patterning the conductive metal films on PVA is a challenge. Previous studies reported some techniques including lithography and ion etching thin films onto PVA [20–22]. These technologies provide patterned films with high accuracy but there are problems such as complicated processes and high cost associated with cleanroom facilities. In addition, greater demands were being placed on the processing equipment, the operator and operating environment. In our technology, metal film and PVA film were directly patterned by laser processing and then bonding with each other for the stickiness of wetting PVA. Experimental studies and finite element analyses (FEA) revealed the important features of our technology and their dependence on key design parameters to predict the durability of the electronic components for a periodic deformation. Different interconnect patterns and polymer matrix materials were applied to evaluate the universality of this new encapsulation and it was proved to be valid. Remarkably, even when stretched with 50% strain, the electrode was still able to recover its original shape and maintain good electrical conductivity. An appropriate dimension and lubricating fluid for the tunnel which enclose the interconnects are critical. In particular, fixing-pillar was proposed to strengthen local stiffness in the arc regions of the serpentine, which could strongly enhance the lifetime of the interconnects. Overall, we provided experimental and simulation results that can fabricate a high-performance electronic device in encapsulation with excellent stretchability and durability. The fundamental difference between the direct encapsulation and tunnel encapsulation is the introduction of the three-dimensional tunnel. For direct encapsulation, interconnect embedded in polymer will encounter severe stress concentration when stretched. The severe stress concentration results in the limited stretchability and lifetime of the interconnect. For tunnel encapsulation, stress concentration is alleviated when stretched at the same degree of elongation for the introduction of tunnel space. Meanwhile, the interconnect would slide and buckle in the tunnel to release stress. With resizing the tunnel, the maximum stretchability and fatigue performance of interconnect can be modified. Therefore, the tunnel encapsulation helps to improve the maximum stretchability and lifetime of the system.

2. Materials and Technologies

2.1. Preparation of PVA Membrane

A 9:1 mixture of deionized water and granular Polyvinyl Alcohol (PVA) (87.0∼89.0% hydrolyzed, Mw 130,000 g/mol, Shanghai Macklin Biochemical Co., Ltd., Shanghai, China) was heated in a glass beaker at 75 °C for 1 h with stirring, Then, the mixture was kept at room temperature (25 °C) with vigorous stirring for about 30 min to obtain 10 wt.% PVA aqueous solution. We fabricated pristine PVA membranes by casting the 10 wt.% PVA aqueous solution onto a culture dish and drying the solutions at 75 °C for 2 h.

2.2. Preparation of PDMS

Poly-dimethyl siloxane (PDMS) (Sylgard 186, Dow Corning (Midland, MI, USA)) was chosen as the elastomeric substrate to carry the patterned metal on top. The two-part liquid components (PDMS base and curing agent) were mixed with weight ratio of 10:1 in a plastic cup successively and mixed together manually for 5 min with a glass rod. Prior to putting into use, vacuumization (10 min) and refrigeration (about −18 °C for 1 h) were used to eliminate the air bubbles.

2.3. Fabrication of Stretchable Electronics in Tunnel Encapsulation

A schematic of the tunnel encapsulation process is shown in Figure 1. Firstly, PDMS was poured on top of a ceramic carrier at 90 °C for 10 min to form the donor substrate. In our demonstration, a commercial electrolytic copper foil (EQ-bccf-9u, thinness: 9 μm, Shenzhen Kejing Star Tech, Shenzhen, China) was adhered to the donor substrate. Similarly, two solidified PVA membranes (the width was 1000 μm and the thickness was 50 μm; we dyed the PVA solution with black ink for visibility) were adhered to another two donor substrates. Afterwards, a UV laser marker (HGL-LSU3/5EI, Wuhan Huagong Laser Engineering Co., Ltd., Wuhan, China) was used to pattern the copper Foil and PVA membrane, respectively [23]. The required pattern and unnecessary parts were cut off by UV laser. The patterns of copper and PVA are shown in Figure S1. After tearing off the obsolete PVA membrane and copper foil (Figure 1a), a patterned PVA film was transferred from a donor to a receiver substrate (made at 90 °C for 4 min) for the different stickiness of substrate. Then, the required copper circuit was wetted with deionized water (place the copper over a bottle filled with 100 °C deionized water about 15 s) and put into contact with the copper and PVA without external pressure for 5 s. Afterwards, the copper circuit was sticking to the patterned PVA due to stickiness increasing of wetting PVA (Figure 1b). In other words, the copper was transferred from the donor to the patterned PVA which was on the receiver substrate. Same as above procedure, another donor substrate covered with patterned PVA was wetted with deionized water, and the patterned PVA was transferred from the donor to the top of the double layer. After stacking up, the water would dissolve the interface of patterned PVA. The copper was embedded in PVA when constructing the PVA/copper/PVA sandwich structure (Figure 1c). Then, the PDMS was poured to directly encapsulate the sandwich structure and cured for 2 h at 75 °C (Figure 1d). The PDMS/sandwich structure/PDMS laminate was cut into strips, each strip has one patterned meander metal interconnect encapsulated in the stretchable substrate. The edges of strips were cut to expose the entrance and exit. At the final step, these strips were placed in deionized water for 24 h at 75 °C to dissolve the PVA which enclosed those copper circuits (Figure 1e). The residual PVA was washed away by an injection needle filled with deionized water. From a practical point of view, the entrance and exit of these strips should be sealed to prevent the corrosion of the circuits. The critical point of the fabrication is the intimate bonding of PVA and copper, and it can be assured by vacuuming when pouring the upper PDMS before curing. Nevertheless, it may affect the shaping of tunnel.

2.4. Stretching and Electrical Test

All the strips were clamped by two pieces of thin PDMS on both sides, and a high-frequency fatigue testing machine (E1000, Instron, Boston, MA, USA) stretched the sample to a specified displacement specified through program for cyclic stretching at 2 Hz. The ends of the interconnects were welded to connect the test electrode; a digital multimeter (34,461A, KeySight Technologies, Santa Rosa, CA, USA) was used for electrical test.

2.5. Finite Element Analysis

ABAQUS commercial software (ABAQUS6.14, ABAQUS Inc., Palo Alto, CA, USA) was used to study the mechanics response of tunnel encapsulation and direct encapsulation. PDMS was modeled by the hexahedron element (C3D8R), while the interconnect was modeled by the shell element (S4R). 22,749 elements for silicone and 1419 elements for interconnects were used to conduct FEA modeling after grid independence testing. Displacement boundary conditions were applied to both edges of the system to apply different levels of stretching and refined meshes were adopted to ensure the accuracy. ABAQUS/Explicit was applied to analyze the deformation and normal stress distribution of interconnects with tunnel encapsulation and direct encapsulation.

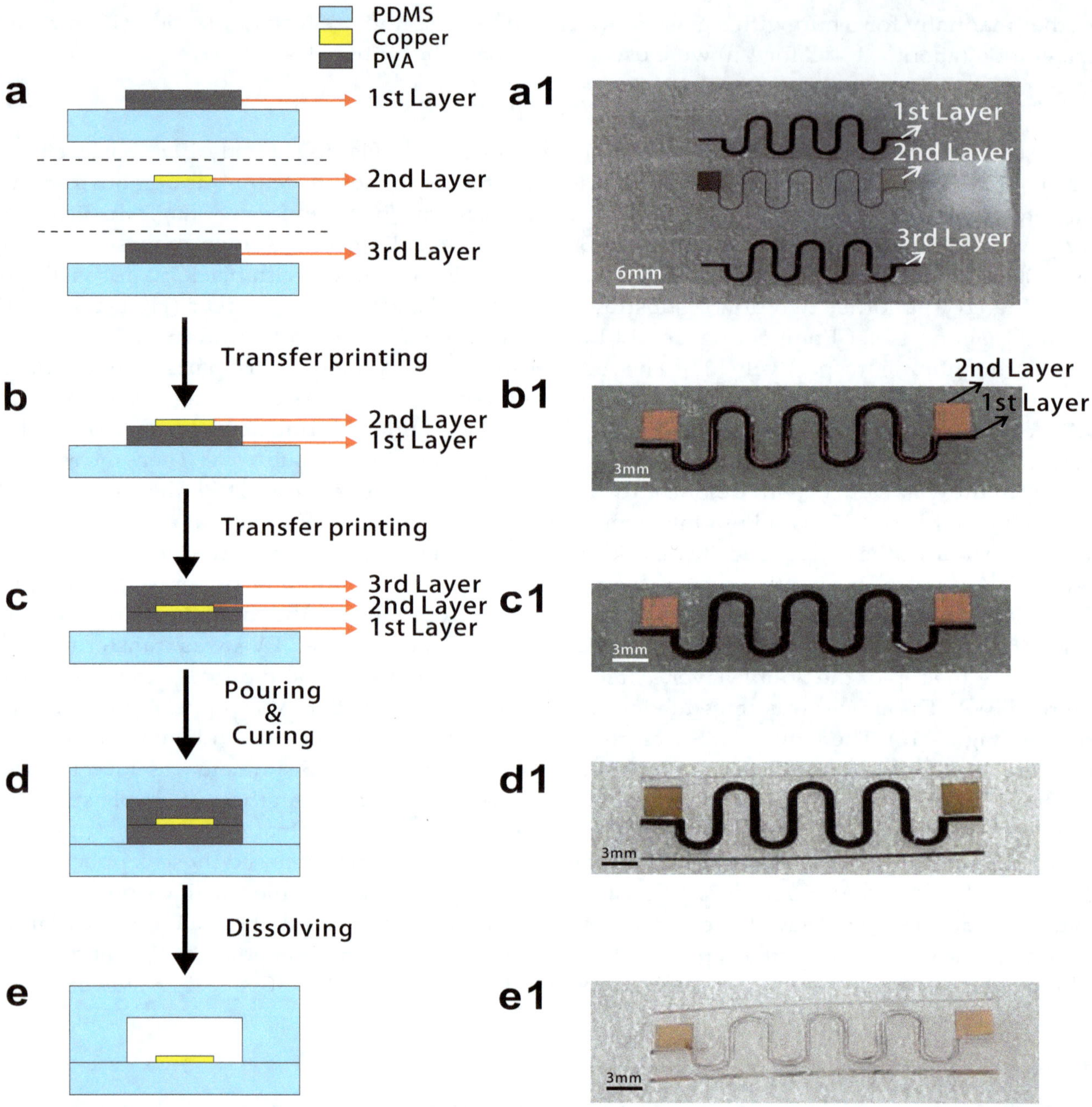

Figure 1. Schematic diagram of the tunnel encapsulation fabrication process. (**a**) Patterned copper foil and PVA membrane, where the first layer and the third layer have the same dimension. (**a1**) Optical image of patterned copper film and PVA. (**b**) Metal circuit was transferred to the patterned PVA. (**b1**) Optical image of the double layer on the receiver substrate. (**c**) Another patterned PVA was transferred to the top of the double layer structure. (**c1**) Optical image of the sandwich structure. (**d**) PDMS was poured to direct encapsulate the sandwich structure and cured at 75 °C for 2 h. (**d1**) Optical image of the sandwich structure encapsulated with PDMS. (**e**) Both ends of the laminate strip were cut to expose the entrance and exit for dissolving the PVA around the Copper circuit. (**e1**) Optical image of the sample after dissolving the PVA.

3. Results and Discussion

3.1. *The Performance during Stretching*

Figure 2a shows three different patterns of serpentine interconnects which consist of two periodic unit cells of two straight lines and two arc of circles, and the straight lines are tangent to the arc. R is the outer radius, S is the space of two unit cells, L is the center distance of two arc along the longitudinal axis, and W is the width of the copper trace. The center distance of these patterns was fixed to set a limit for the dimensions of these circuits. Therefore, R is the critical parameter of these pattern, which controls the shape of the pattern. We obtained horseshoe design (Pattern-A, R = 1.95 mm, S = 6 mm, L = 4 mm, W = 100 μm), semicircle design (Pattern-B, R = 1.55 mm, S = 6 mm, L = 4 mm, W = 100 μm) and the snakelike design (Pattern-C, R = 1.15 mm, S = 6 mm, L = 4 mm, W = 100 μm). Let Ht denote the height of the tunnel made of PVA; Wt the width of the tunnel; Ws the width of the strip; and Ts the thickness of the strip. Through the thickness, the strip has a layered structure (Figure 2b) from bottom (substrate/serpentine/encapsulation layer) to top. A sample (Ht = 100 μm, Wt = 1000 μm, Ws = 1 cm, Ts = 500 μm) with serpentine designed Pattern-B is presented sequentially to illustrate the characteristic of tunnel encapsulation. Figure 2c shows the original shape of the interconnect with tunnel encapsulation before stretched and the tunnel was filled with green fluid to identify the deformation of tunnel. Figure 2d exhibits the serpentine interconnect at 30% applied strain obtained from dynamic mechanical analysis. Figure 2e exhibits the serpentine interconnect at 50% applied strain. Figure 2g illustrates the interconnects slip and buckle in the tunnel when stretched. It is shown that buckling occurred in the arc region highlight with the yellow frame and it is identical with the FEA result (magnification in Figure 2f). The corresponding FEA predictions on buckling deformation of serpentine interconnects illustrate that the tunnel provides enormous space for the interconnects to relieve stress with slipping and buckling.

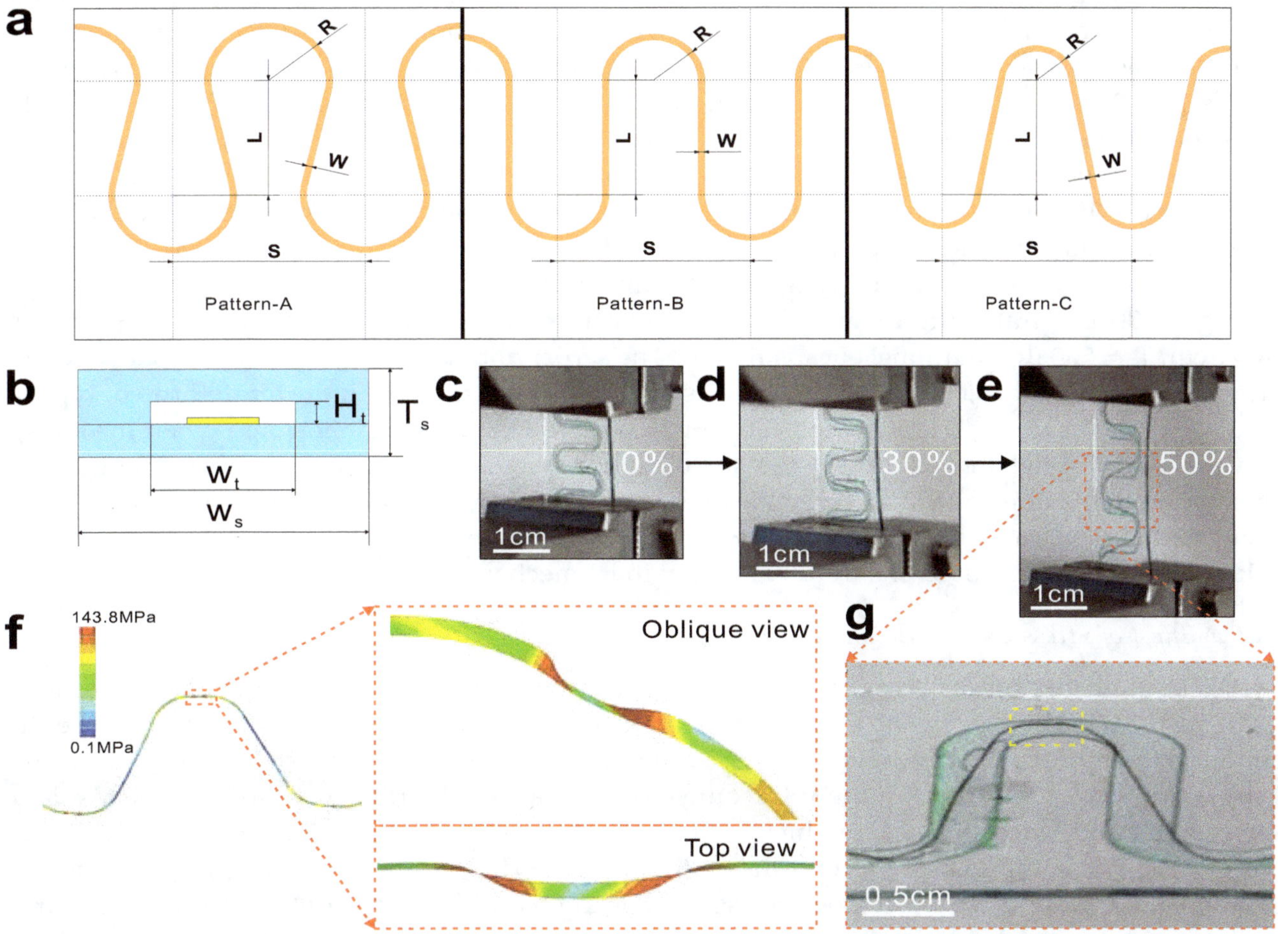

Figure 2. (**a**) Three different patterns of serpentine interconnects. (**b**) Cross-sectional illustration of representative layers in a hollow structure with embedded stretchable interconnects which collapse in the tunnel. (**c**–**e**) Optical images of the sample before stretched, morphology at 30% and 50% applied strain, respectively. (**f**) FEA results of morphology at 50% strain. (**g**) Magnification of slipping and buckling region.

We performed elongation test with both encapsulation technologies to investigate the effect of encapsulation technology on maximum stretchability. Three serpentines of different patterns were encapsulated in PDMS with both technologies to confirm the effect of patterns on maximum stretchability. In addition, to evaluate the effect of thickness of encapsulation layer, we made samples with three different thicknesses (300 μm, 500 μm, and 700 μm). The results of the experiment are shown in Figure 3a, where the direct encapsulation and tunnel encapsulation are abbreviated to DE and TE, respectively. For direct encapsulation without further processing, the maximum elongation is limited by localized fracture of the copper embedded in silicone, and the interconnect encounters severe stress concentration originating from constraints of surrounding PDMS (Figures S2 and S3). The FEA result in Figure 3e illustrates the stress of interconnect with direct encapsulation increase dramatically (stress at 80–110% strain exceed the maximum stress, hence values are not shown in the graph). For tunnel encapsulation, the serpentine interconnect was freestanding in the tunnel except the two ends. Upon stretching of the whole system, the interconnect is dragged from the two ends, geometrical opening in arc segment increases to accommodate the elongation, the straight part and the arc segment undergoes both stretching and bending. As the applied strain exceeds a critical value, lateral buckling occurs to reduce the strain energy, which helps to alleviate stress concentration of interconnects. Figure 3e shows that the stress of interconnects with tunnel encapsulation is considerably lower than interconnect with direct encapsulation. Besides, the platform period of stress with tunnel encapsulation in Figure 3e proved that the outer stretching is accommodated by the sliding and buckling behavior of interconnect in tunnel (Figures S4–S6). It is shown that Pattern-C owns the best stretchability in both encapsulation technologies. A larger curvature in the arc region become closer to a straight line. That means Pattern-C serpentine has a larger redundancy to deform with both technologies as the curvature of Pattern-C is the smallest. The stretchability of the system decreases as the thickness of the strip increases when encapsulated directly (Figure 3a) since the interconnects suffer severer mechanical constraints at elongation. On the contrary, the trend of the curve of tunnel encapsulation is opposite to the direct encapsulation. In tunnel encapsulation, PDMS around the serpentine broke first before the copper rupture as the tunnel would decrease the strength of the strip. As a result, with the same dimension of tunnel, the degradation of a thinner strip will become worse.

Figure 3b illustrates the reversibility of serpentine (Pattern-B, Ht = 100 μm, Wt = 1000 μm, Ws = 1 cm, Ts = 500 μm) in tunnel encapsulation. The serpentine can recover its original shape roughly under 50% strain after 20,000 cycles at 2 Hz. As the interconnects hang in the air deformed, typically by out-of-plane buckling and in-plane bending, they will lose the original shape when the stress is relieved. With tunnel encapsulation, the surrounding silicone matrix will mechanically constrain deformation of the interconnects in the case of the serpentine coming loose in the tunnel space. In the experiment, as long as the applied strain was under maximum strain for all kinds of patterns and thicknesses, the system could ensure reversibility in its mechanics.

3.2. The Effect of Stress Concentration on Durability

We studied the influence of our technology on the durability with the same pattern design (Pattern-B). We performed cyclic loading experiments with a focus on the lifetime of the serpentine, defined as the number of stretching cycles to failure in the interconnect. Figure 3c shows the number of stretching cycles to failure in the interconnect versus applied strain for different encapsulation technologies and different system thicknesses (300 μm, 500 μm, and 700 μm). For the direct encapsulation, the differences on lifetime with corresponding elongation are caused by the fact that the encapsulation would further constrain the interconnects from deforming out of plane if elongation is increased. Meanwhile, stress concentration occurs in the arc region of the serpentine and the phenomenon becomes rather obvious with increasing elongation. The FEA result (Figure 3e,f) also verifies the hypothesis. Remarkably, the lifetime of serpentine improved tremendously by the tunnel encapsulation. Likewise, the differences on lifetime with corresponding elongation are caused by further constraints and severer friction contact. Note that the outstanding improvement on lifetime with tunnel encapsulation stem from the introduction of tunnel space. It offers the activity space for the interconnects to slide and buckle in the space to relieve stress at elongation. In addition, the tunnel isolates the interconnect from surrounding PDMS and helps to decrease the friction of interconnect when stretching. The result in Figure 3c reveals that a thinner system would improve the lifetime of the serpentine as the constraints of a thinner encapsulation layer are weaker. It can be concluded from those results that the lifetime of the serpentine decreases significantly with increasing system elongation. Besides, both the tunnel encapsulation technology and a thinner encapsulation layer help to improve the lifetime of the interconnects.

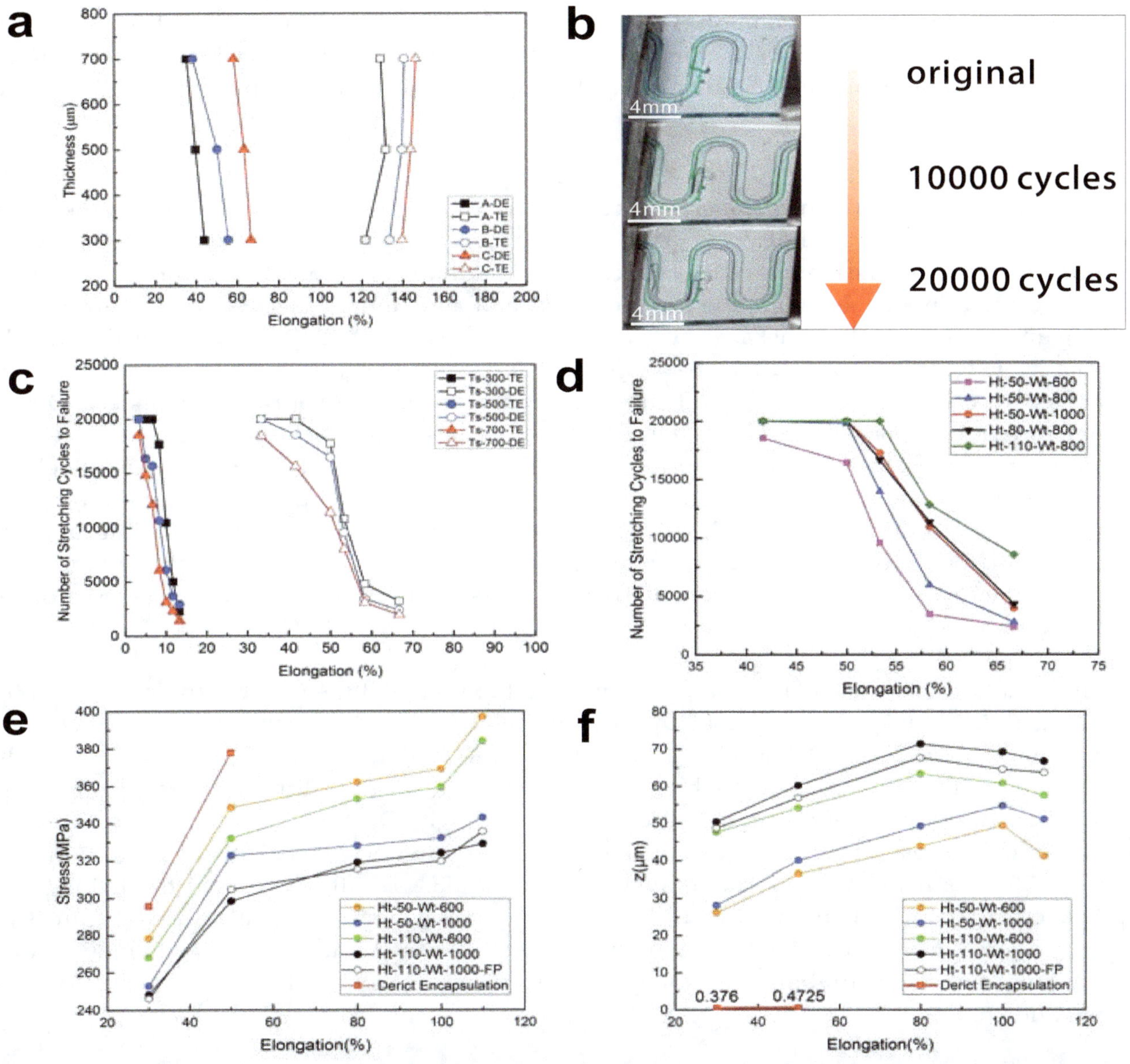

Figure 3. (**a**) Maximum stretchability when the serpentine (different patterns) encapsulated in different technologies with different sample thicknesses Ts from 300 to 700 μm and the width of these samples is 600 μm. (**b**) Reversibility illustration of serpentine in tunnel encapsulation (Ht = 100 μm, Wt = 1000 μm, Ws = 1 cm, Ts = 500 μm). (**c**) Number of stretching cycles to failure when the serpentine encapsulated in different technologies with different sample thicknesses Ts from 300 to 700 μm and the width of these samples is 600 μm. (**d**) Experimental results for different width and height of tunnel in system (Ts = 500 μm). (**e**,**f**) FEA predictions on stress and relative displacement in z direction of serpentine interconnects under stretching.

The available space depends on the width and the height of the tunnel. Thus, a comparison of different width tunnel encapsulation is presented in Figure 3d. It is obvious that the lifetime increase with a wider tunnel. It might be attributed to a broader space enabling a greater degree of slipping and buckling of the interconnects (see green and black line in Figure 3f). In addition, the result depicts the relationship between lifetime of interconnects and the height of tunnel. The lifetime increased as the height increased, which gives rise to a larger space to buckle for the serpentine in the tunnel under stretching, which is a desirable characteristic for cycling stretch. The FEA result (see orange and green line, blue and black line in Figure 3f) shows that the greater heights enable the interconnects to have a larger displacement in z direction, which means the stress relieved when buckling in the direction (see orange and green line, blue and black line in Figure 3e). The displacement in z direction would decrease after a specified value of strain for the severer buckling (Supplementary Materials, Figure S7). Therefore, the durability of the interconnects can be improved by increasing the width and height of the tunnel.

The tunnel was filled with a silicone oligomer (Sylgard 184, without curing agent) using an injection needle to lubricate the contact surface and reduce the friction between interconnects and silicon. The effect of injected fluid was quantitatively studied by experimental tests, as shown in Figure 4a. It is clear that the lifetime of the interconnects was improved as the lubricating fluid reduced the nonspecific adhesion between the interconnects and the tunnel. Besides, it is hydrophobic, expelling moisture from the package, and is optically transparent. It is remarkable that all the experiments were done at once after dissolving. It can be seen that the lifetime of interconnect in water-filled tunnel declines rapidly and the water may corrode the copper interconnect. On the contrary, interconnect in silicone oligomer-filled tunnel sustains excellent fatigue performance without the risk of corrosion. The contact between the interconnects and the wall of the tunnel will definitely influence the lifetime of the serpentine. Apparently, the height of the tunnel plays an important role in optimization as described above. Taking the interconnect itself into account, diminishing the out-of-plane displacement when stretched may contribute much to the lifetime of the system. Here, we made several flexible fixing-pillars in the tunnel to fix the position of the interconnect.

The results are shown in Figure 4b. It is important to note that the experiment is based on the lubricating fluid. It was observed that the lifetime increased enormously with introduction of fixing-pillars. We reduced the local stiffness of the arc region of the serpentine so that the deformation in the arc part would be alleviated, and the probability of contacting between interconnects and the wall of the tunnel diminished accordingly. Of course, the pillars are located in the minimum deformation part of the interconnects. For the patterns we used, it was the middle part of the straight segments of the serpentine structure. The FEA result shows that the fixing-pillar help to decrease displacement in z direction while maintaining similar degree of stress (Figure 3e,f). Besides, it just needs to modify the pattern of the PVA a little bit to be an effective technology that is able to improve the durability of interconnects when applying this new technology of encapsulation.

3.3. Electrical Performance

The stretchable serpentine shows excellent cycling stability in Figure 4c. The parameters of the sample are Ts = 500 μm, Wt = 100 μm, and Ht = 100 μm, after 20,000 cycles at 50% applied strain, the interconnects still exhibited an almost constant value. Note that no visible cracks formed in the interconnects when inspected under an optical microscope, and the system can maintain stable electrical performance under repetitively large deformation.

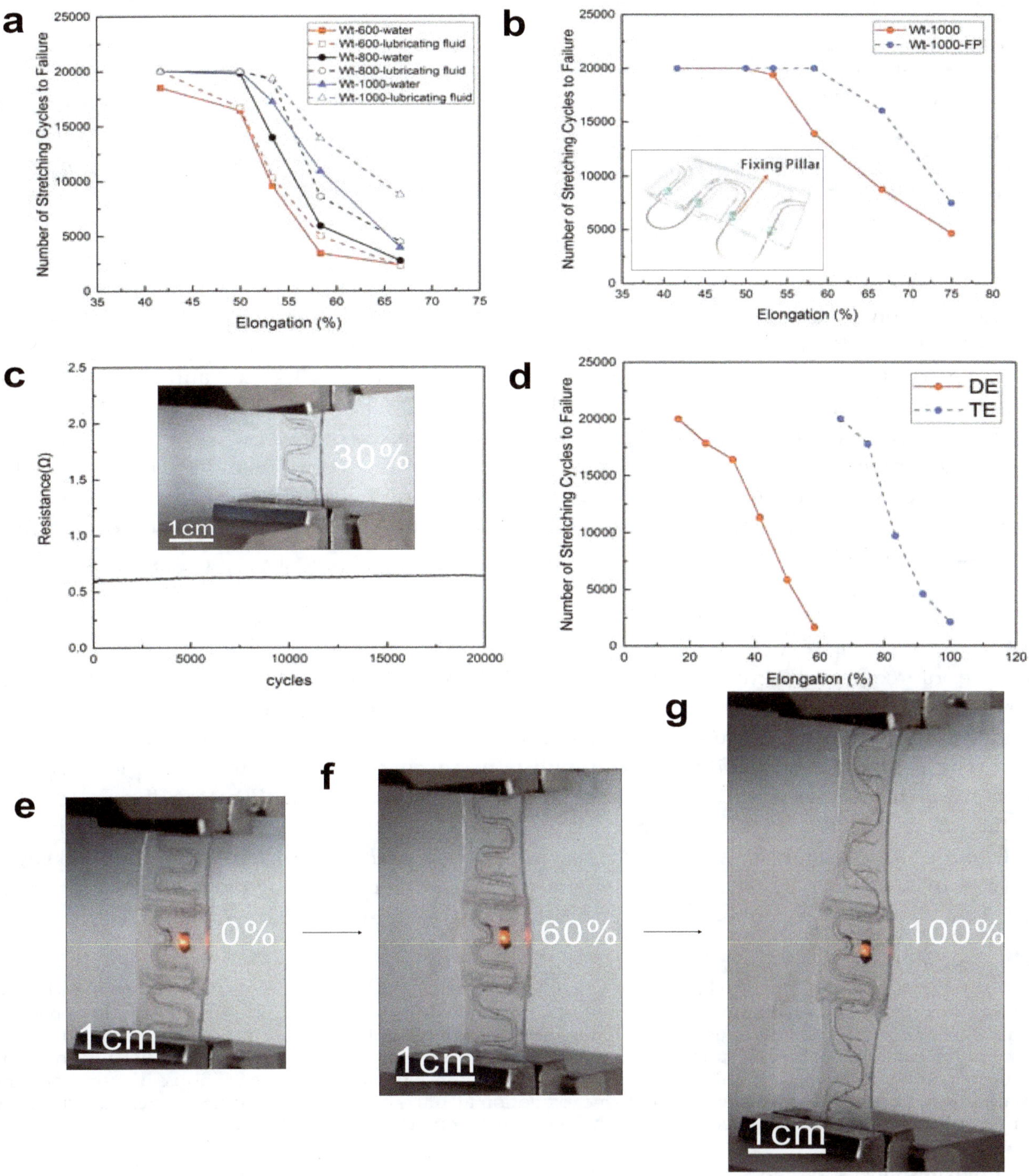

Figure 4. (**a**) Experimental results for effect of lubricating fluid on lifetime of samples with different tunnel widths (sample thickness Ts = 500 μm and height of the tunnel Ht = 100 μm). (**b**) The corresponding improvement in lifetime when adopt fixing-pillars (300 μm in diameter) in the tunnel filled with lubrication fluid (Ht = 100 μm, Wt = 1000 μm, Ws = 1 cm, Ts = 500 μm) and schematic illustration of fixing-pillars (insets). (**c**) Experimental results of resistance change with fixing-pillars in tunnel (Ht = 100 μm, Wt = 1000 μm, Ws = 1 cm, Ts = 500 μm) and optical images of deformation with pillars in tunnel at 30% strain. (**d**) Experimental results of serpentine with direct encapsulation and tunnel encapsulation in Ecoflex. (**e**–**g**) Optical images of device before stretched and morphology at 60% and 100% applied strain, respectively.

3.4. *The Universality of Tunnel Encapsulation Technology*

To study the universal property of this new encapsulation, Ecoflex was selected as encapsulation material to test the durability of the system in Figure 4d. The lifetime of the serpentine encapsulated with Ecoflex improved enormously compared with PDMS, and it is suitable for tunnel encapsulation. In addition, it was observed that the interconnects reached its rupture strain before the Ecoflex broke. The stretchability is up to 200% for the serpentine structure of the metal, but not the ultimate limit for interconnects which are not encapsulated. The interconnects in the tunnel can recover the original shape practically even at the maximum elongation for the constraints of the tunnel.

4. Device Display

A long serpentine interconnect encapsulated in Ecoflex with tunnel encapsulation is presented in Figure 4e–g. An LED was integrated in the device, and fixing-pillars were applied in the tunnel. The parameters of the device are Ts = 500 μm, Wt = 1000 μm, and Ht = 150 μm, and the interconnect with tunnel encapsulation can sustain electrical performance when stretched to reach 100% strain. In this device, a piece of PDMS was attached over the central part to protect the connecting of the LED chip and interconnects. When the system was elongated, the interconnects of both end endure the whole elongation while the middle part would sustain the original shape. Considering integration of microchip, it is worth exploring placing a microchip in the three-dimensional tunnel to further alleviate stress concentration on the chip.

5. Conclusions

In summary, we report a technology of tunnel encapsulation to improve the lifetime of stretchable interconnects A tunnel is formed from dissolution of PVA in deionized water which can help alleviate the stress concentration by provide the space for the sliding and buckling of interconnects. Our tunnel encapsulation confers superior properties compared to direct encapsulation, namely: (1) exceptional stretchability and durability, our approach offers a three-dimensional activity space for the interconnects to buckle, twist and stretch, the stretchability can be up to 200% when the interconnects encapsulated in Ecoflex; (2) excellent stability of electronics conductivity, the interconnects can sustain excellent electrical performance after 20,000 cycles at 50% applied strain; and (3) ease of patterning, the pattern of the tunnel can be readily changed by modifying the pattern of PVA. This new encapsulation has a good application prospect for consumer wearable electronics.

Supplementary Materials
s1, Figure S1: (a) Pattern of copper; (b) Pattern of Polyvinyl Alcohol (PVA), Figure S2: Stress distributions and strain contours of the interconnects with direct encapsulation at 30% elongation, Figure S3: Stress distributions and strain contours of the interconnects with direct encapsulation at 50% elongation, Figure S4: Stress distributions and strain contours of the interconnects with tunnel encapsulation at 30% elongation, Figure S5: Stress distributions and strain contours of the interconnects with tunnel encapsulation at 50% elongation, Figure S6: Stress distributions and strain contours of the interconnects with tunnel encapsulation at 100% elongation, Figure S7: Stress distributions and strain contours of the interconnects with tunnel encapsulation at 110% elongation.

Author Contributions: K.W. and K.L. conceived the concept and designed the experiments. K.L. conducted the devices fabrication, testing and FEA modeling. K.L., K.W., and G.C. conducted the data analysis. Z.W. supervised and supported the work. All authors contributed to the manuscript and revision.

Acknowledgments: This work was supported by the National Key Research and Development Program of China (2017YFB1303103), National Natural Science Foundation of China (No. U1613204) and China Postdoctoral Science Foundation (2018M632833). Wu and Guo thank the support from Chinese central government through its Thousand Youth Talents program.

References

1. Sun, Y.; Kumar, V.; Adesida, I.; Rogers, J.A. Buckled and wavy ribbons of GaAs for high-performance electronics on elastomeric substrates. *Adv. Mater.* **2006**, *18*, 2857–2862. [CrossRef]

2. Sun, Y.; Rogers, J.A. Structural forms of single crystal semiconductor nanoribbons for high-performance stretchable electronics. *J. Mater. Chem.* **2007**, *17*, 832–840. [CrossRef]
3. Kim, D.H.; Kim, Y.S.; Wu, J.; Liu, Z.; Song, J.; Kim, H.S.; Huang, Y.Y.; Hwang, K.C.; Rogers, J.A. Ultrathin silicon circuits with strain-isolation layers and mesh layouts for high-performance electronics on fabric, vinyl, leather, and paper. *Adv. Mater.* **2009**, *21*, 3703–3707. [CrossRef]
4. Zhao, Y.; Huang, X. Mechanisms and materials of flexible and stretchable skin sensors. *Micromachines* **2017**, *8*, 69. [CrossRef]
5. Gonzalez, M.; Axisa, F.; Bossuyt, F.; Hsu, Y.Y.; Vandevelde, B.; Vanfleteren, J. Design and performance of metal conductors for stretchable electronic circuits. *Circuit World* **2009**, *35*, 22–29. [CrossRef]
6. Kim, D.H.; Liu, Z.; Kim, Y.S.; Wu, J.; Song, J.; Kim, H.S.; Huang, Y.; Hwang, K.C.; Zhang, Y.; Rogers, J.A. Optimized structural designs for stretchable silicon integrated circuits. *Small* **2009**, *5*, 2841–2847. [CrossRef] [PubMed]
7. Li, Z.; Wang, Y.; Xiao, J. Mechanics of curvilinear electronics and optoelectronics. *Curr. Opin. Solid State Mater. Sci.* **2015**, *19*, 171–189. [CrossRef]
8. Li, Z.; Wang, Y.; Xiao, J. Mechanics of bioinspired imaging systems. *Theor. Appl. Mech. Lett.* **2016**, *6*, 11–20. [CrossRef]
9. Li, Z.; Xiao, J. Mechanics and optics of stretchable elastomeric microlens array for artificial compound eye camera. *J. Appl. Phys.* **2015**, *117*, 014904. [CrossRef]
10. Sun, Y.; Choi, W.M.; Jiang, H.; Huang, Y.Y.; Rogers, J.A. Controlled buckling of semiconductor nanoribbons for stretchable electronics. *Nat. Nanotechnol.* **2006**, *1*, 201–207. [CrossRef] [PubMed]
11. Yeo, W.H.; Kim, Y.S.; Lee, J.; Ameen, A.; Shi, L.; Li, M.; Wang, S.; Ma, R.; Jin, S.H.; Kang, Z.; et al. Multifunctional epidermal electronics printed directly onto the skin. *Adv. Mater.* **2013**, *25*, 2773–2778. [CrossRef] [PubMed]
12. Meitl, M.A.; Zhu, Z.T.; Kumar, V.; Lee, K.J.; Feng, X.; Huang, Y.Y.; Adesida, I.; Nuzzo, R.G.; Rogers, J.A. Transfer printing by kinetic control of adhesion to an elastomeric stamp. *Nat. Mater.* **2006**, *5*, 33–38. [CrossRef]
13. Carlson, A.; Bowen, A.M.; Huang, Y.; Nuzzo, R.G.; Rogers, J.A. Transfer printing techniques for materials assembly and micro/nanodevice fabrication. *Adv. Mater.* **2012**, *24*, 5284–5318. [CrossRef] [PubMed]
14. Hsu, Y.Y.; Gonzalez, M.; Bossuyt, F.; Axisa, F.; Vanfleteren, J.; De Wolf, I. The effects of encapsulation on deformation behavior and failure mechanisms of stretchable interconnects. *Thin Solid Films* **2011**, *519*, 2225–2234. [CrossRef]
15. Jablonski, M.; Lucchini, R.; Bossuyt, F.; Vervust, T.; Vanfleteren, J.; De Vries, J.; Vena, P.; Gonzalez, M. Impact of geometry on stretchable meandered interconnect uniaxial tensile extension fatigue reliability. *Microelectron. Reliab.* **2015**, *55*, 143–154. [CrossRef]
16. Xu, S.; Zhang, Y.; Jia, L.; Mathewson, K.E.; Jang, K.I.; Kim, J.; Fu, H.; Huang, X.; Chava, P.; Wang, R.; et al. Soft microfluidic assemblies of sensors, circuits, and radios for the skin. *Science* **2014**, *344*, 70–74. [CrossRef] [PubMed]
17. Hussain, A.M.; Ghaffar, F.A.; Park, S.I.; Rogers, J.A.; Shamim, A.; Hussain, M.M. Metal/polymer based stretchable antenna for constant frequency far-field communication in wearable electronics. *Adv. Funct. Mater.* **2015**, *25*, 6565–6575. [CrossRef]
18. Pan, T.; Pharr, M.; Ma, Y.; Ning, R.; Yan, Z.; Xu, R.; Feng, X.; Huang, Y.; Rogers, J.A. Experimental and theoretical studies of serpentine interconnects on ultrathin elastomers for stretchable electronics. *Adv. Funct. Mater.* **2017**, *27*, 1702589. [CrossRef]
19. Zhang, Y.; Wang, S.; Li, X.; Fan, J.A.; Xu, S.; Song, Y.M.; Choi, K.J.; Yeo, W.H.; Lee, W.; Nazaar, S.N.; et al. Experimental and theoretical studies of serpentine microstructures bonded to prestrained elastomers for stretchable electronics. *Adv. Funct. Mater.* **2014**, *24*, 2028–2037. [CrossRef]
20. Le Borgne, B.; De Sagazan, O.; Crand, S.; Jacques, E.; Harnois, M. Conformal Electronics Wrapped Around Daily Life Objects Using an Original Method: Water Transfer Printing. *ACS Appl. Mater. Interfaces* **2017**, *9*, 29424–29429. [CrossRef] [PubMed]
21. Rogel, R.; Borgne, B.L.; Mohammed-Brahim, T.; Jacques, E.; Harnois, M. Spontaneous Buckling of Multiaxially Flexible and Stretchable Interconnects Using PDMS/Fibrous Composite Substrates. *Adv. Mater. Interfaces* **2017**, *4*, 1600946. [CrossRef]

22. Jin, S.H.; Kang, S.K.; Cho, I.T.; Han, S.Y.; Chung, H.U.; Lee, D.J.; Shin, J.; Baek, G.W.; Kim, T.i.; Lee, J.H.; et al. Water-soluble thin film transistors and circuits based on amorphous indium–gallium–zinc oxide. *ACS Appl. Mater. Interfaces* **2015**, *7*, 8268–8274. [CrossRef] [PubMed]
23. Peng, P.; Wu, K.; Lv, L.; Guo, C.F.; Wu, Z. One-Step Selective Adhesive Transfer Printing for Scalable Fabrication of Stretchable Electronics. *Adv. Mater. Technol.* **2018**, *3*, 1700264. [CrossRef]

9

Ultrasonic Spray-Coating of Large-Scale TiO_2 Compact Layer for Efficient Flexible Perovskite Solar Cells

Peng Zhou [1], Wangnan Li [2], Tianhui Li [1], Tongle Bu [1], Xueping Liu [1], Jing Li [1], Jiang He [1], Rui Chen [1], Kunpeng Li [1], Juan Zhao [3,*] and Fuzhi Huang [1,*]

1 State Key Laboratory of Advanced Technology for Materials Synthesis and Processing, Wuhan University of Technology, Wuhan 430070, China; whutzhp@gmail.com (P.Z.); litianhui@whut.edu.cn (T.L.); tonglebu@whut.edu.cn (T.B.); xuepingliu@whut.edu.cn (X.L.); lijing123@whut.edu.cn (J.L.); milijiang@whut.edu.cn (J.H.); Rui-Chen@whut.edu.cn (R.C.); likunpeng@whut.edu.cn (K.L.)

2 Hubei Key Laboratory of Low Dimensional Optoelectronic Materials and Devices, Hubei University of Arts and Science, Xiangyang 441053, China; wangnan.li@yahoo.com

3 School of Automotive Engineering, Wuhan University of Technology, Wuhan 430070, China

* Correspondence: juan.zhao@whut.edu.cn (J.Z.); fuzhi.huang@whut.edu.cn (F.H.)

Academic Editors: Seung Hwan Ko, Daeho Lee and Zhigang Wu

Abstract: Flexible electronics have attracted great interest in applications for the wearable devices. Flexible solar cells can be integrated into the flexible electronics as the power source for the wearable devices. In this work, an ultrasonic spray-coating method was employed to deposit TiO_2 nanoparticles on polymer substrates for the fabrication of flexible perovskite solar cells (PSCs). Pre-synthesized TiO_2 nanoparticles were first dispersed in ethanol to prepare the precursor solutions with different concentrations (0.5 mg/mL, 1.0 mg/mL, 2.0 mg/mL) and then sprayed onto the conductive substrates to produce compact TiO_2 films with different thicknesses (from 30 nm to 150 nm). The effect of the different drying processes on the quality of the compact TiO_2 film was studied. In order to further improve the film quality, titanium diisopropoxide *bis*(acetylacetonate) (TAA) was added into the TiO_2-ethanol solution at a mole ratio of 1.0 mol % with respect to the TiO_2 content. The final prepared PSC devices showed a power conversion efficiency (PCE) of 14.32% based on the indium doped tin oxide coated glass (ITO-glass) substrate and 10.87% on the indium doped tin oxide coated polyethylene naphthalate (ITO-PEN) flexible substrate.

Keywords: ultrasonic spray; titanium oxide; flexible perovskite solar cells; low temperature; large area

1. Introduction

With the global energy consumption increasing, cheap, green and environmentally friendly energy sources are in urgent demand due to the reduction of fossil fuels. Photovoltaic technology is an ideal solution to alleviate the energy crisis and environmental pollution problems. Organic-inorganic lead halide perovskite solar cells (PSCs) have attracted great interest due to their rapid increase in power conversion efficiencies (PCE) from 3.8% to 22.1% within only seven years [1–3]. These great improvements are mainly attributed to the excellent photo-electronic properties of perovskite materials, such as high light absorption properties, direct bandgaps, high charge-carrier mobility and a long electron–hole exciton transport distance (more than 1 μm) [4–6]. Compared to the commercial silicon-based solar cells, PSCs show great advantages with a simplified architecture and a low-cost solution-processed technology, which give them great potential for the future photovoltaic industry [7].

Organic-inorganic lead halide perovskite ($CH_3NH_3PbI_3$) was first used as a sensitizer in dye-sensitized solar cells (DSSCs) by Kojima in 2009 [8] with a PCE of 3.8%, but the performance decreased rapidly due to the dissolution of the perovskite in the liquid electrolytes. Two years later, the solid-state hole-transporting material 2,2′,7,7′-tetrakis-(N,N-di-p-methoxypheny-lamine) 9,9′-spirobifluorene (spiro-OMeTAD) was introduced by Park and Grätzel, and achieved a reported PCE of 9.7% [9]. Since the all solid-state-type PSCs were fabricated, the photo-electronic performance improved rapidly with the use of different electronic/hole transporting semi-conductive materials, such as TiO_2, ZnO, SnO, PCBM, P3HT, etc. [10]. Depending on the different architectures, the PSCs can be generally classified into mesoscopic, meso-superstructured and planar heterojunction types [11]. Among these types, the planar PSCs have a simplified architecture and are easily produced by a solution process. In a planar solar cell, the photoactive layer, $CH_3NH_3PbI_3$, is sandwiched between an electron-transporting layer (ETL) and a hole-transporting layer (HTL), which is suitable for large-scale commercial manufacturing layer-by-layer. The ETL plays an important role as it allows the transport of electrons while blocking the holes. Thus, it influences the carriers' injection, collection, transportation, recombination, and then the overall performance of the PSCs [12]. Anatase TiO_2 is the most widely used ETL for planar PSCs, but it still requires a high temperature sintering of the TiO_2 compact layer to achieve a high efficiency. In general, this compact layer is prepared by spin-coating or spray pyrolysis of a TiO_2 precursor solution with subsequent sintering at 500 °C to transform the amorphous oxide layer into the crystalline phase (anatase), which provides good charge transport properties [13]. The involvement of the high temperature sintering process of TiO_2 has limited the development of flexible PSCs fabricated on plastic substrates, such as polyethylene terephthalate (PET) and polyethylene naphthalate (PEN).

Since the first flexible perovskite solar cell was reported [14], more efforts have been devoted to it and have achieved efficiencies of over 10% on polymer substrates [15,16], demonstrating that efficient perovskite solar cells can be fabricated at low temperatures with a "regular" design. Atomic layer deposition [17,18], microwave sintering [19] and inductively coupled plasma (ICP)-assisted DC magnetron sputtering [20] have been used for deposition of the TiO_2 compact layer at low temperatures. A photonic-cured compact TiO_2 layer has been used on a PET substrate with a high efficiency of 11.2% by Xiao [21]. Snaith et al. realized flexible PSC with an efficiency of 15.9% on a low-temperature processed TiO_2 compact layer by spin-coating the TiO_2 precursor on fluorine doped tin oxide coated glass (FTO-glass) [22]. Yang et al. developed a process to fabricate a very dense amorphous TiO_2 using DC magnetron sputtering at room temperature and achieved 15.07% PCE based on a flexible PET substrate [23]. Sanjib et al. used a combination of ultrasonic spray-coating and a low-thermal-budget photonic curing technology for the first time to fabricate a flexible PSC with an efficiency of 8.1% [24]. Supasai et al. fabricated compact layers of crystalline TiO_2 thin films using aerosol spray pyrolysis on FTO-glass for PSCs, achieving the best efficiency of 6.24% [25]. Kim et al. fabricated a mesoporous TiO_2 ETL with a large area of 10 cm × 10 cm using electro-spray deposition (ESD) for the first time, which resulted in an optimized PCE of 15.11%, higher than that (13.67%) of the PSC with spin-coated TiO_2 films [26]. Both the aerosol spraying and ESD methods demonstrated great potential in the large-scale fabrication of PSCs. However, the TiO_2 films deposited by the above two spraying methods require high temperature sintering, which is not suitable for the preparation of flexible PSCs. The use of these technologies has allowed for the development of all-low-temperature processed PSCs on flexible substrates, but at the same time, has made the process more complicated, resulting in the increase of the manufacturing cost.

Here, we report a low temperature fabrication (<150 °C) of a compact layer composed of highly crystalline small nanoparticles of anatase TiO_2 (diameter <10 nm) dispersed in ethanol. The ultrasonic spray-coating method was employed for the deposition of the TiO_2 compact layer, demonstrating the capability to precisely and reliably deposit thin and uniform layers. Various parameters have to be considered in the process of ultrasonic spraying, such as the flow rate of the TiO_2 ethanol solution, the gas flow pressure which carries the sprayed droplets to the substrate, the distance between the spray nozzle and the substrate, as well as the moving speed of the nozzle during spraying [27,28]. Through the optimization of these parameters, two technological regimes named "wet-film" (w-film) and "dry-film" (d-film) are compared in our work. The former approach results in a dense TiO_2 layer of different thicknesses (30 nm to 150 nm) by changing the concentration of the dispersion. In addition, a small amount of titanium diisopropoxide *bis*(acetylaceto-nate) (TAA) (1.0 mol % with respect to the TiO_2 content) is added to the colloidal TiO_2 dispersion, resulting in the highest PCE of 14.32% based on ITO-glass and 10.87% based on ITO-PEN, both with an active area of 0.4 cm × 0.4 cm. Through the same method, a large-area flexible PSC of 5 cm × 5 cm is fabricated.

2. Materials and Methods

2.1. Materials

Unless specified otherwise, all the chemicals were purchased from either Alfa Aesar or Sigma-Aldrich and used as received. 2,2′,7,7′-tetrakis-(N,N-di-p-methoxypheny-lamine) 9,9′-spirobifluorene (spiro-OMeTAD) was purchased from Shenzhen Feiming Science and Technology Co., Ltd. (Shenzhen, China), and MAI (CH_3NH_3I) was purchased from Lumtec, Taiwan.

2.2. Substrates

Indium doped tin oxide coated glass (ITO-Glass) and polyethylene naphthalate (ITO-PEN) were etched by zinc powder and hydrochloric acid, then followed by ultrasonic cleaning in detergent, pure water and ethanol for 15 min, respectively. The substrate (ITO-Glass or ITO-PEN) was cut into suitable size and plasma cleaned for 5 min to remove any organic material on the surface. Especially, the flexible ITO-PEN substrate was mounted onto glass micro-slides before using.

2.3. Synthesis of TiO_2 Nanoparticles

The anatase titanium oxide nanoparticles were synthesized according to previously reported method published elsewhere [29]. Briefly, 0.5 mL of anhydrous $TiCl_4$ (99.9%, Aladdin, Shanghai, China) was added dropwise into 2 mL of anhydrous ethanol (Sigma-Aldrich, Shanghai, China), stirring for 5 min till the mixed yellow liquid was cooled down to room temperature. Then the whole content was transferred into a vial containing 10 mL anhydrous benzyl alcohol (Aladdin, Shanghai, China), with the color of the clear solution changing to light yellow. The solution was heated to 80 °C and reacted for 9 h. After the reaction the solution was cooled down to the room temperature, and a translucent dispersion of very fine TiO_2 nanoparticles was obtained. Then add 36 mL of diethyl ether to 4 mL of the above solution to precipitate the TiO_2 which was centrifuged at 4000 rpm for 5 min, washed with ethanol and diethyl ether. The above steps were repeated three times and the final precipitation was redispersed in 10 mL of anhydrous ethanol, resulting in a colloidal solution of approximately 12 mg TiO_2/mL ethanol but easily aggregated. The TiO_2 ethanol solution was further diluted into 0.5 mg/mL, 1.0 mg/mL and 2.0 mg/mL respectively for the late ultrasonic spraying. In order to disperse the TiO_2 nanoparticles and enhance the adhesion as well as the connection between nanoparticles, a small amount of TAA was added into the diluted dispersion with a mole ratio of 1.0 mol %, 2.0 mol % and 3.0 mol % respectively. The solution was left to stand for at least 2 h before spraying, but could be stable for months.

2.4. Deposition of the TiO_2 Compact Layer

The low-temperature processed TiO_2 compact layer was deposited by ultrasonic spraying the colloidal dispersion of anatase particles in anhydrous ethanol, formulated with TAA, followed by treating at 135 °C for 1 h to remove ethanol solvent. Figure 1a shows the schematic representation of setup for the spray coating. Figure 1b showed the spray nozzle moving path during the spray. Spray coating was carried out in an Exacta Coat Ultrasonic Spraying System (Sono-Tek Corpration, Milton,

NY, USA) equipped with an AccuMist nozzle. The thickness of the compact layer was tuned by the concentration of TiO_2 nanoparticles (0.5–2.0 mg/mL).

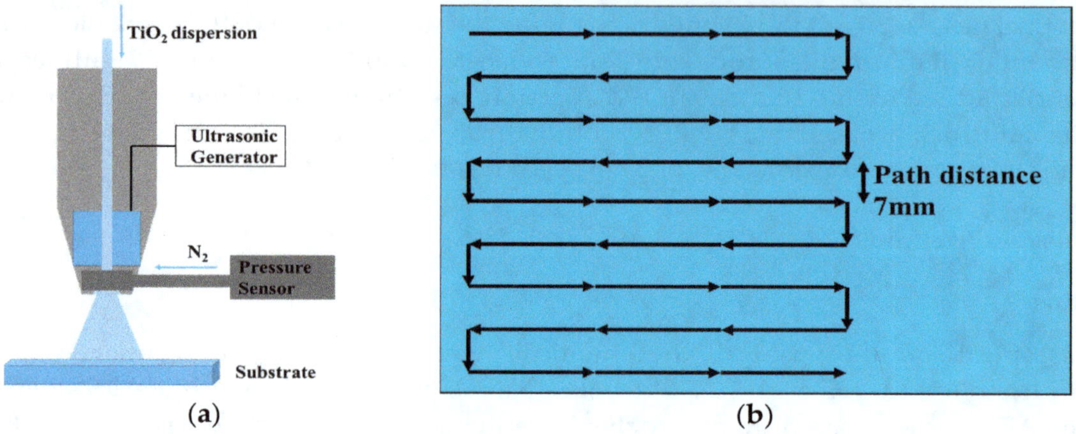

Figure 1. Schematic representation of (**a**) setup for the spray coater and (**b**) the spray nozzle moving path.

2.5. Device Fabrication

After deposition of the TiO_2 compact layer, the substrate with appropriate size (1.5 cm × 1.5 cm for small devices and 5.0 cm × 5.0 cm for large area devices) was transferred into a N_2 filled glove box. The perovskite and hole conductor solutions were prepared in a N_2 glove box before deposition. The perovskite was deposited by spin coating 25 μL of a 1.25 mol/L solution of CH_3NH_3I and PbI_2 (molar ratio 1:1) in DMF at 6500 rpm for 30 s using the gas-assisted method [30], meanwhile a 60 psi dry N_2 gas stream was blown onto the film for 8 s from the third second of spinning. For the fabrication of flexible devices with large area, 200 μL perovskite solution was coated on TiO_2 compact layer and spin-coated at 5000 rpm for 30 s, followed by dropping 200 μL chlorobenzene (CBZ) at the fifth second of spinning. The films were subsequently annealed on a hot plate at 100 °C for 10 min in the glove box. After letting the films cool for 5 min, 20 μL spiro-OMeTAD/CBZ solution (68 mM spiro-OMeTAD, 150 mM tert-butylpyridine and 25 mM lithium *bis*(*tri*-fluoromethanesulphonyl)imide) was spin-coated at 3000 rpm for 30 s. Promptly after the hole transport material deposition, a gold counter electrode (60 nm) was evaporated under high vacuum to complete the device.

2.6. Characterizations

The morphologies and microstructures of the prepared low-temperature processed TiO_2 compact and the cross-sectional structure of the perovskite solar cells were investigated using a field-emission scanning electron microscopy (FE-SEM, Zeiss Ultra Plus). To examine the surface roughness, the films were characterized by BY2000 atomic force microscopy (AFM). The TiO_2 anatase phase was tested by an X-ray diffractometer (XRD, D8 Advance). The photocurrent density-voltage curves of the PSCs were measured using a solar simulator (Oriel 94023A, 300 W). The intensity (100 mW/cm^2) was calibrated using a standard Si-solar cell (Oriel, VLSI standards). All the devices were tested under AM 1.5 G sun light (100 mW/cm^2) using a metal mask of 0.16 cm^2 with a scan rate of 0.01 V/s. Some parameters would be mentioned to evaluate the performance of PSCs, such as open circuit voltage (V_{oc}), short circuit current (J_{sc}), fill fact (FF), and power conversion efficiency (PCE).

3. Results and Discussion

Anatase-type TiO_2 nanoparticles were synthesized through a method published elsewhere [29], using titanium tetrachloride ($TiCl_4$) as the precursor, and anhydrous ethanol and benzyl alcohol as solvents. The TiO_2 nanoparticles synthesized via this method are highly crystalline, shown in Figure 2a,b. In the XRD pattern of the TiO_2 powder prepared by drying the as-synthesized nanoparticles at 135 °C to evaporate the ethanol solvent, three typical diffraction peaks occur at 25.5°, 38.5° and 48°, which respectively belong to the (101), (004) and (200) planes of the anatase TiO_2 crystal. The as-synthesized TiO_2 nanoparticles could easily be dispersed in ethanol, which can be used for ultrasonic spraying directly. After the deposition of the TiO_2 layer, the films had to be treated at 135 °C for 1 h, replacing the high-temperature sintering process. The size of the TiO_2 nanoparticles was around 5 nm, as shown in the TEM image (Figure 2a), which can also be calculated by the Scherrer equation using the (101) diffraction peak through the formula reported elsewhere [31]. Figure 2c shows the XRD patterns of ITO-glass with/without the TiO_2 layer, in which just a weak diffraction peak of TiO_2 occurs at 25.5° due to the thin thickness of the TiO_2 layer.

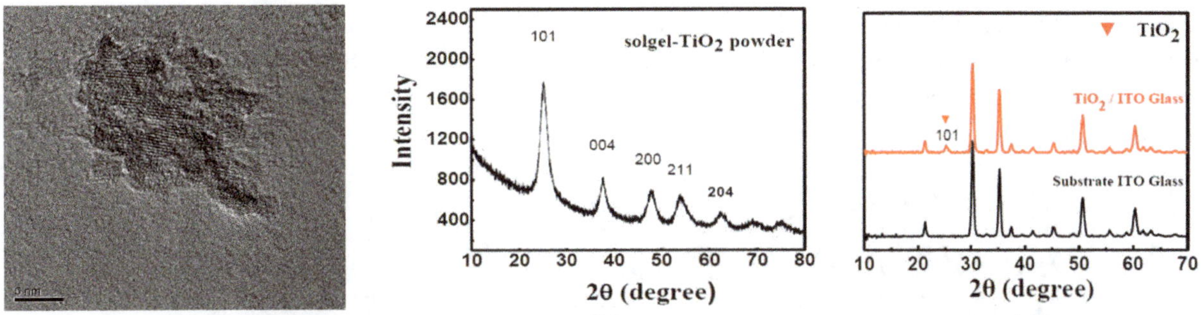

(a) (b) (c)

Figure 2. (**a**) Transmission electron microscopy (TEM) image of the TiO_2 nanoparticles dispersed in ethanol at a concentration of 1.0 mg/mL; (**b**) XRD pattern of the TiO_2 powder drying at 135 °C; and (**c**) XRD patterns of ITO-glass with/without the TiO_2 compact layer coated by the ultrasonic spraying.

In order to obtain uniform and dense films of TiO_2 compact layers, two technological regimes of ultrasonic spraying were built. During ultrasonic spraying, many parameters influence the uniformity, roughness, and coverage of the TiO_2 film, such as the flow rate of the TiO_2 ethanol solution, the gas flow pressure which carries the sprayed droplets to the substrate, the distance between the spray nozzle and the substrate, as well as the moving speed of the nozzle. The ultrasonic spraying technology was employed in the deposition of polymer films such as polyvinylpyrrolidone (PVP) by Sanjukta Bose et al. in 2013 [27]. Parts of the parameters have been discussed, resulting in uniform organic films with small roughness compared to the thickness of the films. Based on the work mentioned above, we optimized the process of spraying, and employed it to deposit inorganic TiO_2 film. Figure 1a shows the schematic representation of the setup for the spray-coating. Figure 1b shows the ultrasonic spraying trajectory over the substrate. Snaith et al. realized low-temperature processed PSCs with power convention efficiencies (PCE) of up to 15.9% using a spin-coating method, demonstrating the optimum thickness of TiO_2 to be approximately 45 nm [22], while the thickness of TiO_2 prepared by high-temperature sintering at 500 °C is around 30 nm [7]. Based on these reports, we tuned the thickness of ultrasonic spraying TiO_2 compact layer ranging from 30 nm to 100 nm by changing the spraying parameters and the concentration of the TiO_2 ethanol solution.

During the spraying, two different technological regimes were termed "wet-film" (w-film) and "dry-film" (d-film), depending on the drying type of the ethanol solvent. The former regime means that the ethanol solvent evaporates after the droplets are sprayed onto the substrate, forming a very uniform thin liquid membrane layer before the ethanol evaporates. A few seconds later, the solvent begins to evaporate, resulting in a uniform thin compact layer of TiO_2. The latter regime means that the processing of ethanol evaporation occurs mainly before the droplets are sprayed onto the substrate, without forming the liquid membrane. The TiO_2 compact layer prepared by the former regime was influenced by the "coffee-ring" effect, which is described by Deegan et al. as the result of capillary flow [32], where liquid evaporates faster from the pinned contact line of a deposited solution and is replenished by additional liquid from inside; however, this did not happen in the latter regime. During the spraying of the w-film regime, the pressure of the N_2 flow was set to 0.3 psi, the flow rate of solution was set to 1.2 mL/min, and the speed of the moving nozzle was set to 50 mm/s, which can result in a thin liquid membrane on the surface of the substrate before the solvent evaporates; during the d-film regime, the three parameters were respectively set to 1.0 psi, 0.3 mL/min, and 20 mm/s. Figure 3 showed the AFM surface images of the TiO_2 compact layers by ultrasonic spraying using 1.0 mg/mL of the TiO_2 (without TAA) ethanol solution. The TiO_2 film prepared by the w-film regime presented a root-mean-square (RMS) surface roughness of around 3.3 nm, while it increased to 44.6 nm rapidly with the d-film regime. It was attributed to the big holes between the TiO_2 nanoparticles prepared through the d-film regime. In the film of the d-film regime, the connection between the particles was so weak that big holes appeared because of the evaporation of ethanol before the droplets were deposited on the substrate. In addition, more spraying cycles were needed to obtain the optimum thickness of the TiO_2 film through the d-film regime, which increased the roughness of film, while just one cycle was enough to achieve the thickness with the w-film regime. Thus, it is better to fabricate the TiO_2 compact layer using the w-film regime.

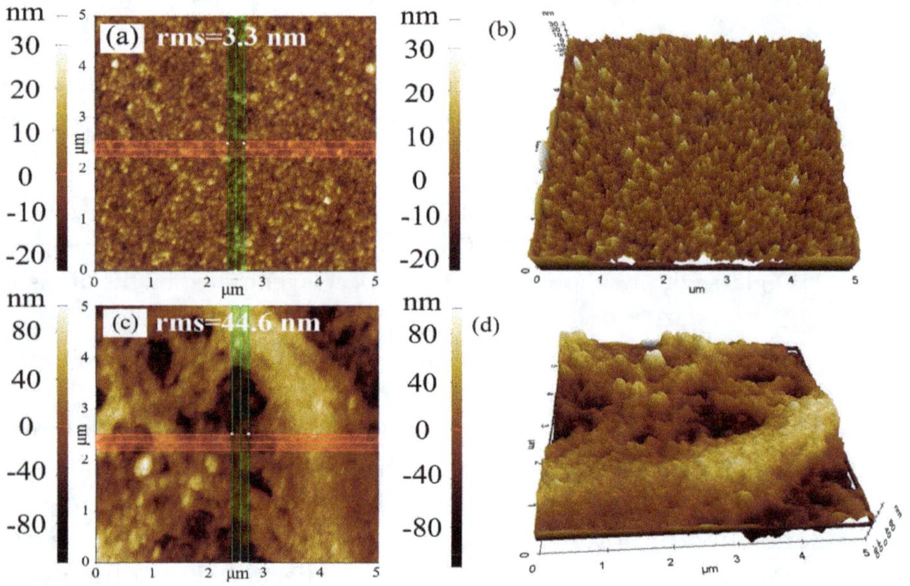

Figure 3. Atomic force microscope (AFM) images of the surface of the TiO_2 layers sprayed by (**a**,**b**) the w-film; (**c**,**d**) the d-film using the TiO_2 ethanol solution without TAA.

In order to discuss the influence of the thickness of the ultrasonically sprayed TiO_2 compact layer on the performance of PSC devices, ITO-glass substrates coated with TiO_2 films of different thicknesses (30–100 nm) were made into complete planar PSC devices. The thickness of the TiO_2 film was tuned by changing the concentration. Table 1 shows the photovoltaic parameters obtained from the current-voltage measurements of devices with different compact layers. It is obvious that the optimal concentration for the performance was 1.0 mg/mL, with a thickness around 60 nm. Figure 4 shows the surface image of different functional layers, and a dense TiO_2 compact layer was obtained (Figure 4b). Each functional layer can be clearly distinguished in the cross-section scanning electron microscopy (SEM) image (Figure 4d). The clear interface between the TiO_2 layer and perovskite layer confirmed that the TiO_2 layer was not utilized as a mesoscopic scaffold. It can be concluded that the uniform TiO_2 compact layer with low roughness is suitable for the preparation of perovskite (Figure 4c), and the optimal thickness ranges around 60 nm. Figure 5a shows the current-voltage curves of PSCs prepared with a TiO_2 solution of 1.0 mg/mL using the w-film regime, with an average PCE of 10.79%, and the highest PCE of 11.23%.

Table 1. The performance of devices fabricated based on the w-film ultrasonically sprayed TiO_2 with different thicknesses tuned by the concentration of TiO_2 dispersion.

Concentration of TiO_2 (mg/mL)	Average Thickness of TiO_2 Layer (nm)	V_{oc} (mV)	J_{sc} (mA/cm^2)	FF (%)	PCE (%)
0.5	30	654	18.74	0.38	4.61
1.0	60	896	17.57	0.69	10.82
2.0	100	838	15.72	0.59	7.81

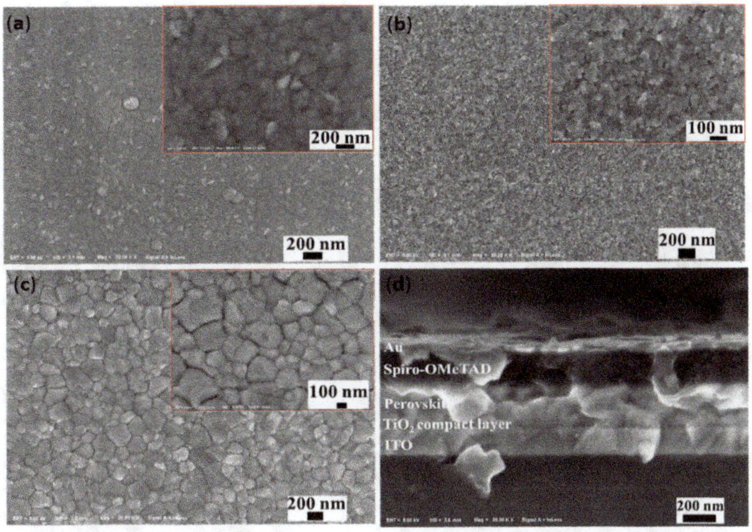

Figure 4. Scanning electron microscopy (SEM) image of (**a**) PEN-ITO substrate; (**b**) TiO_2 compact layer; (**c**) perovskite prepared by the gas-assisted method; and (**d**) cross-section SEM image of the PSCs.

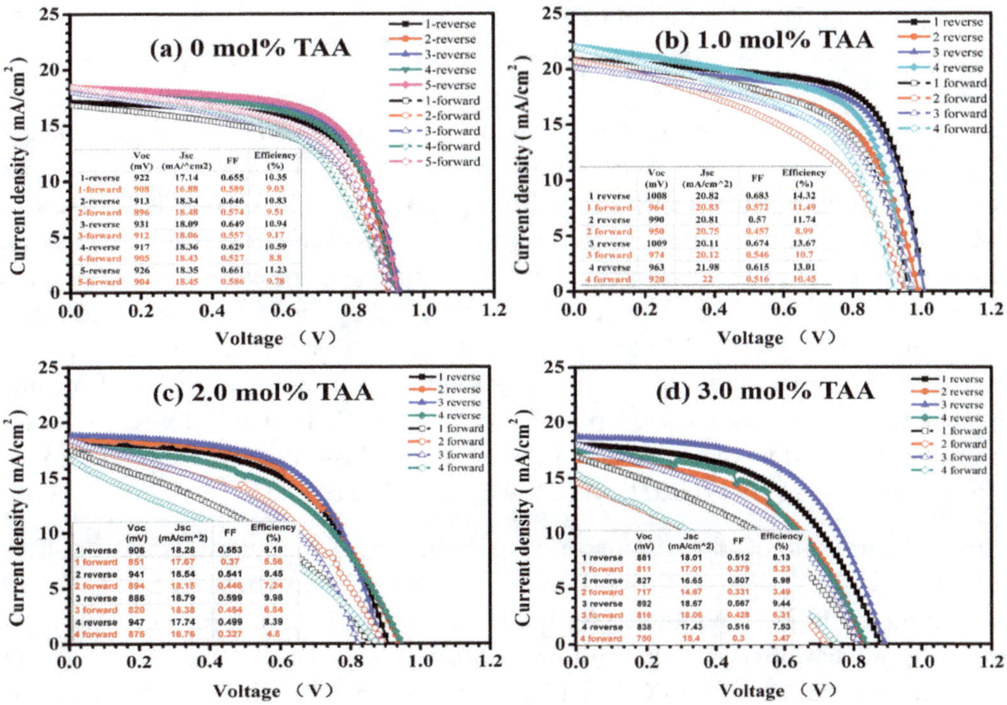

Figure 5. J-V characteristics of the gas-assisted processed PSCs based on TiO_2 compact layer deposited by the ultrasonic spray-coating of TiO_2 solution with the addition of TAA (**a**) 0 mol %; (**b**) 1.0 mol %; (**c**) 2.0 mol %; (**d**) 3.0 mol % with respect to the TiO_2 content.

In addition, more works have been done to improve the performance of TiO_2 by adding a small amount of TAA (1.0 mol %, 2.0 mol %, 3.0 mol % with respect to the TiO_2 content), with almost a negligible influence on the thickness of the TiO_2 film. The solutions with different amounts of TAA were ultrasonically sprayed on the substrates but with a similar thickness of the TiO_2 compact layers, around 60 nm. The RSM increased to 7.6 nm from 3.3 nm without a significant change of the roughness of the bare ITO after 1.0 mol % TAA was added to a TiO_2 solution of 1.0 mg/mL (Figure 6). It was found that the RMS increased with the additional increase of TAA. Furthermore, the increasing size of the large aggregates on the surface of the TiO_2 film was observed in the AFM images (Figure 6).

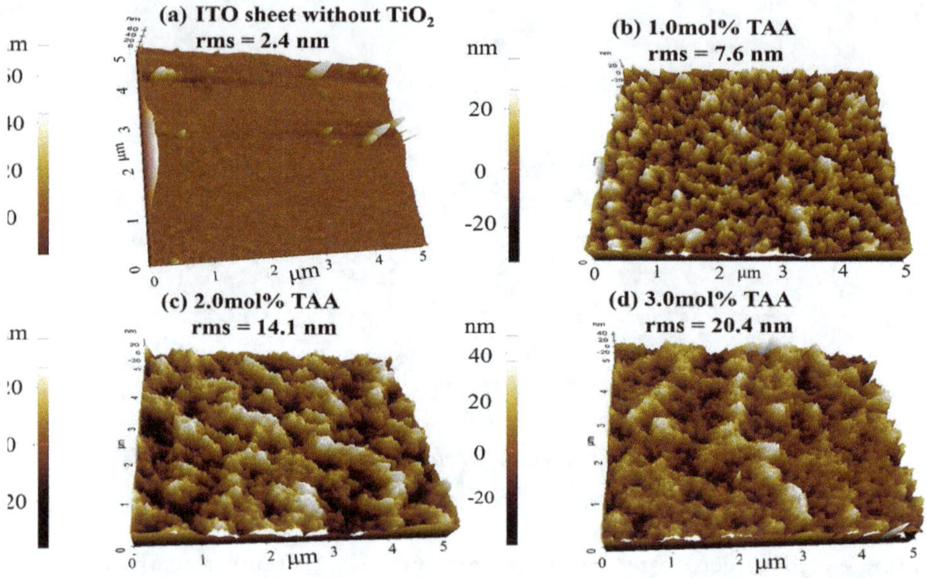

Figure 6. AFM images of the surface of the TiO_2 layers sprayed by the TiO_2 ethanol solution with the addition of TAA, (**a**) ITO sheet without TiO_2 film; (**b**) 1.0 mol % TAA; (**c**) 2.0 mol % TAA; and (**d**) 3.0 mol % TAA with respect to the TiO_2 content.

The devices with the highest performance were prepared based on the TiO_2 compact layer sprayed using the TiO_2 solution with 1.0 mol % of TAA. The highest PCE of 14.32% was achieved, with a V_{oc} of 1.008 mV, a J_{sc} of 20.82 mA/cm^2, and a FF of 68.3%. The improvement of the performance was attributed to the addition of TAA, which enhanced the connection of the TiO_2 nanoparticles. However, the addition of TAA increased the roughness of the film, which had a negative influence on the preparation of the perovskite. It was also observed that the hysteresis of the PSCs increased with the addition of TAA (Figure 5). To study the stability of PSCs based on ultrasonically sprayed TiO_2, a typical PSC on ITO-glass was placed in a N_2-filled glove box for long-duration stability testing. The PSC showed a good stability, as shown in Figure 7. The PCE of the device increased after two days, which can be attributed to the increase of the FF, caused by the enhancement of the connection between the TiO_2 particles and the oxidation of the Spiro-OMeTAD.

The optimized process of the ultrasonic spraying of 1.0 mg/mL TiO_2 solution with an addition of 1.0 mol % TAA was employed to fabricate PSCs on the flexible ITO-PEN substrate. A photographic image of the flexible device with a large area of 5.0 cm × 5.0 cm is shown in Figure 8a. The best-performing cell exhibited an outstanding photovoltaic performance with a V_{oc} of 945 mV, a J_{sc} of 20.68 mA/cm^2, a FF of 55.6% and a PCE of 10.87%, tested under AM 1.5 G sunlight (100 mW/cm^2) using a metal mask of 0.16 cm^2 (Figure 8b), which offers great potential for flexible PSCs with a large area. Figure 9 shows the current-voltage curve of a flexible large-area PSC with an active area of 13.5 cm^2, tested without a mask under standard sunlight. A PCE of 4.33% was achieved for the large-area flexible PSC.

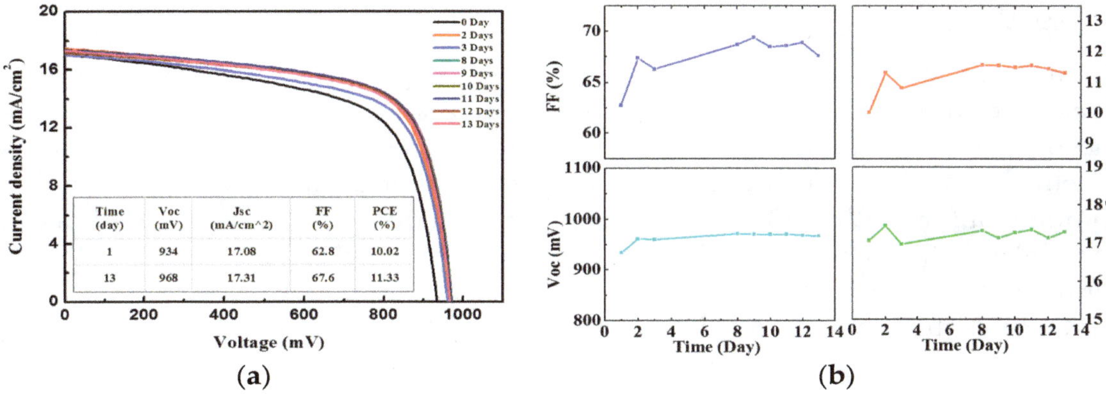

Figure 7. Long-duration stability test of a typical PSC based on ITO-glass, measured under AM 1.5 G sunlight (100 mW/cm^2) using a metal mask of 0.16 cm^2. (**a**) J-V curve of the PSC; and (**b**) FF, PCE, J_{SC}, V_{OC}.

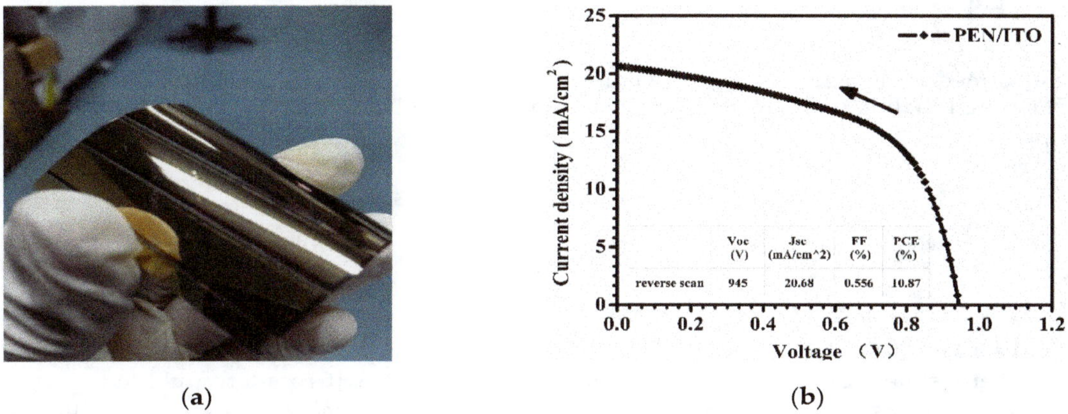

Figure 8. (**a**) Photographic image of the flexible device with a large area of 5.0 cm × 5.0 cm and (**b**) J-V curve of the flexible device fabricated by using the optimized parameters of ultrasonic spraying on polymer ITO-PEN substrate, tested under AM 1.5 G sunlight (100 mW/cm^2) using a metal mask of 0.16 cm^2.

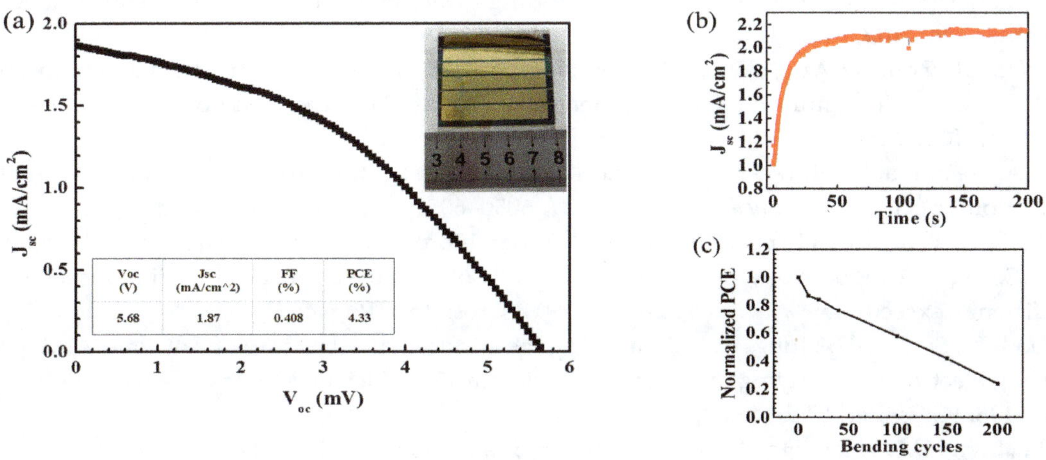

Figure 9. (**a**) J-V characteristics of a typical flexible PSC of 5 cm × 5 cm with an active area of 13.5 cm^2, tested without a mask. Inset photograph is the typical flexible PSC. (**b**) Photocurrent density measured as a function of time for the same flexible PSC held at 2.5 V forward bias. (**c**) Normalized PCE of the large-area flexible PSC after bending tests using 50 mm radii of curvature.

4. Conclusions

In summary, using an ultrasonic spraying technology, we prepared a low-temperature processed TiO_2 compact layer with low roughness compared to the thickness of the TiO_2 film on both rigid and flexible substrates with a large area. Two different drying process-related regimes were studied. The spraying method of the w-film regime showed advantages in preparing low-temperature processed TiO_2, resulting in a film with a RSM of 3.3 nm. A small amount of TAA was added into the TiO_2 dispersion to enhance the connection of the particles, thus improving the performance of the PSCs. The best-performing device was achieved with a PCE of 14.32% based on ITO-glass, and a PCE of 10.87% based on ITO-PEN. This method provides a potential route for the fabrication of efficient large-area flexible PSCs.

Acknowledgments: This work is financially supported by the National Natural Science Foundation of China (NSFC 51672202) and the Fundamental Research Funds for the Central Universities (WUT: 2016IVA085). Hubei Key Laboratory of Low Dimensional Optoelectronic Materials and Devices at Hubei University of Arts and Science was acknowledged for the measurement of AFM. The Analytical and Testing Center of Wuhan University of Technology and the Wuhan National Laboratory for Optoelectronics are also acknowledged for the SEM and TEM characterizations.

Author Contributions: Fuzhi Huang, Juan Zhao, and Peng Zhou conceived and designed the experiments; Peng Zhou performed the experiments; Fuzhi Huang and Peng Zhou analyzed the data; Wangnan Li, Tianhui Li, Tongle Bu, Xueping Liu, Jing Li, Jiang He, Rui Chen and Kunpeng Li contributed characterizations; Peng Zhou wrote the paper.

References

1. Park, N.G. Perovskite solar cells: An emerging photovoltaic technology. *Mater. Today* **2015**, *18*, 65–72. [CrossRef]
2. Kim, H.S.; Im, S.H.; Park, N.G. Organolead halide perovskite: New horizons in solar cell research. *J. Phys. Chem. C* **2014**, *118*, 5615–5625. [CrossRef]
3. Hodes, G. Perovskite-Based Solar Cells. *Science* **2013**, *342*, 317–318. [CrossRef] [PubMed]
4. Bai, S.; Wu, Z.; Wu, X.; Jin, Y.; Zhao, N.; Chen, Z.; Mei, Q.; Wang, X.; Ye, Z.; Song, T.; et al. High-performance planar heterojunction perovskite solar cells: Preserving long charge carrier diffusion lengths and interfacial engineering. *Nano Res.* **2014**, *7*, 1749–1758. [CrossRef]
5. Dong, Q.; Fang, Y.; Shao, Y.; Mulligan, P.; Qiu, J.; Cao, L.; Huang, J. Electron-hole diffusion lengths >175 μm in solution-grown $CH_3NH_3PbI_3$ single crystals. *Science* **2015**, *347*, 967–970. [CrossRef] [PubMed]
6. Stranks, S.D.; Eperon, G.E.; Grancini, G.; Menelaou, C.; Alcocer, M.J.P.; Leijtens, T.; Herz, L.M.; Petrozza, A.; Snaith, H.J. Electron-Hole Diffusion Lengths Exceeding 1 Micrometer in an Organometal Trihalide Perovskite Absorber. *Science* **2013**, *342*, 341–344. [CrossRef] [PubMed]
7. Bu, T.; Wen, M.; Zou, H.; Wu, J.; Zhou, P.; Li, W.; Ku, Z.; Peng, Y.; Li, Q.; Huang, F.; et al. Humidity controlled sol-gel Zr/TiO_2 with optimized band alignment for efficient planar perovskite solar cells. *Sol. Energy* **2016**, *139*, 290–296. [CrossRef]
8. Kojima, A.; Teshima, K.; Shirai, Y.; Miyasaka, T. Organometal halide perovskites as visible-light sensitizers for photovoltaic cells. *J. Am. Chem. Soc.* **2009**, *131*, 6050–6051. [CrossRef] [PubMed]
9. Kim, H.S.; Lee, C.R.; Im, J.H.; Lee, K.B.; Moehl, T.; Marchioro, A.; Moon, S.J.; Humphry-Baker, R.; Yum, J.H.; Moser, J.E.; et al. Lead iodide perovskite sensitized all-solid-state submicron thin film mesoscopic solar cell with efficiency exceeding 9%. *Sci. Rep.* **2012**, *2*, 591. [CrossRef] [PubMed]
10. Mahmood, K.; Swain, B.S.; Jung, H.S. Controlling the surface nanostructure of ZnO and Al-doped ZnO thin films using electrostatic spraying for their application in 12% efficient perovskite solar cells. *Nanoscale* **2014**, *6*, 9127–9138. [CrossRef] [PubMed]
11. Heo, J.H.; Song, D.H.; Han, H.J.; Kim, S.Y.; Kim, J.H.; Kim, D.; Shin, H.W.; Ahn, T.K.; Wolf, C.; Lee, T.W.; et al. Planar $CH_3NH_3PbI_3$ perovskite solar cells with constant 17.2% average power conversion efficiency irrespective of the scan rate. *Adv. Mater.* **2015**, *27*, 3424–3430. [CrossRef] [PubMed]

12. Yang, G.; Tao, H.; Qin, P.; Kw, W.; Fang, G. Recent progress in electron transport layers for efficient perovskite solar cells. *J. Mater. Chem. A* **2016**, *4*, 3970–3990. [CrossRef]
13. Peng, B.; Jungmann, G.; Jager, C.; Haarer, D.; Schmidt, H.W.; Thelakkat, M. Systematic investigation of the role of compact TiO_2 layer in solid state dye-sensitized TiO_2 solar cells. *Coord. Chem. Rev.* **2004**, *248*, 1479–1489. [CrossRef]
14. Kumar, M.H.; Yantara, N.; Dharahi, S.; Graetzel, M.; Mhaisalkar, S.; Boix, P.P.; Mathews, N. Flexible, low-temperature, solution processed ZnO-based perovskite solid state solar cells. *Chem. Commun.* **2013**, *49*, 11089–11091. [CrossRef] [PubMed]
15. Liu, D.; Yang, J.; Kelly, T.L. Compact layer free perovskite solar cells with 13.5% efficiency. *J. Am. Chem. Soc.* **2014**, *136*, 17116–17122. [CrossRef] [PubMed]
16. Kim, B.J.; Kim, D.H.; Lee, Y.Y.; Shin, H.W.; Han, G.S.; Hong, J.S.; Mahmood, K.; Ahn, T.K.; Joo, Y.C.; Hong, K.S.; et al. Highly efficient and bending durable perovskite solar cells: Toward a wearable power source. *Energy Environ. Sci.* **2015**, *8*, 916–921. [CrossRef]
17. Jiang, C.Y.; Koh, W.L.; Leung, M.Y.; Chiam, S.Y.; Wu, J.S.; Zhang, J. Low temperature processing solid-state dye sensitized solar cells. *Appl. Phys. Lett.* **2012**, *100*, 113901. [CrossRef]
18. Munoz-Rojas, D.; Sun, H.; Iza, D.C.; Weickert, J.; Chen, L.; Wang, H.; Schmidt-Mende, L.; MacManus-Driscoll, J.L. High-speed atmospheric atomic layer deposition of ultra thin amorphous TiO_2 blocking layers at 100 °C for inverted bulk heterojunction solar cells. *Prog. Photovolt. Res. Appl.* **2013**, *21*, 393–400.
19. Hart, J.N.; Menzies, D.; Cheng, Y.B.; Simon, G.P.; Spiccia, L. Microwave processing of TiO_2 blocking layers for dye-sensitized solar cells. *J. Sol-Gel Sci. Technol.* **2006**, *40*, 45. [CrossRef]
20. Kim, H.-J.; Jeon, J.-D.; Kim, D. Y.; Lee, J.-J.; Kwak, S.-Y. Improved performance of dye-sensitized solar cells with compact TiO_2 blocking layer prepared using low-temperature reactive ICP-assisted DC magnetron sputtering. *J. Ind. Eng. Chem.* **2012**, *18*, 1807–1812. [CrossRef]
21. Das, S.; Gu, G.; Joshi, P.C.; Yang, B.; Aytug, T.; Rouleau, C.M.; Geohegan, D.B.; Xiao, K. Low thermal budget, photonic-cured compact TiO_2 layers for high-efficiency perovskite solar cells. *J. Mater. Chem. A* **2016**, *4*, 9685. [CrossRef]
22. Wojciechowski, K.; Saliba, M.; Leijtens, T.; Abate, A.; Snaith, H.J. Sub-150 degrees C processed meso-superstructured perovskite solar cells with enhanced efficiency. *Energy Environ. Sci.* **2014**, *7*, 1142–1147. [CrossRef]
23. Yang, D.; Yang, R.; Zhang, J.; Yang, Z.; Liu, S.; Li, C. High efficiency flexible perovskite solar cells using superior low temperature TiO_2. *Energy Environ. Sci.* **2015**, *8*, 3208. [CrossRef]
24. Sanjib, D.; Yang, B.; Gu, G.; Joshi, P.C.; Lvanov, L.N.; Rouleau, C.M.; Aytug, T.; Grohegan, D.B.; Xiao, K. High-performance flexible perovskite solar cells by using a combination of ultrasonic spray-coating and low thermal budget photonic curing. *ACS Photonics* **2015**, *2*, 680–686.
25. Supasai, T.; Henjongchom, N.; Tang, I.-M.; Deng, F.; Rujisamphan, N. Compact nanostructured TiO_2 deposited by aerosol spray pyrolysis for the hole-blocking layer in a $CH_3NH_3PbI_3$ perovskite solar cell. *Solar Energy* **2016**, *136*, 515–524. [CrossRef]
26. Kim, M.-C.; Kim, B.J.; Yoon, J.; Lee, J.-W.; Suh, D.; Park, N.-G.; Choi, M.; Jung, H.S. Electro-spray deposition of a mesoporous TiO_2 charge collection layer: Toward large scale and continuous production of high efficiency perovskite solar cells. *Nanoscale* **2015**, *7*, 20725. [CrossRef] [PubMed]
27. Bose, S.; Keller, S.S.; Alstrom, T.S.; Boisen, A.; Almdal, K. Process optimization of ultrasonic spray coating of polymer films. *Langmuir* **2013**, *29*, 6911–6919. [CrossRef] [PubMed]
28. Tait, J.G.; Rand, B.P.; Heremans, P. Concurrently pumped ultrasonic spray coating for donor: Acceptor and thickness optimization of organic solar cells. *Org. Electron.* **2013**, *14*, 1002–1008. [CrossRef]
29. Wang, J.; Polleux, J.; Lim, J.; Dunn, B. Pseudocapacitive contributions to electrochemical energy storage in TiO_2 (Anatase) nanoparticles. *J. Phys. Chem. C* **2007**, *2*, 14925–14931. [CrossRef]
30. Huang, F.Z.; Dkhissi, Y.; Huang, W.; Xiao, M.; Benesperi, I.; Rubanov, S.; Zhu, Y.; Lin, X.; Jiang, L.; Zhou, Y. Gas-assisted preparation of lead iodide perovskite films consisting of a monolayer of single crystalline grains for high efficiency planar solar cells. *Nano Energy* **2014**, *10*, 10–18. [CrossRef]

31. Li, X.; Bi, D.; Yi, C.; Décoppet, J.D.; Luo, J.; Zakeeruddin, S.M.; Hagfeldt, A.; Grätzel, M. A vacuum flash-assisted solution process for high-efficiency large-area perovskite solar cells. *Science* **2016**, *353*, 58–62. [CrossRef] [PubMed]
32. Deegan, R.; Bakajin, O.; Dupont, T. Capillary flow as the cause of ring stains from dried liquid drops. *Nature* **1997**, *389*, 827–829. [CrossRef]

10

Tunable Adhesion for Bio-Integrated Devices

Zhaozheng Yu [1] and Huanyu Cheng [1,2,*]

[1] Department of Engineering Science and Mechanics, The Pennsylvania State University, University Park, State College, PA 16802, USA; zqy5106@psu.edu
[2] Materials Research Institute, The Pennsylvania State University, University Park, State College, PA 16802, USA
* Correspondence: huanyu.cheng@psu.edu

Abstract: With the rapid development of bio-integrated devices and tissue adhesives, tunable adhesion to soft biological tissues started gaining momentum. Strong adhesion is desirable when used to efficiently transfer vital signals or as wound dressing and tissue repair, whereas weak adhesion is needed for easy removal, and it is also the essential step for enabling repeatable use. Both the physical and chemical properties (e.g., moisture level, surface roughness, compliance, and surface chemistry) vary drastically from the skin to internal organ surfaces. Therefore, it is important to strategically design the adhesive for specific applications. Inspired largely by the remarkable adhesion properties found in several animal species, effective strategies such as structural design and novel material synthesis were explored to yield adhesives to match or even outperform their natural counterparts. In this mini-review, we provide a brief overview of the recent development of tunable adhesives, with a focus on their applications toward bio-integrated devices and tissue adhesives.

Keywords: bio-integrated devices; tissue adhesives; tunable adhesion; dry/wet conditions; soft biological tissue

---

1. Introduction

Although adhesion has long been studied, early efforts focused on the contact between stiff materials [1]. Due to the emerging interest in reconfigurable systems [2] and bio-integrated devices [3,4], adhesion that involves a soft material with different levels of adhesion strength or even a tunable range started attracting attention. Soft materials of interest range from synthetic polymers to biological tissues [5]. Adhesion involving soft materials could be affected by the surface structure/morphology, the deformation of soft materials, and wet/dry conditions, among many others [6–8]. The strategies to design and achieve various levels of adhesion strength can be achieved through structural designs or material innovations. The rapid development in both these classes is greatly promoted by bio-inspiration from several marvelous animals (e.g., gecko, octopus, and mussel), which shed light on the effects of surface roughness, the directionality of the adhesive, and surface chemistry [6,9–11]. Although adhesion is greatly modulated by the properties of the adhesive layer, it is also affected by the target substrates due to interaction at the adhesive–substrate interface; thus, the adhesive has to be specifically designed for each application [12].

When it comes to adhesion to biological tissues, tunable adhesion is of great importance. For instance, strong adhesion to the wound edge is expected in a tissue adhesive to suture the wound [13]. Upon completion of wound healing, a weak adhesion is then desirable for easy removal of the adhesive. Tunable adhesion is also one essential step for realizing repeatable use, as easily removed adhesive could be sanitized and prepared for further use. Current commercial tissue adhesives such as Dermabond® [14] are designed for one-time use. However, when combined with tunable properties, tissue adhesives can be used as an alternative to surgical sutures in clinical practice to eliminate the need for stitch removal. As a decrease in adhesion is observed following multiple uses of the adhesive, strategies to minimize such a decrease need to be explored [15].

When integrated with functional sensing components and actuators, soft adhesives could provide a natural interface to enable bio-integrated devices (e.g., flexible displays [16–18], wellness monitors [19–23], and therapeutic devices [24,25]). The soft nature of the adhesive could minimize discomfort, while strong adhesion improves signal transfer through an enhanced signal-to-noise ratio [26]. Reversible adhesion also allows multiple uses of bio-integrated electronics. Moreover, the capability to modulate interfacial adhesion finds application in the development of transfer printing [27–30], which is widely used for the assembly of heterogeneous materials [31] with applications from wearable devices [3,32–34] to biodegradable sensors [35–39].

In this mini-review, we firstly provide a brief overview of the structural design for adhesives with applications mostly in the dry environment. As extensive review articles exist for dry adhesives [8,40–42], only selected key developments are highlighted here. Next, we discuss material innovations for using adhesives in the wet environment, which are largely based on bio-inspiration from mussels. As special considerations have to be given to the application of adhesives on biological tissue surfaces, we then highlight several recently developed techniques for such applications.

2. Structural Design for Dry Adhesion

Due to its remarkable ability to climb rapidly up a variety of vertical surfaces (Figure 1A (i)), the gecko inspired researchers to uncover the underlying mechanisms behind its significant enhanced, highly robust and repeatable, and reversible adhesion. Observation of the pad area (Figure 1A (ii)) shows nearly 500,000 keratin setae (pillars) (Figure 1A (iii)), with each seta consisting of branches of spatulas that are approximately 200 nm in diameter and 20–60 μm in length (Figure 1A (iv)) [11]. Experimental evidence confirmed that the dry adhesion of gecko setae results from van der Waals forces rather than mechanisms associated with a high surface polarity such as capillary adhesion [43], which indicates that the exceptional adhesion is merely a result of the size and shape of the setae tips. The direct observation of the van der Waals interaction indicates that the adhesion is not affected by the surface chemistry, and repeatable use is possible [40]. In order to reveal the role of the van der Waals interaction on the enhanced adhesion observed in the gecko pad [41,44,45], an array of biomimetic microscopic fibrils on an elastic support was created [41,46]. In direct contrast to a flat surface that only has limited contact to the target substrate with a microscale surface roughness, the array of fibrils with a high aspect ratio in a dense arrangement [47] was observed to form intimate contact with the target substrate due to its low effective Young's modulus and increased effective contact area, especially when a preload was applied (Figure 1B) [48,49]. The principle of contact mechanics was further applied to illustrate that contact splitting (i.e., reducing the radius of the fibril) yielded substantially improved adhesion, and the scaling was found to be applicable to animals differing in weight by six orders of magnitude (Figure 1C) [45,47,50]. The use of soft polymers in most biomimetic systems helps increase the adhesion, but their tacky nature also makes them more susceptible to particulate fouling; thus, a hydrophobic surface with the capacity for self-cleaning is desired. In fact, a fibrillar adhesive can partially transfer particles in a certain size range from its surface to the clean substrate and recover ca. one-third of its shear adhesion [51], as observed in gecko setae [52]. In the practical application where defects commonly exist, the fibrillar structure also localizes the contact failure at individual fibrils and minimizes the effect on contact adhesion, thereby increasing the defect tolerance. Moreover, the adhesion is also affected by the underlying supports. By peeling a polydimethylsiloxane (PDMS) substrate patterned with different hexagonal arrays of cylindrical pillars (to mimic fibrils) from an acrylic adhesive, the enhancement in the adhesion was shown to be more than the increase of the contact area, and this was attributed to the deformation of the underlying support [6]. The fibrillary structure can also be used directly beneath the existing viscoelastic adhesive film (e.g., pressure-sensitive adhesives) to change the dissipative crack trapping and the stress field in the viscoelastic layer for enhanced adhesion [53].

Due to the need for locomotion [54], a reversible adhesion is desirable, and animals such as geckos are observed to use direction to switch from strong to weak adhesion [55]. This direction-dependent adhesion is attributed to the angled fibrils on the gecko's foot, as evidenced by 20° cryo-SEM imaging [56]. While several methods were explored to fabricate vertical structures (e.g., e-beam lithography [57], nano-molding [58], constructing polymers from stiff thermoplastic [59], nano-drawing of stretched polymers [60], and growth of carbon nanotubes [61,62]), it is a challenge to obtain angled structures with high resolution and high aspect ratio, though several attempts were made (e.g., directional exposure in the lithography [63,64], deforming the shape memory polymer of vertical structures from soft lithography [65], post directional e-beam exposure of Pt-coated vertical polyurethane acrylate (PUA) nanohairs [66], and direct laser writing [67]).

In another effort to address this challenge, an angled etching technique was developed, where a Faraday cage introduced in the conventional plasma etching system allows vertical movement of ions to induce an angled etching to the silicon substrate that is placed on an inclined stage [9]. Curing the polymer (e.g., polyurethane acrylate resin) in the etched Si master yields slanted structures with the designed angle and aspect ratio (Figure 1D). Taken together with the ultraviolet (UV)-assisted capillary force lithography, the etched Si master can be further used to create two-level hierarchical PUA hairs for enhanced robustness to a rough surface (<20 μm). In addition to the hierarchical structure [9,68,69], the shape of the tip was also found to have an influential role on adhesion strength [70] (e.g., mushroom-like and spatula tips [50,64,71,72] were shown to have higher adhesion than flat and round tips [73]).

On a separate route to structural design for enhanced adhesion, octopus suckers that reversibly adhere to wettable surfaces provided another source of inspiration [74–77]. The strong adhesion in both dry and wet environments results from the lower pressure in the octopus suckers than that of the environment. Using an external control (e.g., suctioning system [78], vacuum pump [79], dielectric elastomer actuator [80], or magnetic actuated film [81]), the biomimetic system can be easily created. The miniaturization of the system was also achieved through the use of lithographic processes [82,83].

In one attempt to create nanoscale suction cups [84], a non-close-packed self-assembled silica nanoparticle array served as an etching mask to prepare mushroom-like structures consisting of polymer stems and silica caps (Figure 2A (i)). Drying and peeling a polyvinyl alcohol (PVA) film from the etched structure created a replica with embedded silica nanoparticles (Figure 2A (ii)), which formed a mold to yield silicone polymer with nano-sucker structures (Figure 2A (iii)). By controlling the meniscus of a liquid precursor from applied pressure, a simple molding process from the mold with different surface energies could yield artificial micro-suckers with well-controlled cross-sectional profiles [85].

The adhesion in both dry and underwater environments was shown to increase as the curvature of the cross-sectional profile increased, due to the increased contact area from the preload (Figure 2B). In the wet environment, a model that combined the suction effect and capillary interaction [86] captured the experimental observation. When the elastomeric PDMS film with suction-cup structures was covered by thermoresponsive hydrogel of poly(*N*-isopropylacrylamide)(pNIPAM), a resulting smart adhesive pad could respond to temperature change, with an increased temperature inducing an increased volume and decreased pressure in the suction cup due to the deformation of the pNIPAM layer (Figure 2C) [87].

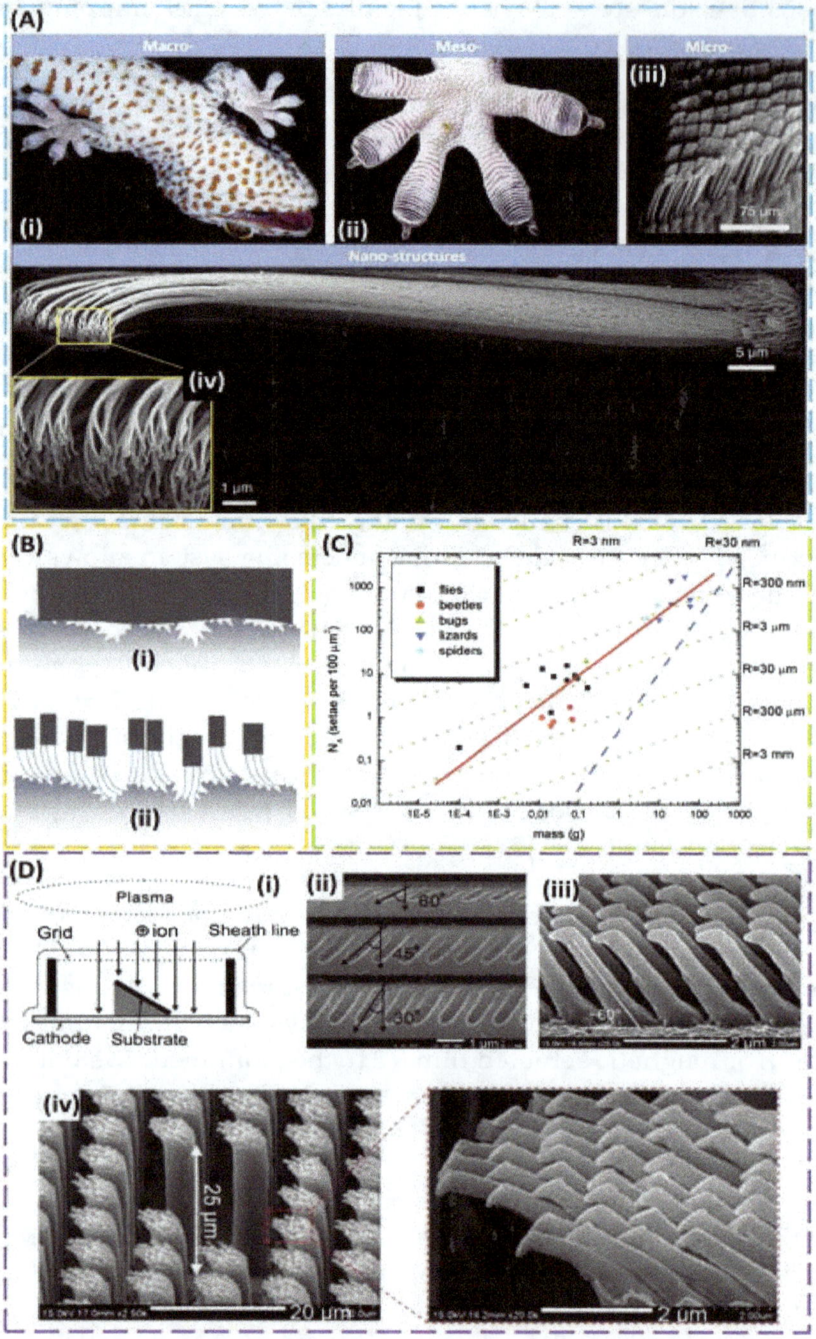

Figure 1. Gecko-inspired dry adhesive. (**A**)(**i**) A close-up view of a gecko climbing a glass wall and (**ii**) mesostructure of gecko toe pad with (**iii**) microscale array of high-aspect-ratio setae structure. (**iv**) Cryo-SEM view of a single seta (ca. 110 µm in length and 4.2 µm in diameter) that branches at the tips into 100–1000 more structures known as spatulas. Reproduced with permission from Reference [56]; (**B**) Schematic illustrations of adhesives with (**i**) a relatively flat surface and (**ii**) an array of fibrils in contact with a rough surface. Reproduced with permission from Reference [49]; (**C**) Dependence of the terminal element density (N_A) of the attachment pads on the body mass (m) in hairy-pad systems of diverse animal groups. The red line that fits all data corresponds to the self-similarity criterion with a slope of ~0.67. The green lines (slope of ~0.33) correspond to the model with a curvature invariance of contacts with radius *R*. The blue line shows the approximate limit of maximum contact for such attachment devices. Reproduced with permission from Reference [47]; (**D**)(**i**) A schematic showing the mechanism of a plasma sheath with a Faraday cage, where

ions are incident on the substrate surface in a direction normal to the grid plane for an angled etching in the silicon substrate. (**ii**) SEM images of polySi etch profiles with angles of 30°/45°/60°, and (**iii**) polyurethane acrylate (PUA) nano-hairs formed on a poly(ethylene terephthalate) (PET) film substrate with a slanted angle of 60°. (**iv**) SEM image of two-level hierarchical PUA hairs with well-defined high aspect ratio over a large area using two-step ultraviolet (UV)-assisted capillary force lithography; the inset shows angled nano-hairs on the tip. Reproduced with permission from Reference [9].

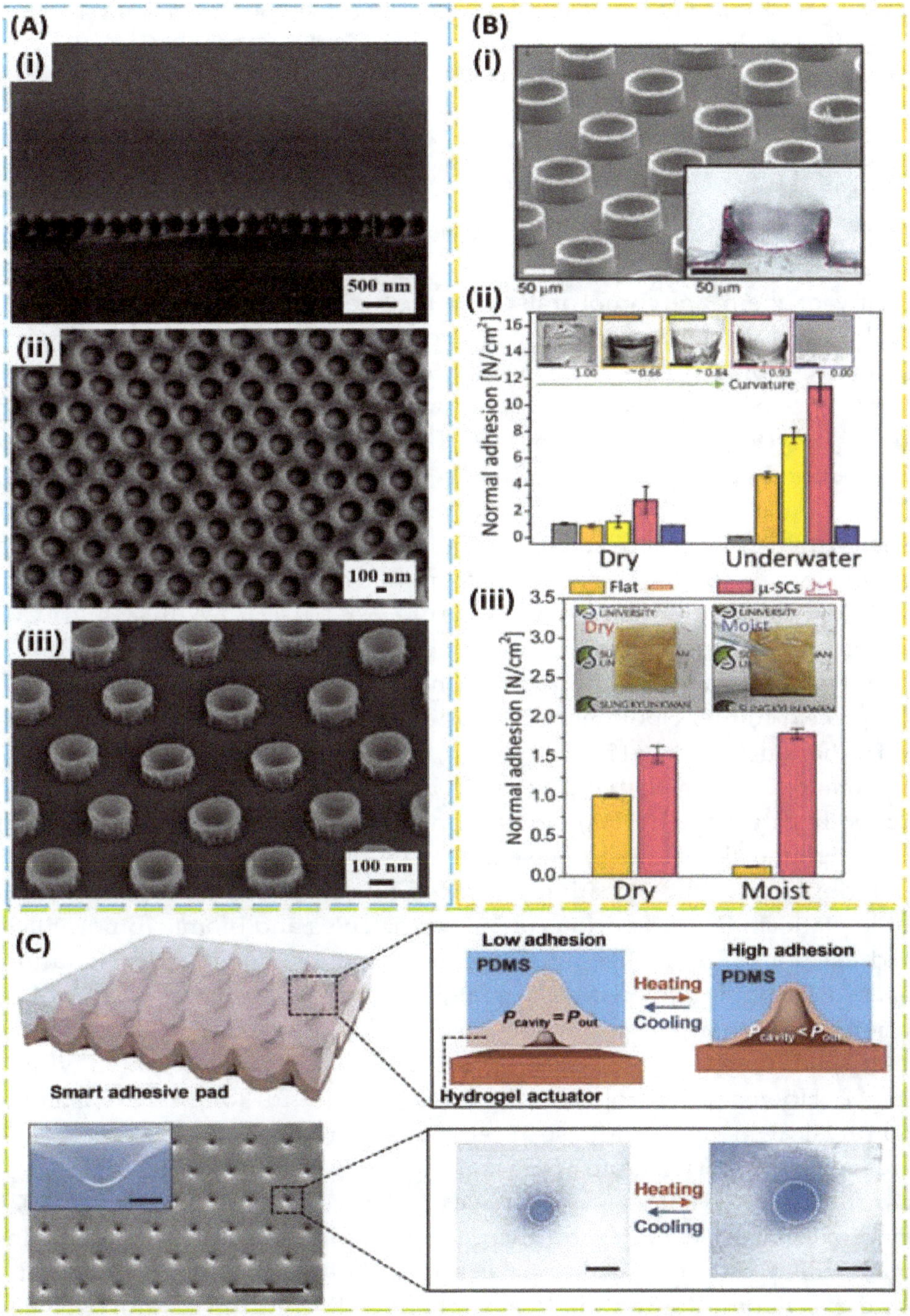

Figure 2. Octopus-inspired adhesive. (**A**) SEM images of (**i**) mushroom-like structures consisting of silica caps and polymer stems on a silicon wafer prepared by plasma etching; (**ii**) a polyvinyl alcohol (PVA) composite film embedding silica nanoparticles; (**iii**) a polydimethylsiloxane (PDMS) nano-sucker array obtained from the PVA composite mold. Reproduced with permission from Reference [84];

(**B**) (**i**) SEM image and cross-sectional optical image of micro-suckers (100 μm in diameter and 75 μm in height); (**ii**) dry/wet adhesion strength increases as the curvature of the cross-sectional profile increases; (**iii**) pull-off strength for PDMS-based micro-suckers and a non-patterned PDMS patch against a rough pigskin, with dry and moist pigskins shown in the inset. Reproduced with permission from Reference [85]; (**C**) Schematic representation and SEM image (scale bar: 10 μm) of micro-cavity arrays within a smart adhesive pad and its corresponding switchable adhesion mechanism (i.e., temperature-dependent hydrogel actuation in response to the environmental temperature). The inset in the bottom right shows SEM images of a single hole on the smart adhesive pad at a low temperature (left, T ≈ 3 °C) and a high temperature (right, T ≈ 40 °C) (scale bar: 500 nm). Reproduced with permission from Reference [85].

3. Design of the Material for Use in Wet Conditions

Though mechanical properties (e.g., the previously discussed structural designs and Young's modulus of the structure [88–90]) showed significant effects on the strength of adhesion in the dry environment, many of them are compromised in the wet environment. In order to address the challenge, several bio-inspired materials [91–93] and their integration with structural designs [7,94] were explored. As a celebrated biological model for wet adhesion [95], mussels were shown to attach virtually all types of inorganic and organic surfaces, including classically adhesion-resistant materials such as poly(tetrafluoroethylene) (PTFE). Clues to this versatility may lie in the amino-acid composition of the specialized adhesive proteins that contain the catecholic amino acid 3,4-dihydroxy-L-phenylalanine (DOPA) and lysine [96]. DOPA and other catechol components perform well as adhesives. With inspiration from both geckos and mussels, a flexible organic nano-adhesive "geckel" was created by dip-coating the gecko-foot-mimetic PDMS pillar array in an ethanol solution of mussel-adhesive-protein-mimetic polymer (Figure 3A (i)) [7]. With a high catechol content, the adhesive monomer, dopamine methacrylamide (DMA), was used in a free-radical polymerization to synthesize poly(dopamine methacrylamide-co-methoxyethyl acrylate) (p(DMA-co-MEA)) as the mussel-adhesive-protein-mimetic polymer. The addition of a p(DMA-co-MEA) coating on the pillars enhanced the wet adhesion by nearly 15 times (Figure 3A (ii)), and this geckel nanoadhesive maintained its adhesive performance for over 1000 contact cycles in both dry and wet environments (Figure 3A (iii)).

Containing both catechol (DOPA) and amine (lysine) functional groups, dopamine as a simple-molecule compound also shows promise to achieve adhesion to a wide spectrum of materials [97]. The adherent polydopamine (PDA) coating produced by self-polymerization of dopamine can also serve as a versatile platform to graft various organic molecules and biomacromolecules for secondary surface-mediated reactions (Figure 3B). Taken together with its biocompatibility and hydrophilicity, polydopamine-based materials demonstrated great potential toward biomedical applications, ranging from cell adhesion/encapsulating/patterning to tissue engineering and re-endothelialization of vascular devices [98,99]. As a versatile building block, PDA was also integrated with other materials such as a hydrogel. However, hydrogel is often associated with long-term instability from water evaporation and physical changes from use at relatively extreme temperatures [100,101]. In order to provide hydrogel with long-term stable operation in a wide temperature window, a glycerol–water (GW) mixture as the binary solvent was used in hydrogel development (Figure 3C) [102], as glycerol is a well-known nontoxic anti-freezing agent. Incorporating PDA-decorated carbon nanotubes (CNTs) as conductive nano-fillers into hydrogel imparts good conductivity (~8 S/m), enhanced toughness (~2000 J/m^2), and excellent adhesion (57 kPa to porcine skin) to the resulting GW hydrogel. Due to its advantages of good adhesiveness and anti-heating or anti-freezing properties, GW hydrogel demonstrated its capability to protect skin from damage in harsh environments (e.g., during frostbite or burn) by serving as an excellent wearable dressing. Other challenges of hydrogel also include foreign body response and poor mechanical properties (i.e., toughness and stretchability). The former could be attenuated by encapsulating mesenchymal stem cells within the hydrogel, such as poly(ethylene glycol) (PEC) [103], and the latter is addressed by the design of tough hydrogel [104–106], discussed in Section 4.

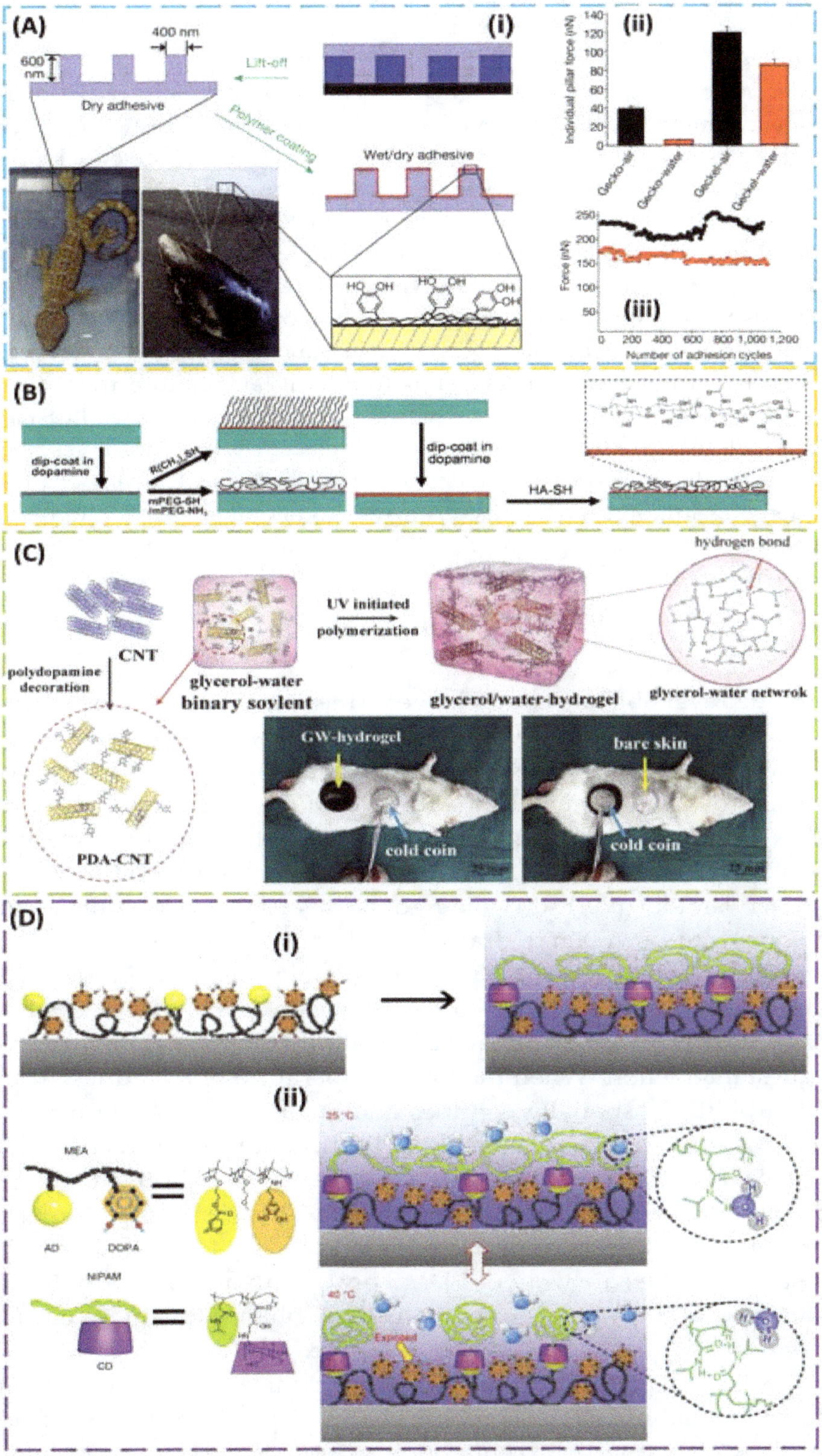

Figure 3. Mussel-inspired wet adhesives. (**A**) (**i**) PDMS casting onto the master is followed by curing, and lift-off results in gecko-foot-mimetic nano-pillar arrays. Next, the fabricated nanopillars are coated with a mussel-adhesive-protein-mimetic polymer that contains catechols, a key component of wet adhesive proteins found in mussel holdfasts. (**ii**) Adhesion force per pillar for gecko and geckel in water (red)

and air (black). (**iii**) Performance of geckel adhesive during multiple contact cycles in water (red) and air (black) with error bars to represent the standard deviation. Reproduced with permission from Reference [7]; (**B**) Schematic illustrations of alkanethiol monolayer and poly(ethylene glycol) (PEG) polymer grafting, as well as the glycosaminoglycan hyaluronic acid (HA) conjugation to polydopamine (PDA)-coated surfaces. Reproduced with permission from Reference [7]; (**C**) Polydopamine decorating carbon nanotubes (PDA-CNTs) in water results in PDA-CNT dispersion. Copolymerizing acrylamide (AM) and acrylic acid (AA) monomers in the PDA-CNT dispersion forms a glycerol–water (GW) hydrogel in a glycerol–water binary solvent. Frostbite and burn models on rats' back skin demonstrate anti-freezing and anti-burning properties of the adhesive GW-hydrogel. Reproduced with permission from Reference [102]; (**D**) (**i**) Schematic diagram showing the specific procedure to prepare the wet adhesive, where dip-coating an adhesive guest copolymer poly(3,4-dihydroxy-L-phenylalanine) (pDOPA)/adamantine (AD)/methoxyethyl acrylate (MEA) on a clean Si substrate is followed by the self-assembly of host copolymer poly(*N*-isopropylacrylamide) (pNIPAM)/cyclodextrin (CD) using host–guest molecular recognition. (**ii**) Schematic drawing showing the tunable wet adhesion that responds to a local temperature trigger. When the local temperature of the adhesive is below lower critical solution temperature (LCST), the pNIPAM easily forms intermolecular hydrogen bonding with adjacent water molecules, and the infused water layer transforms the pNIPAM side chains to a swelling layer, which spatially stabilizes and confines the underlying adhesive moiety, DOPA. On the other hand, heating above the LCST leads to a phase transition and the collapse of pNIPAM-CD chains to form numerous agglomerates, exposing the adhesive group. Reproduced with permission from Reference [10].

In order to provide reversible and tunable wet adhesion that responds to a local temperature trigger in an on-demand manner, a mussel-inspired guest-adhesive copolymer was combined with a thermoresponsive host copolymer [10]. The guest copolymer pDOPA/adamantine (AD)/methoxyethyl acrylate (MEA) consists of a mussel-inspired adhesive DOPA polymer, a guest motif adamantine (AD), and a methoxyethyl acrylate (MEA) monomer as a hydrophobic matrix to enhance the wet adhesion of DOPA (Figure 3D (i)). In the host copolymer pNIPAM/cyclodextrin (CD), the poly(*N*-isopropylacrylamide) (pNIPAM) undergoes a reversible lower critical solution temperature (LCST) phase transition from a swollen hydrated state to a shrunken dehydrated state when heated above the LCST and β-cyclodextrin (β-CD) is the host molecule providing selective binding with the AD moiety in the guest copolymer. Dip-coating the as-prepared guest copolymer on the target substrate surface (e.g., Si, Ti, Al, glass, PTFE, or PDMS) allows the self-assembly of the host copolymer through the host–guest interaction. When the local temperature of the adhesive is below the LCST, the swollen hydrated pNIPAM spatially confines and stabilizes the underneath adhesive moiety, DOPA, through the host–guest interaction, resulting in a dramatically screened interaction area and reduced adhesion. In contrast, the collapsed pNIPAM exposes the adhesive moiety, DOPA, when the local temperature is above LCST (Figure 3D (ii)). The versatile demonstration of the wet adhesive also goes from inorganic (Si, Ti, Al, and glass) to organic surfaces (PDMS and PTFE). In addition, the gecko-like surface structure (e.g., an array of PDMS posts with a diameter of 5 μm and and height of 10 μm), discussed in Section 2, was explored to further enhance the interfacial adhesion strength, which is in direct contrast with the gecko-like dry adhesive.

Although mussel-like wet adhesion was successfully realized, typical catechol functionalization and solution processing entail complex components and steps. In order to reduce the complexity, synthetic low-molecular-weight catecholic zwitterionic surfactants were developed to adhere to diverse surfaces with very strong adhesion (~50 mJ/m^2) [107]. Based on catechol-modified amphiphilic poly(propylene oxide)/poly(ethylene oxide) (PPO-PEO) block copolymers, a mechanically tough zero- or negative-swelling mussel-inspired surgical adhesive was synthesized, minimizing the weakening mechanism from swelling [108]. The range of zero to −25 % swelling was achieved through a hydrophobic collapse of PPO blocks upon heating to physiological temperature. The lap shear adhesion measurements of decellularized porcine dermis show nearly 50 kPa adhesive strength. Although the single-layer mussel-like adhesion is effective, a layer-by-layer (LbL) assembly may be explored to further enhance the adhesion strength due to the versatile control in the assembly process (e.g., introducing sodium chloride in the assembly process yields an adhesion enhanced by two orders of magnitude [109]).

4. Adhesion to Biological Tissues

When it comes to adhesion to biological tissues such as skin, several additional challenges are encountered, including soft properties, multiscale roughness, and biocompatibility. For instance, adhesives based on chemical bonding may irritate the skin and cause discomfort upon removal due to strong adhesion. Though several commercial adhesives were used in bio-integrated electronics on the skin [110,111], their applications are limited by their given properties, and the adhesion strength was also shown to be dependent on the target tissue. Taking a synthetic tissue adhesive (i.e., Dermabond®, 2-octyl cyanoacrylate) as an example, its adhesion to collagen films was observed to be 40 times that when compared with its adhesion to muscle tissue, due to increased wetting (and the decreased contact angle) of the Dermabond® adhesive on the collagen film [12]. In the two classes of tissue adhesives, biologic (e.g., fibrin glue) and synthetic (e.g., *n*-butyl-2-cyanoacrylate) [112], a variety of different bonding mechanisms were explored (e.g., physical interaction, mechanical interlocking, and chemical bonding) [113,114]. As an extensively used synthetic polymer for tissue engineering, polyethylene glycol (PEG) was used with chondroitin sulfate (CS) to form a biodegradable CS–PEG adhesive hydrogel that can covalently bond to proteins in tissue or to collagen in the extracellular matrix via amide bonds, improving the adhesion strength by ten times that of fibrin glue (Figure 4A) [115]. In a separate effort, a buckypaper (BP) film produced from oxidized multi-walled carbon nanotubes demonstrated enhanced adhesion to the rimmed muscular fascia of the abdominal wall of New Zealand female rabbit during both peeling and shearing tests, due to soft tissue deformation from water suction resulting in water bridge formation and BP–tissue mechanical interlocking, respectively (Figure 4B) [116].

Because of the self-cleaning effect and adaptation capability, the gecko-inspired fibrillar adhesive demonstrated repeatable and restorable adhesion to the skin surface over multiple cycles of use [5]. Though the adhesion does not show a direct correlation with classical roughness parameters, strong adhesion was shown to decrease significantly when surface roughness increased [117–119] and a clear correlation was even observed when a newly integrated roughness parameter was introduced [120]. In order to address these challenges, special considerations have to be given to the design of the adhesion layer. In one effort to utilize the unique advantages of the gecko-inspired fibrillar adhesive, a composite adhesive was designed by coating polymer microfibers with a skin interfacing material (e.g., vinylsiloxane) to form mushroom-shaped tips (Figure 4C) [26]. As a member of the family of silicone rubbers, the biocompatible vinylsiloxane (VS) approved for biomedical applications (e.g., forming dental impression) can fully cross-link and form covalent bonds with the PDMS microfibers within a few minutes at room temperature, enabling its direct cross-linking on and conformal contact to the skin surface with multiscale roughness. Due to the high flexibility and strong attachment to the skin surface, the adhesive layer shows an efficient strain signal to the strain sensor integrated on top, thereby significantly increasing the signal-to-noise ratio when compared with medical tape or fibrillar adhesive film fully immersed into a flat vs. film. The adhesion strength of the composite adhesive to a wet skin surface was also shown to be comparable to that of a dry environment. In contrast to the gecko-inspired adhesive, the octopus-inspired adhesive that relies on the pressure difference was less affected by the surface roughness. Such an adhesive also showed comparable adhesion strength to pigskin in moist conditions (40% of area covered with droplets) to that in dry conditions, even when hairs were present [85]. When integrated with physiological sensors and drug-delivery actuators, the octopus-inspired adhesive allowed sensitive biometric measurements and transdermal drug delivery through tight skin coupling (Figure 4D) [83].

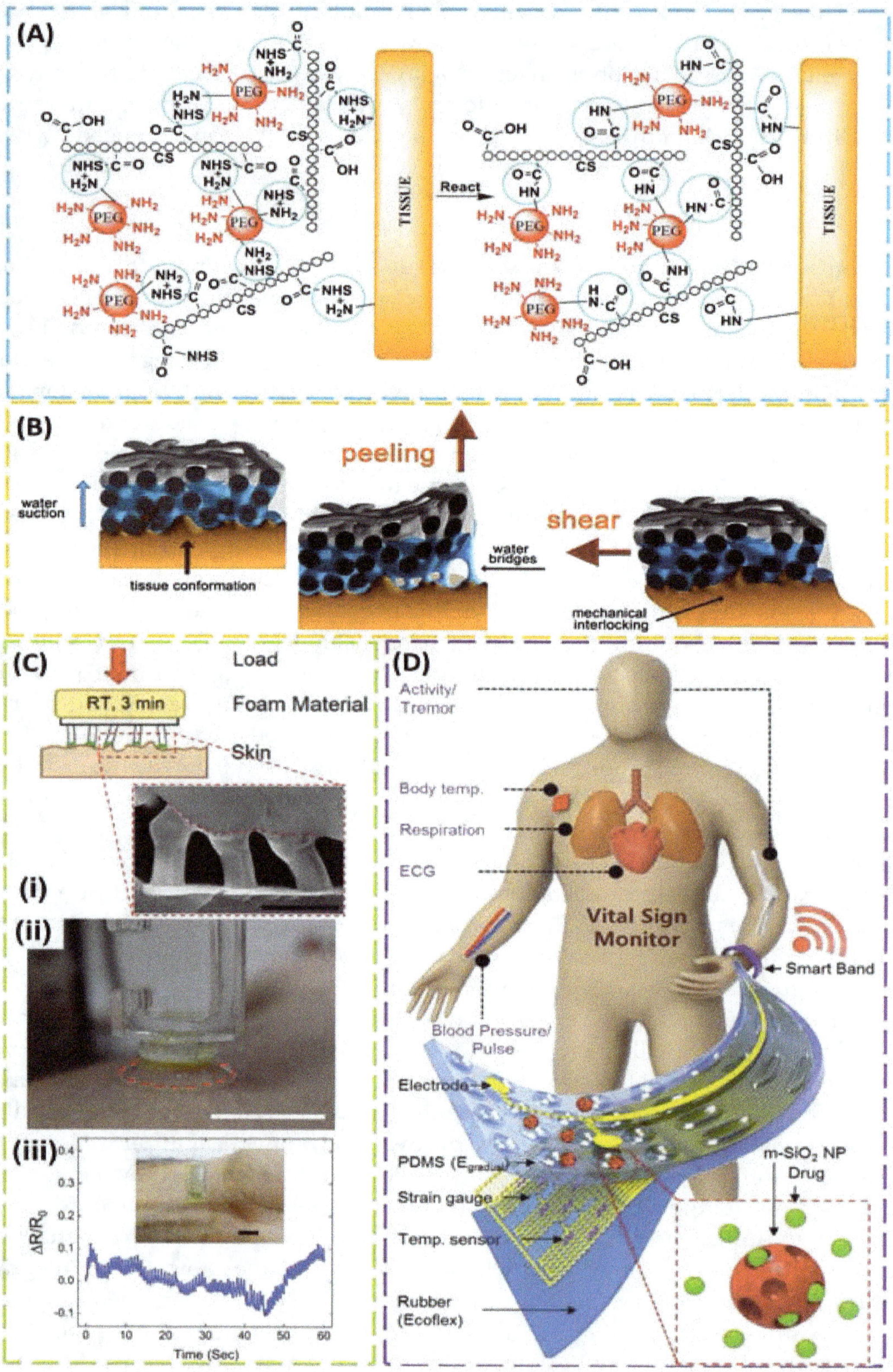

Figure 4. Adhesives on soft tissues. (**A**) Chondroitin sulfate *N*-succinimidyl succinate (CS–NHS) reacts with primary amines of both six-arm polyethylene glycol amine PEG–$(NH_2)_6$ and proteins of the tissue to form a covalently bonded hydrogel to the tissue. Reproduced with permission from Reference [115]; (**B**) When the buckypaper (BP) comes into contact with the wet biological tissue, water suction leads to deformation of the soft tissue and conformal contact of the BP to the tissue.

The resulting water bridge formation and BP–tissue mechanical interlocking help enhance the adhesion during peeling and shearing tests. Reproduced with permission from Reference [116]; (**C**) (**i**) Inking vinylsiloxane (VS) precursor onto uniformly shaped cylindrical microfibers followed by curing vs. directly onto the skin surface. The inset shows the cross-sectional SEM image of an adhesive film attached to an artificial skin replica, where the red dashed line indicates the interface between the skin-adhesive film and artificial skin. (**ii**) Optical image of a skin-adhesive film attached to the human forearm during the retraction cycle of the adhesion experiment (dashed red line indicates the interfacing border between the adhesive film and the skin; scale bar: 1 cm). (**iii**) The output signal from a microfibrillar strain sensor mounted onto the radial artery of the wrist; the inset shows a photograph of the strain sensor attached to the wrist through the inking and printing process. Reproduced with permission from Reference [26]; (**D**) Schematic illustration of the multifunctional electronic patch integrated onto an octopus-inspired adhesive layer with miniaturized suction cups (mSCs). The multifunctional electronic patch is capable of monitoring vital signs from electrodes, strain gauges, and temperature sensors, as well as transdermally delivering drugs loaded in mesoporous silica nanoparticles via iontophoresis. The smart band (connected to the mSC patch) is used to provide wireless functionalities and a power source. Reproduced with permission from Reference [83].

Relying on relatively weak physical interactions, existing tissue adhesives (including mussel-inspired adhesives) are associated with low adhesive energy on the order of 10 J/m^2 [121], which is far from ideal, especially when compared with the example in nature. For instance, cartilage bonds to bones with an adhesion energy of 800 J/m^2 [122]. In order to achieve high adhesion, a synergy from an adhesive surface layer and a dissipative matrix was explored (Figure 5A) [105]. The design was inspired by a sticky and tough secretion from slug *Arion subfuscus* [123], which may arise from two interpenetrating networks of polymers [124]. The strong adhesion from the adhesive surface layer to the tissue substrate can be achieved through electrostatic interactions, covalent bonds, or physical interpenetration. Meanwhile, the energy dissipation through hysteresis in the matrix amplifies the effective adhesion energy. As the surface of tissues or cells is negatively charged, a bridging polymer that bears positively charged primary amine groups enables covalent binding via electrostatic attraction. In the case of a permeable target surface, the bridging polymer penetrating into the target forms a physical entanglement and a chemical anchor for the adhesive. As for the dissipative matrix, a substrate that can dissipate energy is used. By exploiting the synergy of these two factors, a class of tough adhesives demonstrated high adhesion energy (~1000 J/m^2) on wet surfaces. In vivo demonstrations included strong adhesion to a beating porcine heart in the presence of blood, heart sealants to prevent liquid leakage, and hemostatic dressing for a deep wound (Figure 5B). This simple yet effective strategy opens up a wide range of applications, including tissue adhesives, wound dressing, and tissue repair.

In order to provide a reversible adhesion that can respond to the external stimuli, a responsive polymer was explored to diffuse into and form an entangled network with two polymer networks of two wet materials for enhanced adhesion upon the trigger of one signal (e.g., in one pH range), while being soluble and separating both wet materials upon trigger removal (e.g., in the other pH range) (Figure 5C) [125]. In the demonstration, several stitching polymers were identified to cover the full range of pH (e.g., cellulose forms a network for pKa < 13, alginate for pKa < 3.5, chitosan for pKa > 6.5, and poly(4-aminostyrene) for pKa > 4.5). Adhesion energy as high as 1000 J/m^2 could be achieved when the stitching polymer introduced hysteresis in the wet hydrogel materials. The demonstrated strong adhesion also went beyond the hydrogel to various porcine tissues (e.g., skin, liver, heart, artery, and stomach) and the skin was shown to exhibit a relatively high adhesion energy (100 J/m^2) due to its relatively high toughness.

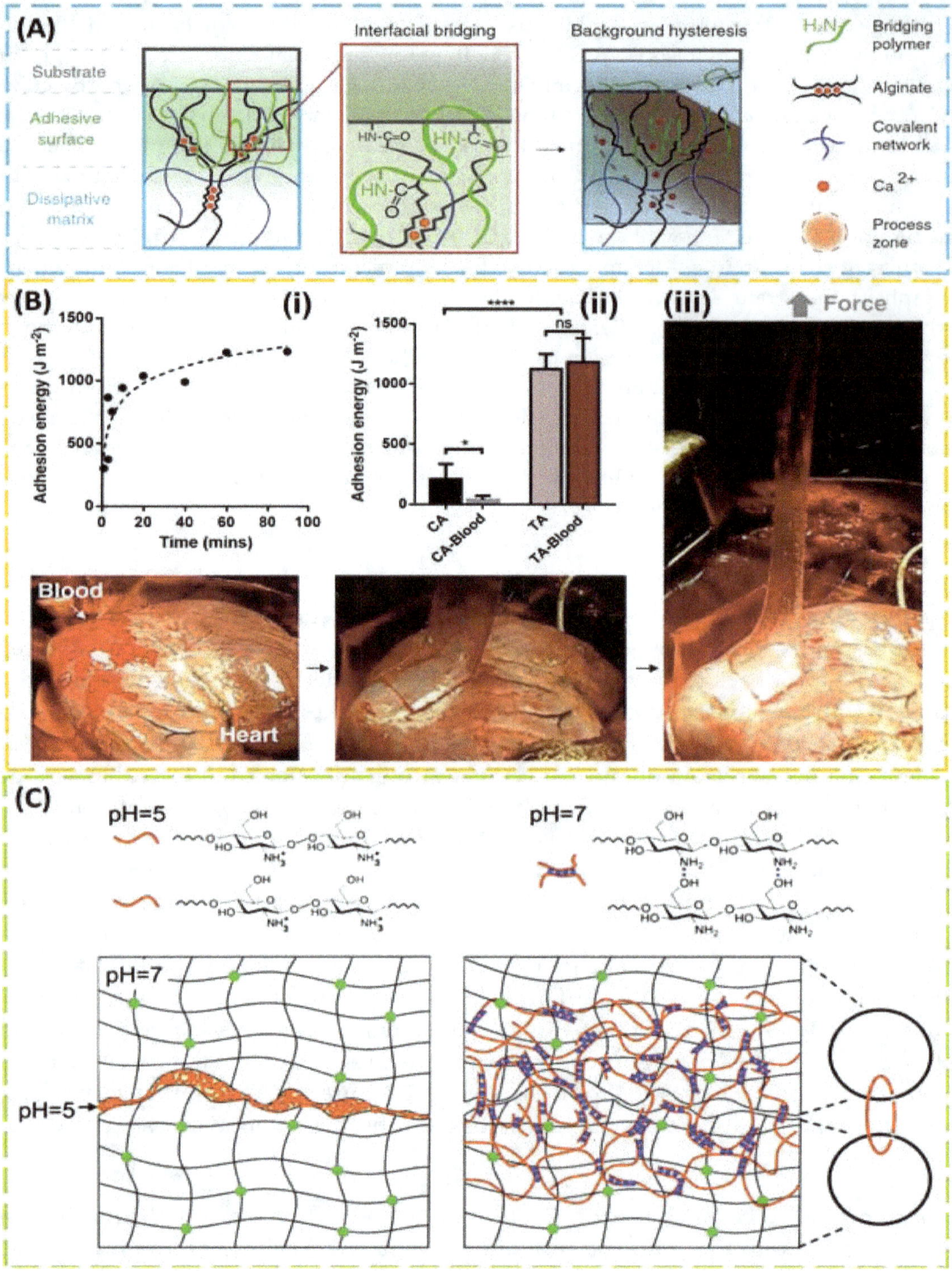

Figure 5. Tough and topological adhesives. (**A**) Tough adhesives (TAs) consist of a dissipative matrix (light-blue square), made of a hydrogel containing both ionically (calcium; red circles) cross-linked and covalently cross-linked polymers (black and blue lines), and an adhesive surface that contains a bridging polymer with primary amines (green lines). The bridging polymer can penetrate into the TA and the substrate (light-green region) to facilitate covalent-bond formation. In the presence of a crack, the process zone (orange area) dissipates significant amounts of energy as ionic bonds between alginate chains and calcium ions break. (**B**) (**i**) Tough adhesives exhibit a rapid increase in adhesion energy to porcine skin over time. (**ii**) Comparing with cyanoacrylate (CA), TAs showed strong adhesion even when the porcine skin was exposed to blood in the in vitro experiment. n = 4–6. (**iii**) The TAs were

further tested in an in vivo experiment on a beating porcine heart with blood exposure. (**A**,**B**) Reproduced with permission from References [105]; (**C**) Chitosan chains dissolve in water at pH 5 and form a network in water at pH 7. Placing an aqueous solution of chitosan of pH 5 between two hydrogels (or biological tissues) of pH 7 is followed by the diffusion of chitosan chains into the two hydrogels, forming a network that topologically entangles with the networks of both hydrogels. Reproduced with permission from Reference [125].

5. Conclusions and Future Perspectives

In order to robustly adhere bio-integrated devices to soft biological tissues, an adhesive layer with tunable adhesion is of great interest. The capability to switch between strong and weak adhesion would allow the use of strong adhesion to efficiently transfer vital signals to the device for accurate measurement, followed by the use of weak adhesion for easy removal. However, this tunable adhesion to the tissue surface has long been a challenge due to multiscale roughness, wet conditions, biocompatibility, and natural motion, among many others. Thanks to the recent developments that shed light on the underlying mechanisms of the remarkable adhesion observed in several animal species, great strides were made, and effective strategies ranging from structural design for dry adhesion to novel material synthesis for wet conditions were explored to yield adhesives that can match or even outperform those from nature. The importance of the developed adhesives also goes beyond bio-integrated devices to cell culture [126,127] and to tissue glues that can potentially replace sutures in clinical practice.

When combined with the tunable properties of the adhesive, tissue glues would promise repeatable use, which can dramatically reduce the cost and pave the way for commercialization. Despite great strides made in the field of tissue adhesives, several challenges still exist, including fabricating high-aspect-ratio fibrillar structures with diameters down to submicron scales [40], long-term reliability of the tissue adhesives to diverse wet surfaces, tunable properties in the adhesives to accommodate dynamic changes in target tissues, and integration with multifunctional electronics for real-time sensing and closed-loop control [128,129].

In the burgeoning field of tissue adhesives, different testing methods and tissue models are used to evaluate the adhesive properties of newly developed structures and materials. Thus, it is a bit challenging to directly compare the results reported by different research groups. It would be desirable to have standardized testing procedures and tissue models in place to allow for direct comparison among the newly developed tissue adhesives. Nevertheless, the challenges simply represent a small fraction of the great opportunities for future development, which may require the collective wisdom of material scientists, chemists, mechanical engineers, and clinicians, among many others.

Author Contributions: Z.Y. and H.C. wrote the manuscript and assembled the figures. H.C. led the preparation of the manuscript and contributed to editorial modifications of the overall text.

References

1. Shull, K.R. Contact mechanics and the adhesion of soft solids. *Mater. Sci. Eng. R Rep.* **2002**, *36*, 1–45. [CrossRef]
2. Zhu, J.; Dexheimer, M.; Cheng, H. Reconfigurable systems for multifunctional electronics. *npj Flex. Electron.* **2017**, *1*, 8. [CrossRef]
3. Choi, S.; Lee, H.; Ghaffari, R.; Hyeon, T.; Kim, D.H. Recent Advances in Flexible and Stretchable Bio-Electronic Devices Integrated with Nanomaterials. *Adv. Mater.* **2016**, *28*, 4203–4218. [CrossRef] [PubMed]

4. Rogers, J.A.; Someya, T.; Huang, Y. Materials and mechanics for stretchable electronics. *Science* **2010**, *327*, 1603–1607. [CrossRef] [PubMed]
5. Kwak, M.K.; Jeong, H.E.; Suh, K.Y. Rational design and enhanced biocompatibility of a dry adhesive medical skin patch. *Adv. Mater.* **2011**, *23*, 3949–3953. [CrossRef] [PubMed]
6. Poulard, C.; Restagno, F.; Weil, R.; Léger, L. Mechanical tuning of adhesion through micro-patterning of elastic surfaces. *Soft Matter* **2011**, *7*, 2543–2551. [CrossRef]
7. Lee, H.; Lee, B.P.; Messersmith, P.B. A reversible wet/dry adhesive inspired by mussels and geckos. *Nature* **2007**, *448*, 338–341. [CrossRef] [PubMed]
8. Boesel, L.F.; Cremer, C.; Arzt, E.; Del Campo, A.; Greiner, C.; Arzt, E.; del Campo, A. Gecko-inspired surfaces: A path to strong and reversible dry adhesives. *Adv. Mater.* **2010**, *22*, 2125–2137. [CrossRef] [PubMed]
9. Jeong, H.E.; Lee, J.-K.; Kim, H.N.; Moon, S.H.; Suh, K.Y. A nontransferring dry adhesive with hierarchical polymer nanohairs. *Proc. Natl. Acad. Sci. USA* **2009**, *106*, 5639–5644. [CrossRef] [PubMed]
10. Zhao, Y.; Wu, Y.; Wang, L.; Zhang, M.; Chen, X.; Liu, M.; Fan, J.; Liu, J.; Zhou, F.; Wang, Z. Bio-inspired reversible underwater adhesive. *Nat. Commun.* **2017**, *8*, 2218. [CrossRef] [PubMed]
11. Campolo, D.; Jones, S.; Fearing, R.S. Fabrication of gecko foot-hair like nano structures and adhesion to random rough surfaces. In Proceedings of the 2003 Third IEEE Conference on Nanotechnology, San Francisco, CA, USA, 12–14 August 2003.
12. Bochyńska, A.I.; Hannink, G.; Buma, P.; Grijpma, D.W. Adhesion of tissue glues to different biological substrates. *Polym. Adv. Technol.* **2017**, *28*, 1294–1298. [CrossRef]
13. Bruns, T.B.; Worthington, J.M. Using tissue adhesive for wound repair: A practical guide to Dermabond. *Am. Fam. Physician* **2000**, *61*, 1383–1388. [PubMed]
14. Dermabond, Ò.; Corneal, S.; Leung, G.Y.S.; Peponis, V.; Varnell, E.D.; Lam, D.S.C.; Kaufman, H.E. Preliminary In Vitro Evaluation of 2-Octyl Cyanoacrylate to seal corneal incisions. *Cornea* **2005**, *24*, 998–999.
15. Bae, W.G.; Kim, D.; Kwak, M.K.; Ha, L.; Kang, S.M.; Suh, K.Y. Enhanced Skin Adhesive Patch with Modulus-Tunable Composite Micropillars. *Adv. Healthc. Mater.* **2013**, *2*, 109–113. [CrossRef] [PubMed]
16. Gelinck, G.H.; Huitema, H.E.A.; Van Veenendaal, E.; Cantatore, E.; Schrijnemakers, L.; Van Der Putten, J.B.P.H.; Geuns, T.C.T.; Beenhakkers, M.; Giesbers, J.B.; Huisman, B.H.; et al. Flexible active-matrix displays and shift registers based on solution-processed organic transistors. *Nat. Mater.* **2004**, *3*, 106–110. [CrossRef] [PubMed]
17. Kim, R.H.; Bae, M.H.; Kim, D.G.; Cheng, H.; Kim, B.H.; Kim, D.H.; Li, M.; Wu, J.; Du, F.; Kim, H.S.; et al. Stretchable, transparent graphene interconnects for arrays of microscale inorganic light emitting diodes on rubber substrates. *Nano Lett.* **2011**, *11*, 3881–3886. [CrossRef] [PubMed]
18. Sekitani, T.; Nakajima, H.; Maeda, H.; Fukushima, T.; Aida, T.; Hata, K.; Someya, T. Stretchable active-matrix organic light-emitting diode display using printable elastic conductors. *Nat. Mater.* **2009**, *8*, 494–499. [CrossRef] [PubMed]
19. Liao, X.; Liao, Q.; Yan, X.; Liang, Q.; Si, H.; Li, M.; Wu, H.; Cao, S.; Zhang, Y. Flexible and highly sensitive strain sensors fabricated by pencil drawn for wearable monitor. *Adv. Funct. Mater.* **2015**, *25*, 2395–2401. [CrossRef]
20. Trung, T.Q.; Lee, N.E. Flexible and Stretchable Physical Sensor Integrated Platforms for Wearable Human-Activity Monitoringand Personal Healthcare. *Adv. Mater.* **2016**, *28*, 4338–4372. [CrossRef] [PubMed]
21. Jang, K.I.; Han, S.Y.; Xu, S.; Mathewson, K.E.; Zhang, Y.; Jeong, J.W.; Kim, G.T.; Webb, R.C.; Lee, J.W.; Dawidczyk, T.J.; et al. Rugged and breathable forms of stretchable electronics with adherent composite substrates for transcutaneous monitoring. *Nat. Commun.* **2014**, *5*, 4779. [CrossRef] [PubMed]
22. Webb, R.C.; Bonifas, A.P.; Behnaz, A.; Zhang, Y.; Yu, K.J.; Cheng, H.; Shi, M.; Bian, Z.; Liu, Z.; Kim, Y.S.; et al. Ultrathin conformal devices for precise and continuous thermal characterization of human skin. *Nat. Mater.* **2013**, *12*, 938–944. [CrossRef] [PubMed]
23. Xu, L.; Gutbrod, S.R.; Bonifas, A.P.; Su, Y.; Sulkin, M.S.; Lu, N.; Chung, H.J.; Jang, K.I.; Liu, Z.; Ying, M.; et al. 3D multifunctional integumentary membranes for spatiotemporal cardiac measurements and stimulation across the entire epicardium. *Nat. Commun.* **2014**, *5*, 3329. [CrossRef] [PubMed]
24. Dewire, J.; Calkins, H. State-of-the-art and emerging technologies for atrial fibrillation ablation. *Nat. Rev. Cardiol.* **2010**, *7*, 129–138. [CrossRef] [PubMed]
25. Son, D.; Lee, J.; Qiao, S.; Ghaffari, R.; Kim, J.; Lee, J.E.; Song, C.; Kim, S.J.; Lee, D.J.; Jun, S.W.; et al. Multifunctional wearable devices for diagnosis and therapy of movement disorders (Support Information). *Nat. Nanotechnol.* **2014**, *9*, 397–404. [CrossRef] [PubMed]

26. Drotlef, D.M.; Amjadi, M.; Yunusa, M.; Sitti, M. Bioinspired Composite Microfibers for Skin Adhesion and Signal Amplification of Wearable Sensors. *Adv. Mater.* **2017**, *29*, 1–8. [CrossRef] [PubMed]
27. Carlson, A.; Bowen, A.M.; Huang, Y.; Nuzzo, R.G.; Rogers, J.A. Transfer printing techniques for materials assembly and micro/nanodevice fabrication. *Adv. Mater.* **2012**, *24*, 5284–5318. [CrossRef] [PubMed]
28. Yang, S.Y.; Carlson, A.; Cheng, H.; Yu, Q.; Ahmed, N.; Wu, J.; Kim, S.; Sitti, M.; Ferreira, P.M.; Huang, Y.; et al. Elastomer surfaces with directionally dependent adhesion strength and their use in transfer printing with continuous roll-to-roll applications. *Adv. Mater.* **2012**, *24*, 2117–2122. [CrossRef] [PubMed]
29. Kim, S.; Carlson, A.; Cheng, H.; Lee, S.; Park, J.K.; Huang, Y.; Rogers, J.A. Enhanced adhesion with pedestal-shaped elastomeric stamps for transfer printing. *Appl. Phys. Lett.* **2012**, *100*, 171909. [CrossRef]
30. Carlson, A.; Kim-Lee, H.J.; Wu, J.; Elvikis, P.; Cheng, H.; Kovalsky, A.; Elgan, S.; Yu, Q.; Ferreira, P.M.; Huang, Y.; et al. Shear-enhanced adhesiveless transfer printing for use in deterministic materials assembly. *Appl. Phys. Lett.* **2011**, *98*, 264104. [CrossRef]
31. Gao, Y.; Cheng, H. Assembly of Heterogeneous Materials for Biology and Electronics: From Bio-Inspiration to Bio-Integration. *J. Electron. Packag.* **2017**, *139*, 020801. [CrossRef]
32. Wang, T.; Ramnarayanan, A.; Cheng, H. Real time analysis of bioanalytes in healthcare, food, zoology and botany. *Sensors* **2018**, *18*, 5. [CrossRef] [PubMed]
33. Cheng, H.; Yi, N. Dissolvable tattoo sensors: From science fiction to a viable technology. *Phys. Scr.* **2017**, *92*, 13001. [CrossRef]
34. Stoppa, M.; Chiolerio, A. Wearable electronics and smart textiles: A critical review. *Sensors* **2014**, *14*, 11957–11992. [CrossRef] [PubMed]
35. Hwang, S.W.; Tao, H.; Kim, D.H.; Cheng, H.; Song, J.K.; Rill, E.; Brenckle, M.A.; Panilaitis, B.; Won, S.M.; Kim, Y.S.; et al. A physically transient form of silicon electronics. *Science* **2012**, *337*, 1640–1644. [CrossRef] [PubMed]
36. Hwang, S.W.; Park, G.; Cheng, H.; Song, J.K.; Kang, S.K.; Yin, L.; Kim, J.H.; Omenetto, F.G.; Huang, Y.; Lee, K.M.; et al. 25th anniversary article: Materials for high-performance biodegradable semiconductor devices. *Adv. Mater.* **2014**, *26*, 1992–2000. [CrossRef] [PubMed]
37. Hwang, S.W.; Song, J.K.; Huang, X.; Cheng, H.; Kang, S.K.; Kim, B.H.; Kim, J.H.; Yu, S.; Huang, Y.; Rogers, J.A. High-performance biodegradable/transient electronics on biodegradable polymers. *Adv. Mater.* **2014**, *26*, 3905–3911. [CrossRef] [PubMed]
38. Cheng, H. Inorganic dissolvable electronics: Materials and devices for biomedicine and environment. *J. Mater. Res.* **2016**, *31*, 2549–2570. [CrossRef]
39. Cheng, H.; Vepachedu, V. Recent development of transient electronics. *Theor. Appl. Mech. Lett.* **2016**, *6*, 21–31. [CrossRef]
40. Eisenhaure, J.; Kim, S. A review of the state of dry adhesives: Biomimetic structures and the alternative designs they inspire. *Micromachines* **2017**, *8*, 125. [CrossRef]
41. Li, Y.; Krahn, J.; Menon, C. Bioinspired Dry Adhesive Materials and Their Application in Robotics: A Review. *J. Bionic Eng.* **2016**, *13*, 181–199. [CrossRef]
42. Zhou, M.; Pesika, N.; Zeng, H.; Tian, Y.; Israelachvili, J. Recent advances in gecko adhesion and friction mechanisms and development of gecko-inspired dry adhesive surfaces. *Friction* **2013**, *1*, 114–129. [CrossRef]
43. Autumn, K.; Sitti, M.; Liang, Y.A.; Peattie, A.M.; Hansen, W.R.; Sponberg, S.; Kenny, T.W.; Fearing, R.; Israelachvili, J.N.; Full, R.J. Evidence for van der Waals adhesion in gecko setae. *Proc. Natl. Acad. Sci. USA* **2002**, *99*, 12252–12256. [CrossRef] [PubMed]
44. Gao, H.; Yao, H. Shape insensitive optimal adhesion of nanoscale fibrillar structures. *Proc. Natl. Acad. Sci. USA* **2004**, *101*, 7851–7856. [CrossRef] [PubMed]
45. Varenberg, M.; Pugno, N.M.; Gorb, S.N. Spatulate structures in biological fibrillar adhesion. *Soft Matter* **2010**, *6*, 3269–3272. [CrossRef]
46. Jagota, A. Mechanics of Adhesion through a Fibrillar Microstructure. *Integr. Comp. Biol.* **2002**, *42*, 1140–1145. [CrossRef] [PubMed]
47. Arzt, E.; Gorb, S.; Spolenak, R. From micro to nano contacts in biological attachment devices. *Proc. Natl. Acad. Sci. USA* **2003**, *100*, 10603–10606. [CrossRef] [PubMed]
48. Aksak, B.; Murphy, M.P.; Sitti, M. Gecko inspired micro-fibrillar adhesives for wall climbing robots on micro/nanoscale rough surfaces. In Proceedings of the 2008 IEEE International Conference on Robotics and Automation, Pasadena, CA, USA, 19–23 May 2008.

49. Kamperman, M.; Kroner, E.; del Campo, A.; McMeeking, R.M.; Arzt, E. Functional Adhesive Surfaces with "Gecko" Effect: The Concept of Contact Splitting. *Adv. Eng. Mater.* **2010**, *12*, 335–348. [CrossRef]
50. Gorb, S.; Varenberg, M.; Peressadko, A.; Tuma, J. Biomimetic mushroom-shaped fibrillar adhesive microstructure. *J. R. Soc. Interface* **2007**, *4*, 271–275. [CrossRef] [PubMed]
51. Lee, J.; Fearing, R.S. Contact self-cleaning of synthetic gecko adhesive from polymer microfibers. *Langmuir* **2008**, *24*, 10587–10591. [CrossRef] [PubMed]
52. Hansen, W.R.; Autumn, K. Evidence for self-cleaning in gecko setae. *Proc. Natl. Acad. Sci. USA* **2005**, *102*, 385–389. [CrossRef] [PubMed]
53. Shahsavan, H.; Zhao, B. Biologically inspired enhancement of pressure-sensitive adhesives using a thin film-terminated fibrillar interface. *Soft Matter* **2012**, *8*, 8281–8284. [CrossRef]
54. Yao, H.; Rocca, G.D.; Guduru, P.R.; Gao, H. Adhesion and sliding response of a biologically inspired fibrillar surface: Experimental observations. *J. R. Soc. Interface* **2008**, *5*, 723–733. [CrossRef] [PubMed]
55. Lundberg, D. Flow Conditioners. *Control Eng.* **2006**, *53*. [CrossRef]
56. Autumn, K.; Majidi, C.; Groff, R.E.; Dittmore, A.; Fearing, R. Effective elastic modulus of isolated gecko setal arrays. *J. Exp. Biol.* **2006**, *209*, 3558–3568. [CrossRef] [PubMed]
57. Geim, A.K.; Dubonos, S.V.; Grigorieva, I.V.; Novoselov, K.S.; Zhukov, A.A.; Shapoval, S.Y. Microfabricated adhesive mimicking gecko foot-hair. *Nat. Mater.* **2003**, *2*, 461–463. [CrossRef] [PubMed]
58. Mahdavi, A.; Ferreira, L.; Sundback, C.; Nichol, J.W.; Chan, E.P.; Carter, D.J.D.; Bettinger, C.J.; Patanavanich, S.; Chignozha, L.; Ben-Joseph, E.; et al. A biodegradable and biocompatible gecko-inspired tissue adhesive. *Proc. Natl. Acad. Sci. USA* **2008**, *105*, 2307–2312. [CrossRef] [PubMed]
59. Majidi, C.; Groff, R.E.; Maeno, Y.; Schubert, B.; Baek, S.; Bush, B.; Maboudian, R.; Gravish, N.; Wilkinson, M.; Autumn, K.; et al. High friction from a stiff polymer using microfiber arrays. *Phys. Rev. Lett.* **2006**, *97*, 076103. [CrossRef] [PubMed]
60. Jeong, H.E.; Lee, S.H.; Kim, P.; Suh, K.Y. Stretched polymer nanohairs by nanodrawing. *Nano Lett.* **2006**, *6*, 1508–1513. [CrossRef] [PubMed]
61. Ge, L.; Sethi, S.; Ci, L.; Ajayan, P.M.; Dhinojwala, A. Carbon nanotube-based synthetic gecko tapes. *Proc. Natl. Acad. Sci. USA* **2007**, *104*, 10792–10795. [CrossRef] [PubMed]
62. Qu, L.; Dai, L.; Stone, M.; Xia, Z.; Wang, Z.L. Carbon nanotube arrays with strong shear binding-on and easy normal lifting-off. *Science* **2008**, *322*, 238–242. [CrossRef] [PubMed]
63. Aksak, B.; Murphy, M.P.; Sitti, M. Adhesion of biologically inspired vertical and angled polymer microfiber arrays. *Langmuir* **2007**, *23*, 3322–3332. [CrossRef] [PubMed]
64. Murphy, M.P.; Aksak, B.; Sitti, M. Gecko-inspired directional and controllable adhesion. *Small* **2009**, *5*, 170–175. [CrossRef] [PubMed]
65. Reddy, S.; Arzt, E.; Del Campo, A. Bioinspired surfaces with switchable adhesion. *Adv. Mater.* **2007**, *19*, 3833–3837. [CrossRef]
66. Kim, T.I.; Jeong, H.E.; Suh, K.Y.; Lee, H.H. Stooped nanohairs: Geometry-controllable, unidirectional, reversible, and robust Gecko-like dry adhesive. *Adv. Mater.* **2009**, *21*, 2276–2281. [CrossRef]
67. Röhrig, M.; Thiel, M.; Worgull, M.; Hölscher, H. 3D Direct laser writing of nano- and microstructured hierarchical gecko-mimicking surfaces. *Small* **2012**, *8*, 3009–3015. [CrossRef] [PubMed]
68. Murphy, M.P.; Kim, S.; Sitti, M. Enhanced adhesion by gecko-inspired hierarchical fibrillar adhesives. *ACS Appl. Mater. Interfaces* **2009**, *1*, 849–855. [CrossRef] [PubMed]
69. Greiner, C.; Arzt, E.; Del Campo, A. Hierarchical gecko-like adhesives. *Adv. Mater.* **2009**, *21*, 479–482. [CrossRef]
70. Del Campo, A.; Greiner, C.; Arzt, E. Contact shape controls adhesion of bioinspired fibrillar surfaces. *Langmuir* **2007**, *23*, 10235–10243. [CrossRef] [PubMed]
71. Kim, S.; Sitti, M. Biologically inspired polymer microfibers with spatulate tips as repeatable fibrillar adhesives. *Appl. Phys. Lett.* **2006**, *89*, 261911. [CrossRef]
72. Sameoto, D.; Menon, C. A low-cost, high-yield fabrication method for producing optimized biomimetic dry adhesives. *J. Micromech. Microeng.* **2009**, *19*, 115002. [CrossRef]
73. Mengüç, Y.; Yang, S.Y.; Kim, S.; Rogers, J.A.; Sitti, M. Gecko-inspired controllable adhesive structures applied to micromanipulation. *Adv. Funct. Mater.* **2012**, *22*, 1246–1254. [CrossRef]
74. Smith, A.M. Negative Pressure Generated By Octopus Suckers: A Study of the Tensile Strength of Water in Nature. *J. Exp. Biol.* **1991**, *157*, 257–271.

75. Tramacere, F.; Beccai, L.; Kuba, M.; Gozzi, A.; Bifone, A.; Mazzolai, B. The Morphology and Adhesion Mechanism of Octopus vulgaris Suckers. *PLoS ONE* **2013**, *8*, e65074. [CrossRef] [PubMed]
76. Kier, W.M. The Structure and Adhesive Mechanism of Octopus Suckers. *Integr. Comp. Biol.* **2002**, *42*, 1146–1153. [CrossRef] [PubMed]
77. Tramacere, F.; Pugno, N.M.; Kuba, M.J.; Mazzolai, B. Unveiling the morphology of the acetabulum in octopus suckers and its role in attachment. *Interface Focus* **2014**, *5*, 1–5. [CrossRef] [PubMed]
78. Tramacere, F.; Beccai, L.; Mattioli, F.; Sinibaldi, E.; Mazzolai, B. Artificial adhesion mechanisms inspired by octopus suckers. In Proceedings of the IEEE International Conference on Robotics and Automation, Saint Paul, MN, USA, 14–18 May 2012; pp. 3846–3851.
79. Tomokazu, T.; Kikuchi, S.; Suzuki, M.; Aoyagi, S. Vacuum gripper imitated octopus sucker-effect of liquid membrane for absorption. In Proceedings of the IEEE International Conference on Intelligent Robots and Systems, Hamburg, Germany, 28 September–2 October 2015; pp. 2929–2936.
80. Follador, M.; Tramacere, F.; Mazzolai, B. Dielectric elastomer actuators for octopus inspired suction cups. *Bioinspir. Biomim.* **2014**, *9*. [CrossRef] [PubMed]
81. Yu, Q.; Chen, F.; Zhou, H.; Yu, X.; Cheng, H.; Wu, H. Design and Analysis of Magnetic-Assisted Transfer Printing. *J. Appl. Mech.* **2018**, *85*, 101009. [CrossRef]
82. Chang, W.Y.; Wu, Y.; Chung, Y.C. Facile fabrication of ordered nanostructures from protruding nanoballs to recessional nanosuckers via solvent treatment on covered nanosphere assembled monolayers. *Nano Lett.* **2014**, *14*, 1546–1550. [CrossRef] [PubMed]
83. Choi, M.K.; Park, O.K.; Choi, C.; Qiao, S.; Ghaffari, R.; Kim, J.; Lee, D.J.; Kim, M.; Hyun, W.; Kim, S.J.; et al. Cephalopod-Inspired Miniaturized Suction Cups for Smart Medical Skin. *Adv. Healthc. Mater.* **2016**, *5*, 80–87. [CrossRef] [PubMed]
84. Chen, Y.C.; Yang, H. Octopus-Inspired Assembly of Nanosucker Arrays for Dry/Wet Adhesion. *ACS Nano* **2017**, *11*, 5332–5338. [CrossRef] [PubMed]
85. Baik, S.; Kim, J.; Lee, H.J.; Lee, T.H.; Pang, C. Highly Adaptable and Biocompatible Octopus-Like Adhesive Patches with Meniscus-Controlled Unfoldable 3D Microtips for Underwater Surface and Hairy Skin. *Adv. Sci.* **2018**, *5*. [CrossRef] [PubMed]
86. Li, K.; Cai, S. Wet adhesion between two soft layers. *Soft Matter* **2014**, *10*, 8202–8209. [CrossRef] [PubMed]
87. Lee, H.; Um, D.S.; Lee, Y.; Lim, S.; Kim, H.-j.; Ko, H. Octopus-Inspired Smart Adhesive Pads for Transfer Printing of Semiconducting Nanomembranes. *Adv. Mater.* **2016**, *28*, 7457–7465. [CrossRef] [PubMed]
88. Matuda, N.; Baba, S.; Kinbara, A. Internal stress, young's modulus and adhesion energy of carbon films on glass substrates. *Thin Solid Films* **1981**, *81*, 301–305. [CrossRef]
89. Schneider, A.; Francius, G.; Obeid, R.; Schwinté, P.; Hemmerlé, J.; Frisch, B.; Schaaf, P.; Voegel, J.-C.; Senger, B.; Picart, C. Polyelectrolyte Multilayers with a Tunable Young's Modulus: Influence of Film Stiffness on Cell Adhesion. *Langmuir* **2006**, *22*, 1193–1200. [CrossRef] [PubMed]
90. Pan, T.; Pharr, M.; Ma, Y.; Ning, R.; Yan, Z.; Xu, R.; Feng, X.; Huang, Y.; Rogers, J.A. Experimental and Theoretical Studies of Serpentine Interconnects on Ultrathin Elastomers for Stretchable Electronics. *Adv. Funct. Mater.* **2017**, *27*, 1702589. [CrossRef]
91. Pena-Francesch, A.; Akgun, B.; Miserez, A.; Zhu, W.; Gao, H.; Demirel, M.C. Pressure Sensitive Adhesion of an Elastomeric Protein Complex Extracted From Squid Ring Teeth. *Adv. Funct. Mater.* **2014**, *24*, 6227–6233. [CrossRef]
92. Zhao, Q.; Woog Lee, D.; Kollbe Ahn, B.; Seo, S.; Kaufman, Y.; Israelachvili, J.N.; Herbert Waite, J. Underwater contact adhesion and microarchitecture in polyelectrolyte complexes actuated by solvent exchange. *Nat. Mater.* **2016**, *15*, 407–412. [CrossRef] [PubMed]
93. Ahn, B.K. Perspectives on Mussel-Inspired Wet Adhesion. *J. Am. Chem. Soc.* **2017**, *139*, 10166–10171. [CrossRef] [PubMed]
94. Rao, P.; Sun, T.L.; Chen, L.; Takahashi, R.; Shinohara, G.; Guo, H.; King, D.R.; Kurokawa, T.; Gong, J.P. Tough Hydrogels with Fast, Strong, and Reversible Underwater Adhesion Based on a Multiscale Design. *Adv. Mater.* **2018**, *30*, 1801884. [CrossRef] [PubMed]
95. Waite, J.H.; Tanzer, M.L. Polyphenolic substance of Mytilus edulis: Novel adhesive containing L-dopa and hydroxyproline. *Science* **1981**, *212*, 1038–1040. [CrossRef] [PubMed]
96. Waite, J.H.; Qin, X. Polyphosphoprotein from the adhesive pads of *Mytilus edulis*. *Biochemistry* **2001**, *40*, 2887–2893. [CrossRef] [PubMed]

97. Lee, H.; Dellatore, S.M.; Miller, W.M.; Messersmith, P.B. Mussel-inspired surface chemistry for multifunctional coatings. *Science* **2007**, *318*, 426–430. [CrossRef] [PubMed]
98. Liu, Y.; Ai, K.; Lu, L. Polydopamine and its derivative materials: Synthesis and promising applications in energy, environmental, and biomedical fields. *Chem. Rev.* **2014**, *114*, 5057–5115. [CrossRef] [PubMed]
99. Lynge, M.E.; Van Der Westen, R.; Postma, A.; Städler, B. Polydopamine—A nature-inspired polymer coating for biomedical science. *Nanoscale* **2011**, *3*, 4916–4928. [CrossRef] [PubMed]
100. Lee, Y.-Y.; Kang, H.-Y.; Gwon, S.H.; Choi, G.M.; Lim, S.-M.; Sun, J.-Y.; Joo, Y.-C. A Strain-Insensitive Stretchable Electronic Conductor: PEDOT:PSS/Acrylamide Organogels. *Adv. Mater.* **2016**, *28*, 1636–1643. [CrossRef] [PubMed]
101. Yuk, H.; Zhang, T.; Parada, G.A.; Liu, X.; Zhao, X. Skin-inspired hydrogel–elastomer hybrids with robust interfaces and functional microstructures. *Nat. Commun.* **2016**, *7*, 12028. [CrossRef] [PubMed]
102. Han, L.; Liu, K.; Wang, M.; Wang, K.; Fang, L.; Chen, H.; Zhou, J.; Lu, X. Mussel-Inspired Adhesive and Conductive Hydrogel with Long-Lasting Moisture and Extreme Temperature Tolerance. *Adv. Funct. Mater.* **2018**, *28*, 1704195. [CrossRef]
103. Swartzlander, M.D.; Blakney, A.K.; Amer, L.D.; Hankenson, K.D.; Kyriakides, T.R.; Bryant, S.J. Immunomodulation by mesenchymal stem cells combats the foreign body response to cell-laden synthetic hydrogels. *Biomaterials* **2015**, *41*, 79–88. [CrossRef] [PubMed]
104. Han, L.; Lu, X.; Liu, K.; Wang, K.; Fang, L.; Weng, L.-T.; Zhang, H.; Tang, Y.; Ren, F.; Zhao, C.; et al. Mussel-Inspired Adhesive and Tough Hydrogel Based on Nanoclay Confined Dopamine Polymerization. *ACS Nano* **2017**, *11*, 2561–2574. [CrossRef] [PubMed]
105. Li, J.; Celiz, A.D.; Yang, J.; Yang, Q.; Wamala, I.; Whyte, W.; Seo, B.R.; Vasilyev, N.V.; Vlassak, J.J.; Suo, Z.; et al. Tough adhesives for diverse wet surfaces. *Science* **2017**, *357*, 378–381. [CrossRef] [PubMed]
106. Zhao, P.; Wei, K.; Feng, Q.; Chen, H.; Wong, D.S.H.; Chen, X.; Wu, C.-C.; Bian, L. Mussel-mimetic hydrogels with defined cross-linkers achieved via controlled catechol dimerization exhibiting tough adhesion for wet biological tissues. *Chem. Commun.* **2017**, *53*, 12000–12003. [CrossRef] [PubMed]
107. Ahn, B.K.; Das, S.; Linstadt, R.; Kaufman, Y.; Martinez-Rodriguez, N.R.; Mirshafian, R.; Kesselman, E.; Talmon, Y.; Lipshutz, B.H.; Israelachvili, J.N.; et al. High-performance mussel-inspired adhesives of reduced complexity. *Nat. Commun.* **2015**, *6*, 8663. [CrossRef] [PubMed]
108. Barrett, D.G.; Bushnell, G.G.; Messersmith, P.B. Mechanically Robust, Negative-Swelling, Mussel-Inspired Tissue Adhesives. *Adv. Healthc. Mater.* **2013**, *2*, 745–755. [CrossRef] [PubMed]
109. Pettersson, T.; Pendergraph, S.A.; Utsel, S.; Marais, A.; Gustafsson, E.; Wågberg, L. Robust and tailored wet adhesion in biopolymer thin films. *Biomacromolecules* **2014**, *15*, 4420–4428. [CrossRef] [PubMed]
110. Koh, A.; Kang, D.; Xue, Y.; Lee, S.; Pielak, R.M.; Kim, J.; Hwang, T.; Min, S.; Banks, A.; Bastien, P.; et al. A soft, wearable microfluidic device for the capture, storage, and colorimetric sensing of sweat. *Sci. Transl. Med.* **2016**, *8*, 366ra165. [CrossRef] [PubMed]
111. Wang, S.; Xu, J.; Wang, W.; Wang, G.-J.N.; Rastak, R.; Molina-Lopez, F.; Chung, J.W.; Niu, S.; Feig, V.R.; Lopez, J.; et al. Skin electronics from scalable fabrication of an intrinsically stretchable transistor array. *Nature* **2018**, *555*, 83–88. [CrossRef] [PubMed]
112. Panda, A.; Kumar, S.; Kumar, A.; Bansal, R.; Bhartiya, S. Fibrin glue in ophthalmology. *Indian J. Ophthalmol.* **2009**, *57*, 371–379. [CrossRef] [PubMed]
113. Khurana, A.; Parker, S.; Goel, V.; Alderman, P.M. Dermabond wound closure in primary hip arthroplasty. *Acta Orthop. Belg.* **2008**, *74*, 349–353. [PubMed]
114. Agarwal, A.; Kumar, D.A.; Jacob, S.; Baid, C.; Agarwal, A.; Srinivasan, S. Fibrin glue-assisted sutureless posterior chamber intraocular lens implantation in eyes with deficient posterior capsules. *J. Cataract Refract. Surg.* **2008**, *34*, 1433–1438. [CrossRef] [PubMed]
115. Strehin, I.; Nahas, Z.; Arora, K.; Nguyen, T.; Elisseeff, J. A versatile pH sensitive chondroitin sulfate-PEG tissue adhesive and hydrogel. *Biomaterials* **2010**, *31*, 2788–2797. [CrossRef] [PubMed]
116. Martinelli, A.; Carru, G.A.; D'Ilario, L.; Caprioli, F.; Chiaretti, M.; Crisante, F.; Francolini, I.; Piozzi, A. Wet adhesion of buckypaper produced from oxidized multiwalled carbon nanotubes on soft animal tissue. *ACS Appl. Mater. Interfaces* **2013**, *5*, 4340–4349. [CrossRef] [PubMed]
117. Barreau, V.; Hensel, R.; Guimard, N.K.; Ghatak, A.; McMeeking, R.M.; Arzt, E. Fibrillar Elastomeric Micropatterns Create Tunable Adhesion Even to Rough Surfaces. *Adv. Funct. Mater.* **2016**, *26*, 4687–4694. [CrossRef]

118. Bauer, C.T.; Kroner, E.; Fleck, N.A.; Arzt, E. Hierarchical macroscopic fibrillar adhesives: In situ study of buckling and adhesion mechanisms on wavy substrates. *Bioinspir. Biomim.* **2015**, *10*. [CrossRef] [PubMed]
119. Stark, A.Y.; Palecek, A.M.; Argenbright, C.W.; Bernard, C.; Brennan, A.B.; Niewiarowski, P.H.; Dhinojwala, A. Gecko Adhesion on Wet and Dry Patterned Substrates. *PLoS ONE* **2015**, *10*, e0145756. [CrossRef] [PubMed]
120. Kasem, H.; Varenberg, M. Effect of counterface roughness on adhesion of mushroom-shaped microstructure. *J. R. Soc. Interface* **2013**, *10*, 20130620. [CrossRef] [PubMed]
121. Dastjerdi, A.K.; Pagano, M.; Kaartinen, M.T.; McKee, M.D.; Barthelat, F. Cohesive behavior of soft biological adhesives: Experiments and modeling. *Acta Biomater.* **2012**, *8*, 3349–3359. [CrossRef] [PubMed]
122. Moretti, M.; Wendt, D.; Schaefer, D.; Jakob, M.; Hunziker, E.B.; Heberer, M.; Martin, I. Structural characterization and reliable biomechanical assessment of integrative cartilage repair. *J. Biomech.* **2005**, *38*, 1846–1854. [CrossRef] [PubMed]
123. Pawlicki, J.M. The effect of molluscan glue proteins on gel mechanics. *J. Exp. Biol.* **2004**, *207*, 1127–1135. [CrossRef] [PubMed]
124. Wilks, A.M.; Rabice, S.R.; Garbacz, H.S.; Harro, C.C.; Smith, A.M. Double-network gels and the toughness of terrestrial slug glue. *J. Exp. Biol.* **2015**, *218*, 3128–3137. [CrossRef] [PubMed]
125. Yang, J.; Bai, R.; Suo, Z. Topological Adhesion of Wet Materials. *Adv. Mater.* **2018**, *30*, 1–7. [CrossRef] [PubMed]
126. Théry, M.; Pépin, A.; Dressaire, E.; Chen, Y.; Bornens, M. Cell distribution of stress fibres in response to the geometry of the adhesive environment. *Cell Motil. Cytoskelet.* **2006**, *63*, 341–355. [CrossRef] [PubMed]
127. Chen, X.; Cortez-Jugo, C.; Choi, G.H.; Björnmalm, M.; Dai, Y.; Yoo, P.J.; Caruso, F. Patterned Poly(dopamine) Films for Enhanced Cell Adhesion. *Bioconjug. Chem.* **2017**, *28*, 75–80. [CrossRef] [PubMed]
128. Malki, M.; Fleischer, S.; Shapira, A.; Dvir, T. Gold Nanorod-Based Engineered Cardiac Patch for Suture-Free Engraftment by Near IR. *Nano Lett.* **2018**, *18*, 4069–4073. [CrossRef] [PubMed]
129. Liang, S.; Zhang, Y.; Wang, H.; Xu, Z.; Chen, J.; Bao, R.; Tan, B.; Cui, Y.; Fan, G.; Wang, W.; et al. Paintable and Rapidly Bondable Conductive Hydrogels as Therapeutic Cardiac Patches. *Adv. Mater.* **2018**, *30*, 1704235. [CrossRef] [PubMed]

11

Wrinkled Graphene–AgNWs Hybrid Electrodes for Smart Window

Ki-Woo Jun, Jong-Nam Kim, Jin-Young Jung and Il-Kwon Oh *

Creative Research Initiative Centre for Functionally Antagonistic Nano-Engineering, Department of Mechanical Engineering, Korea Advanced Institute of Science and Technology (KAIST), 291 Daehak-ro, Yuseong-gu, Daejeon 34141, Korea; jkw82@kaist.ac.kr (K.-W.J.); jnkim4388@kaist.ac.kr (J.-N.K.); jinyjung@kaist.ac.kr (J.-Y.J.)

* Correspondence: ikoh@kaist.ac.kr

Academic Editors: Seung Hwan Ko, Daeho Lee and Zhigang Wu

Abstract: Over the past few years, there has been an increasing demand for stretchable electrodes for flexible and soft electronic devices. An electrode in such devices requires special functionalities to be twisted, bent, stretched, and deformed into variable shapes and also will need to have the capacity to be restored to the original state. In this study, we report uni- or bi-axially wrinkled graphene–silver nanowire hybrid electrodes comprised of chemical vapor deposition (CVD)-grown graphene and silver nanowires. A CVD-grown graphene on a Cu-foil was transferred onto a bi-axially pre-strained elastomer substrate and silver nanowires were sprayed on the transferred graphene surface. The pre-strained film was relaxed uni-(or bi-)axially to produce a wrinkled structure. The bi-axially wrinkled graphene and silver nanowires hybrid electrodes were very suitable for high actuating performance of electro-active dielectric elastomers compared with the wrinkle-free case. Present results show that the optical transparency of the highly stretchable electrode can be successfully tuned by modulating input voltages.

Keywords: graphene; silver-nanowires; wrinkles; elastomer actuators; tunable transparency

1. Introduction

Smart windows, which can control the level of light transmission, have recently been attracting great interest, because they can be applied to a variety of applications, including vehicle windows, exterior wall windows and skylight windows. Until now, most smart window technology has been based on the redox reaction of a molecular element and a chromic material in response to external stimuli such as light, electricity, or temperature [1–4]. However, such materials are chemically unstable during the conversion processes, and are further disadvantageous in that they are difficult to control [4]. As a result, there is a need for a new smart window that is efficient in manufacturing, simple to implement, durable to operate and has a fast response time.

Among various approaches to smart window technology, there has been recent interest in techniques for modulating transmittance by changing the surface morphology of an elastomer actuator. In order to generate such a morphological change of soft materials, a flexible and stretchable conducting material should be used as compliant electrodes, which can be stretched or shrunk according to the movement of the soft material, such as nanowires, graphene, and hybrid materials [5–12]. In particular, techniques using a soft dielectric elastomer and stretchable electrodes are being studied for controlling surface morphology by applying electric input voltages. Smart window technology based on such soft materials can be applied to micro-lens arrays [13,14], flexible electronic devices [15,16], and variable diffraction gratings [17].

However, since most devices were operated with flat electrodes, the changes in transmittance were not so large, resulting in a limited application. In order to resolve this performance problem, research on smart windows using graphene and silver nanowire hybrid electrodes has been carried out, but a wrinkled structure implemented with stretchable hybrid materials has not been reported yet.

In this study, we developed a wrinkled hybrid electrode, comprised of CVD-grown graphene and silver nanowires, and reported a tunable transmittance by applying electrical fields to the dielectric elastomer actuators sandwiched with the wrinkled graphene–silver nanowire hybrid electrodes. A monolayer of graphene grown by CVD on a Cu-foil was transferred onto a bi-axially pre-strained elastomer substrate and silver nanowires were sprayed on the transferred graphene surface. The pre-strained film was released bi-(or uni-)axially to develop a wrinkled structure. The morphological and electrical characteristics of the wrinkled structures with bi-(or uni-)axially compressive strains were investigated using scanning electron microscopy (SEM). Also, the bi-axially wrinkled graphene and silver nanowire hybrid nanostructures were successfully applied to produce high actuating performance compared with the wrinkle-free case. Using the fabricated electrode, we demonstrated a device whose optical transparency was tunable with input voltages.

2. Preparation and Experiment

2.1. Materials

VHB 4905 film (3M Corp., Maplewood, MN, USA) is an appropriate material for dielectric elastomer actuators because the film is easy to handle, and both sides of the film are sticky enough for coating with electrode materials such as graphene and silver nanowires. In addition, the VHB 4905 is almost transparent, with a thickness of around 5 mm and a density of 960 kg/m^3. The dielectric constant of the VHB 4905 film is 3.21 at 1 kHz. The dissipation factor and dielectric breakdown strength are 0.0214 and 630 V/mm, respectively [18,19].

Silver nanowires (AgNWs) are nanostructures, generally with a diameter of ten to several hundred nanometers and lengths of tens of microns. Silver nanowire material is a kind of grayish powder and is often dispersed in solvents, such as water, ethanol, and isopropanol [20]. The silver nanowires used in this study were synthesized by the polyol process [21–23]. A cleaned two-neck flask was placed in a heating mantle. Polyvinylpyrrolidone (PVP, Mw = 55,000) 5.86 g in 190 mL of glycerol was heated at 150 °C for 2 h to remove moisture in the PVP. The solution was injected into the cleaned two-neck flask and cooled down to 55 °C. A mixture of 0.059 g of sodium chloride, 0.5 mL of deionized water, and 10 mL of glycerol was added into the prepared solution with 1.58 g silver nitrate. The suspension was heated up to 155 °C and left for 20 min for the additional reaction. The final product was washed using glass filters and stored in methanol.

2.2. Fabrication

The fabrication process of the graphene and silver nanowire hybrid electrode with a wrinkled structure is illustrated in Figure 1. In this process, the wrinkled structure of the elastomer surface was formed mainly due to the difference in modulus between the elastomer film and the graphene and silver nanowire electrode.

The first step is the transfer of the CVD-grown graphene onto the surface of the pre-stretched elastomer film. In this step, the copper layer is removed using a wet-etching process to allow the graphene transfer. In order to maintain electrical conductivity between the graphene islands on the elastomer surface, silver nanowires were sprayed on the wrinkled graphene surface. Then, the pre-stretched elastomer film was released, to prepare a hybrid electrode with a wrinkled structure. In order to fabricate a dielectric elastomer actuator with a wrinkled structure, the electrode formation method described above was also applied to the opposite side of the elastomer. Using this method, a wrinkled elastomer actuator whose light transmittance could be controlled was manufactured.

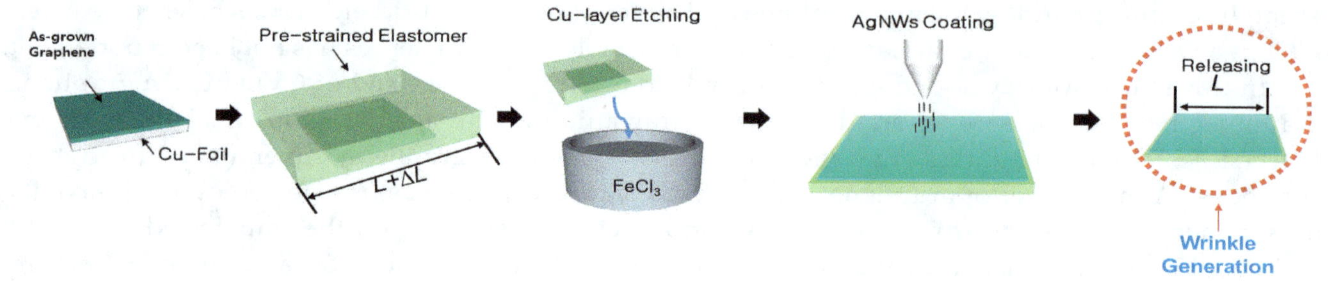

Figure 1. Fabrication process of the wrinkled graphene–AgNWs hybrid electrode.

2.3. *Experiment Method*

To induce the electro-mechanical deformation of the dielectric elastomer actuator, a high input voltage over 1 kV needs to be applied to both electrodes. For this purpose, a high voltage power supply (Trek Model 610E, Lockport, NY, USA) was connected to both the top and bottom electrodes of the actuator. To connect each compliant electrode to a cable from the power supply, conductive copper tape was used instead of the conventional cable-wires. Actuation testing was performed on an experiment table, and video data was acquired using a complementary metal-oxide-semiconductor (CMOS) image sensor to show the areal expansion of the actuator. To observe the variation in optical transmittance under actuating conditions, we placed the fabricated actuator on the sample holder of a UV-VIS spectroscopy. The optical transmittance was then examined, as input voltages were applied to the actuator.

3. Results and Discussions

3.1. *Uni- and Bi-axially Wrinkled Electrode*

Figure 2 shows SEM images of the hybrid electrode, which forms a wrinkled structure in the uni- and bi-axial directions on the elastomer film. Because the as-transferred graphene layer is fragmented, it has low electrical conductivity, and so by itself cannot function as an actuator electrode. In order to solve this problem, we additionally used silver nanowires to act as conductive bridges between the graphene islands. Silver nanowires with a length of about 10 to 20 μm were used, and the synthesized silver nanowires were thermally annealed at a temperature of 120 °C.

In Figure 2b,c, the surface morphology of the wrinkled structure formed along the uni- and bi-directions can be observed. A stretching–releasing method was used to create the wrinkled structure of the hybrid electrode. Pre-stretching was used to elongate the state of the elastomer film, which was used as the elastic body of the actuator. Different wrinkled structures can be formed depending on the pre-stretching direction. In this experiment, we formed wrinkled structures with uni- and bi-directions. The wavelength of the wrinkles is about 5 to 15 μm, as shown in Figure 2b,c.

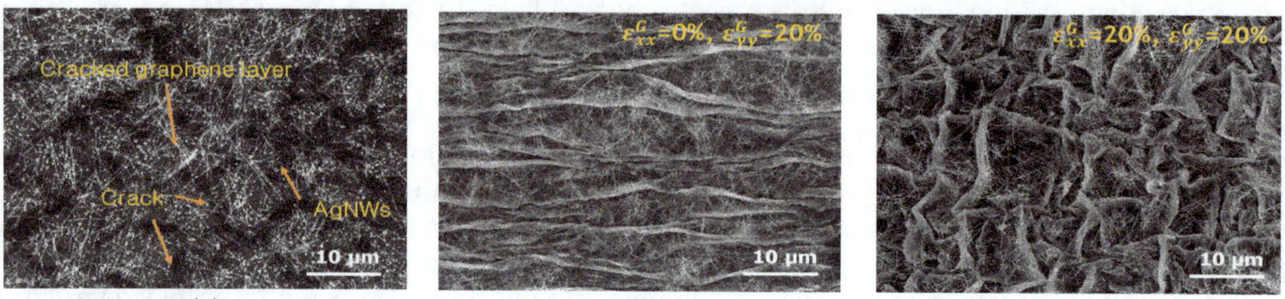

(a) (b) (c)

Figure 2. SEM images of the graphene–AgNWs hybrid electrode: (**a**) flat surface morphology, (**b**) uniaxially compressive strain of 20% and (**c**) biaxially compressive strain of 20%.

3.2. *Morphology Variation with Tensile Strength*

The working principle of the elastomer actuator is based on an attraction force which is generated between the two electrodes. This force causes an expansion of the electrode area in the in-plane direction. In order to examine the change of the surface in the multilayered elastomer-electrodes according to the pulling forces in the in-plane direction, the surface morphology was examined under tensile conditions, as shown in Figure 3. The uni- or bi-axial mechanical forces were applied to make uni- or bi-axially wrinkled patterns. The upper figure shows a wrinkled structure formed in one axis direction, and the lower figure shows a wrinkle structure formed in two axis directions. The surface patterns were observed as 0 to 50% tensile strain applied along the wrinkle direction. As the strain value increases, it can be seen that the wrinkles spread out. In the strained condition, it was confirmed that the interval between the pitches is increased, the height of the wrinkles and the density of electrical networks are lowered.

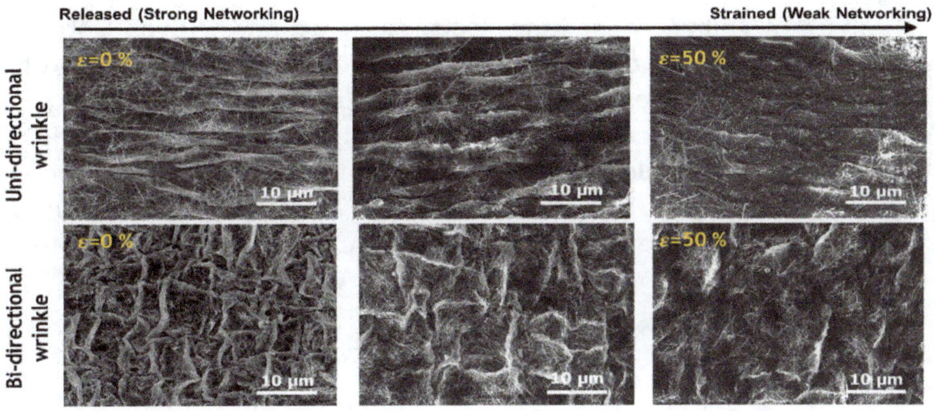

Figure 3. Morphology changes in the electrode surfaces with mechanical tensile strains.

3.3. *Electrical Measurement*

The relative resistance of the electrodes was examined in order to investigate the influence of the wrinkling morphology. Three types of specimens with different compressive strain values were prepared for this experiment. The compressive strain values applied to the specimens were 20%, 33% and 42%, respectively. As shown in Figure 4, the relative resistance ($\Delta R/R_0$) remains stable in the initial section, and as the strain value is increased, the magnitude of the relative resistance value is changed sharply. These results can be explained based on the wrinkled structure of the electrode. As shown in Figure 3, when the value of the tensile strain is increased, the hybrid electrode material, that is, the connectivity between the graphene and the silver nanowires, is weakened. For this reason, it can be understood that the resistance increases sharply in the section where there are fewer wrinkled structures.

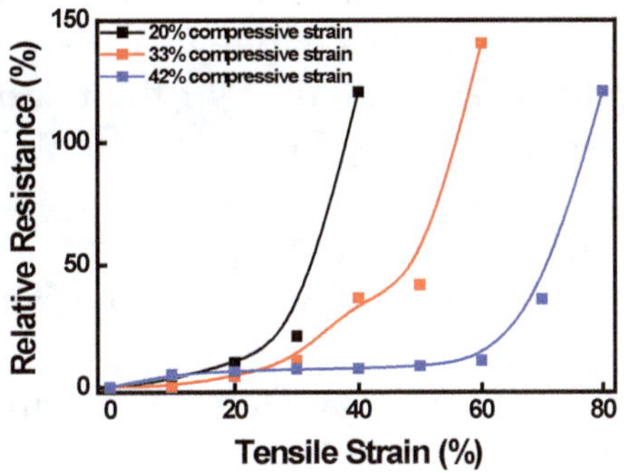

Figure 4. Resistance changes of wrinkled hybrid electrodes according to mechanical tensile strains.

3.4. Electromechanical Transducer with Elastomer Films

The dielectric elastomer actuator has a sandwich structure consisting of an elastomer film with two electrodes on both sides. For application, the elastomer film was fixed in a rigid frame made of polyacetal. Figure 5a illustrates the working principle of the elastomer actuator. When an input voltage is applied to the electrodes, electrons are accumulated on both sides of the wrinkled electrode. At this time, Maxwell stress is generated in the thickness direction of the elastomer film. Deformation occurs in the in-plane direction from the active area where the electrodes are formed.

In order to verify the operating performance of the actuator, we connected the electrode of the actuator to a DC high voltage supply capable of generating 0–10 kV. Then, the applied voltage was gradually increased from 0 to 6 kV, and the change in the area of the actuator was examined. For more precise performance analysis, it is essential to determine the area of the increased actuator under electrical stimuli. For this purpose, the moving images of the actuator were captured frame-by-frame, and analyzed using an image processing program. The actuation of conventional dielectric elastomer actuators using a high input-voltage caused the electrodes to expand with an isotropic areal strain as shown in Figure 5a. Isotropic behaviors of the dielectric elastomer actuator can be differently realized by the morphological configuration of hybrid electrodes. Figure 5b shows the experimental results to confirm the performance difference of the wrinkled and wrinkle-free electrodes when electrical input voltages are applied to two dielectric elastomer actuators. The actuator sample with wrinkled electrodes shows an areal change ($\Delta A/A_0$) of 33%. Within 4 kV, the areal change of the actuator is almost similar in both cases, but there is a remarkable maximum difference of about 6% at 6 kV.

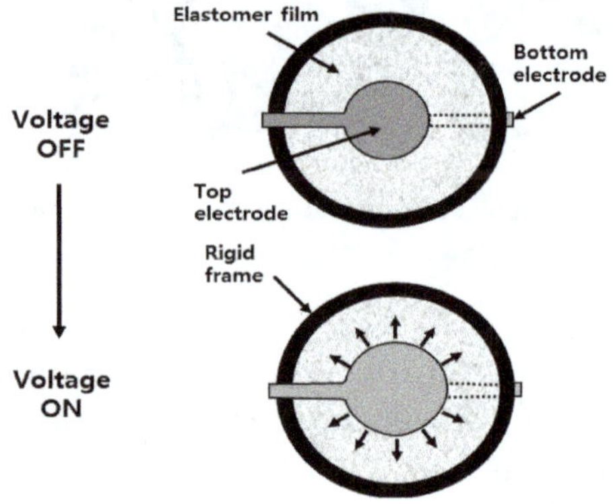

(a)

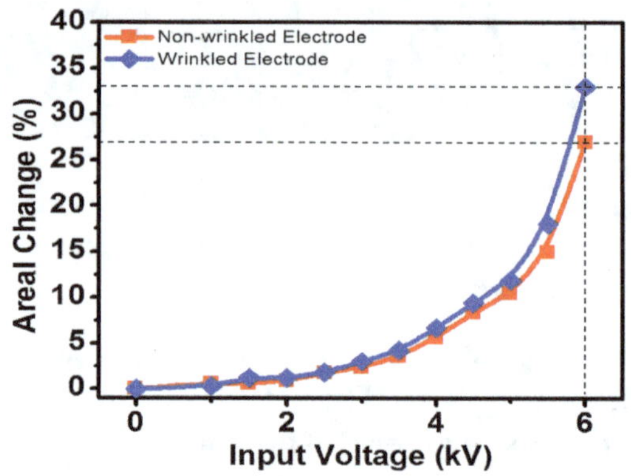

(b)

Figure 5. Actuation behaviors: (**a**) concept of isotropic elastomer actuation, (**b**) comparison of areal changes of wrinkle-free and wrinkled elastomer actuator.

As the applied voltage increases, the strength of the interaction force in the thickness direction increases in the elastomer actuator. This force has the effect of reducing the thickness of the film and increasing the area of the electrode. At this time, the transmittance of the active area is increased due to the reduced thickness of the film and the change in surface morphology on the surface of the elastomer film. Figure 6 is a graphical representation of the relationship between the rate of areal change and transmittance obtained when a voltage is applied to the actuator. Using a sample without the wrinkled electrodes, experimental results confirmed that the areal change rate was about 27% and the transmittance at 550 nm was changed by about 8%.

Using the same method as that used for the wrinkle-free sample, the change in active area of the dielectric elastomer actuator with wrinkled graphene–AgNWs hybrid electrodes was measured with respect to the applied voltages. Then, the transmittance change according to area changes was also examined. The wrinkled electrode structure exhibited about a 33% change in area ratio, and a transmittance change of about 15%. That is, although the wrinkled structure showed the same area change rate, it can be confirmed that the transmittance was about twice as large as that in the case of the wrinkle-free sample. When compared with the structure without wrinkles, the cause of this phenomenon is understood to be the difference in transmittance due to the scattering of light.

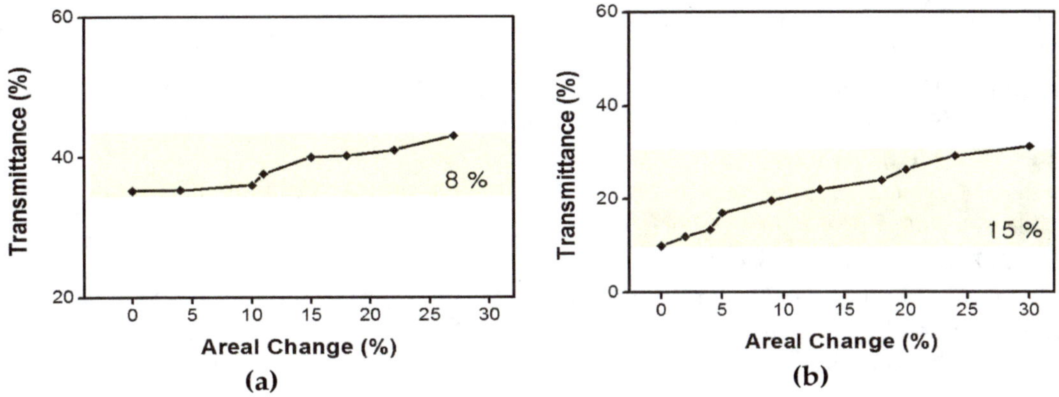

Figure 6. Optical transmittance control of (**a**) a wrinkle-free and (**b**) wrinkled elastomer actuator with electrical input.

4. Conclusions

A novel dielectric elastomer actuator with wrinkled graphene and silver nanowire hybrid electrodes was developed in this study. Graphene and silver nanowires were used to realize the highly conductive and compliant electrodes. The developed dielectric elastomer actuator with the wrinkled surfaces exhibited a large in-plane deformation of up to 33%. As compared with wrinkle-free electrodes, the proposed actuator with wrinkled electrodes exhibited a 10% higher areal change rate because of the synergistic effect of the ultrahigh conductivity of the silver nanowires and the remaining capacitance of the graphene under large deformation. This novel elastomer actuator with wrinkled hybrid electrodes may be a candidate to tune the transmittance of a smart window system.

Acknowledgments: This work was partially supported by the Ministry of Trade, Industry and Energy through Technology Innovation Program (Grant 10044410) and by the Creative Research Initiative program funded by the National Research Foundation of Korea (No. 2015R1A3A2028975). This work was supported by the Center for Advanced Soft Electronics funded by the Ministry of Science, ICT and Future Planning as Global Frontier Project (CASE-2011-0031640).

Author Contributions: K.W.J. performed the experiments and analyzed the data. J.N.K. and J.Y.J. supported the experiments and revised the paper. I.K.O. supervised the research work at the whole stage. All authors read and approved the final manuscript.

References

1. Beaujuge, P.M.; Reynolds, J.R. Color control in π-conjugated organic polymers for use in electrochromic devices. *Chem. Rev.* **2010**, *110*, 268–320. [CrossRef] [PubMed]
2. Bechinger, C.; Ferrer, S.; Zaban, A.; Sprague, J.; Gregg, B.A. Photoelectrochromic windows and displays. *Nature* **1996**, *383*, 608–610. [CrossRef]
3. Granqvist, C.G. Electrochromic tungsten oxide films: Review of progress 1993–1998. *Sol. Energy Mater. Sol. Cells* **2000**, *60*, 201–262. [CrossRef]
4. Lampert, C.M. Chromogenic smart materials. *Mater. Today* **2004**, *7*, 28–35. [CrossRef]
5. An, B.W.; Gwak, E.-J.; Kim, K.; Kim, Y.-C.; Jang, J.; Kim, J.-Y.; Park, J.-U. Stretchable, transparent electrodes as wearable heaters using nanotrough networks of metallic glasses with superior mechanical properties and thermal stability. *Nano Lett.* **2016**, *16*, 471–478. [CrossRef] [PubMed]
6. An, B.W.; Hyun, B.G.; Kim, S.-Y.; Kim, M.; Lee, M.-S.; Lee, K.; Koo, J.B.; Chu, H.Y.; Bae, B.-S.; Park, J.-U. Stretchable and transparent electrodes using hybrid structures of graphene–metal nanotrough networks with high performances and ultimate uniformity. *Nano Lett.* **2014**, *14*, 6322–6328. [CrossRef] [PubMed]

7. He, T.; Xie, A.; Reneker, D.H.; Zhu, Y. A tough and high-performance transparent electrode from a scalable and transfer-free method. *ACS Nano* **2014**, *8*, 4782–4789. [CrossRef] [PubMed]
8. Hsu, P.-C.; Wang, S.; Wu, H.; Narasimhan, V.K.; Kong, D.; Ryoung Lee, H.; Cui, Y. Performance enhancement of metal nanowire transparent conducting electrodes by mesoscale metal wires. *Nat. Commun.* **2013**, *4*, 2522. [CrossRef] [PubMed]
9. Hsu, P.-C.; Wu, H.; Carney, T.J.; McDowell, M.T.; Yang, Y.; Garnett, E.C.; Li, M.; Hu, L.; Cui, Y. Passivation coating on electrospun copper nanofibers for stable transparent electrodes. *ACS Nano* **2012**, *6*, 5150–5156. [CrossRef] [PubMed]
10. Kim, J.; Lee, M.-S.; Jeon, S.; Kim, M.; Kim, S.; Kim, K.; Bien, F.; Hong, S.Y.; Park, J.-U. Highly transparent and stretchable field-effect transistor sensors using graphene–nanowire hybrid nanostructures. *Adv. Mater.* **2015**, *27*, 3292–3297. [CrossRef] [PubMed]
11. Kim, K.; Kim, J.; Hyun, B.G.; Ji, S.; Kim, S.-Y.; Kim, S.; An, B.W.; Park, J.-U. Stretchable and transparent electrodes based on in-plane structures. *Nanoscale* **2015**, *7*, 14577–14594. [CrossRef] [PubMed]
12. Wu, H.; Kong, D.; Ruan, Z.; Hsu, P.-C.; Wang, S.; Yu, Z.; Carney, T.J.; Hu, L.; Fan, S.; Cui, Y. A transparent electrode based on a metal nanotrough network. *Nat. Nanotechnol.* **2013**, *8*, 421–425. [CrossRef] [PubMed]
13. Chan, E.P.; Crosby, A.J. Fabricating microlens arrays by surface wrinkling. *Adv. Mater.* **2006**, *18*, 3238–3242. [CrossRef]
14. Chandra, D.; Yang, S.; Lin, P.-C. Strain responsive concave and convex microlens arrays. *Appl. Phys. Lett.* **2007**, *91*, 251912. [CrossRef]
15. Khang, D.-Y.; Jiang, H.; Huang, Y.; Rogers, J.A. A stretchable form of single-crystal silicon for high-performance electronics on rubber substrates. *Science* **2006**, *311*, 208–212. [CrossRef] [PubMed]
16. Scalisi, R.G.; Paleari, M.; Favetto, A.; Stoppa, M.; Ariano, P.; Pandolfi, P.; Chiolerio, A. Inkjet printed flexible electrodes for surface electromyography. *Org. Electron.* **2015**, *18*, 89–94. [CrossRef]
17. Harrison, C.; Stafford, C.M.; Zhang, W.; Karim, A. Sinusoidal phase grating created by a tunably buckled surface. *Appl. Phys. Lett.* **2004**, *85*, 4016–4018. [CrossRef]
18. Opris, D.M.; Molberg, M.; Walder, C.; Ko, Y.S.; Fischer, B.; Nüesch, F.A. New silicone composites for dielectric elastomer actuator applications in competition with acrylic foil. *Adv. Funct. Mater.* **2011**, *21*, 3531–3539. [CrossRef]
19. Jun, K.W.; Lee, J.M.; Lee, J.Y.; Oh, I.K. Bio-inspired dielectric elastomer actuator with agnws coated on carbon black electrode. *J. Nanosci. Nanotechnol.* **2014**, *14*, 7483–7487. [CrossRef] [PubMed]
20. Tien, H.-W.; Hsiao, S.-T.; Liao, W.-H.; Yu, Y.-H.; Lin, F.-C.; Wang, Y.-S.; Li, S.-M.; Ma, C.-C.M. Using self-assembly to prepare a graphene-silver nanowire hybrid film that is transparent and electrically conductive. *Carbon* **2013**, *58*, 198–207. [CrossRef]
21. Jin, J.; Lee, J.; Jeong, S.; Yang, S.; Ko, J.-H.; Im, H.-G.; Baek, S.-W.; Lee, J.-Y.; Bae, B.-S. High-performance hybrid plastic films: A robust electrode platform for thin-film optoelectronics. *Energy Environ. Sci.* **2013**, *6*, 1811–1817. [CrossRef]
22. Lee, J.; Lee, I.; Kim, T.-S.; Lee, J.-Y. Efficient welding of silver nanowire networks without post-processing. *Small* **2013**, *9*, 2887–2894. [CrossRef] [PubMed]
23. Yang, C.; Gu, H.; Lin, W.; Yuen, M.M.; Wong, C.P.; Xiong, M.; Gao, B. Silver nanowires: From scalable synthesis to recyclable foldable electronics. *Adv. Mater.* **2011**, *23*, 3052–3056. [CrossRef] [PubMed]

12

Stretchability—The Metric for Stretchable Electrical Interconnects

Bart Plovie [1,*], Frederick Bossuyt [1,2] and Jan Vanfleteren [1,2,*]

1 Department of Electronics and Information Systems, Ghent University, Technologiepark 15, 9052 Zwijnaarde, Belgium
2 IMEC vzw, Kapeldreef 75, 3001 Heverlee, Belgium; frederick.bossuyt@imec.be
* Correspondence: bart.plovie@ugent.be (B.P.); jan.vanfleteren@ugent.be (J.V.)

Abstract: Stretchable circuit technology, as the name implies, allows an electronic circuit to adapt to its surroundings by elongating when an external force is applied. Based on this, early authors proposed a straightforward metric: stretchability—the percentage length increase the circuit can survive while remaining functional. However, when comparing technologies, this metric is often unreliable as it is heavily design dependent. This paper aims to demonstrate this shortcoming and proposes a series of alternate methods to evaluate the performance of a stretchable interconnect. These methods consider circuit volume, material usage, and the reliability of the technology. This analysis is then expanded to the direct current (DC) resistance measurement performed on these stretchable interconnects. A simple dead reckoning approach is demonstrated to estimate the magnitude of these measurement errors on the final measurement.

Keywords: stretchability; electronic measurements; stretchable circuits; design metrics; reliability

---

1. Introduction

Stretchable circuit technology is a recent development, having seen the light of day over the span of the last two decades. It is an attractive solution to common design problems; that is, the need for an electronic circuit to cover large areas or to conform to a moving and deformable device, either during production or in use [1,2].

Before the development of stretchable circuit technology, this was usually solved using long spring-loaded cables, stored in an enclosure that fed and retracted the cable as was necessary to cover the distance [3]. This cable-based approach has the advantage of being incredibly reliable if implemented correctly; additionally, it works from direct current (DC) all the way up to the gigahertz range and offers the capability to handle large currents. However, for all these advantages, this technique has a significant drawback: it requires a large volume for even the smallest number of interconnects. An alternative method, which remains popular and still makes a frequent appearance, is the use of coiled flexible circuit boards to cover these transitions [4]. In fact, both methods are surprisingly common in consumer-applications (e.g., devices with retractable power cords and hard drive reading heads).

In comparison, modern-day stretchable electronics use a more elegant approach to provide high-density interconnects over short to medium distances [5]. These methods are often based on thin, flat substrates embedded in soft elastic polymers, and achieve stretchability through careful structuring or speciality materials [1,6]. These techniques are applied on all levels, ranging from pinhead sized semiconductor devices, all the way to circuit boards covering massive composite structures [1,2,4,7–10].

While some exceptions do exist, most technologies have one feature in common: the lumped circuit elements (e.g., resistors, LEDs, and transistors) remain rigid and non-elastic and are relegated to rigid "islands"; meanwhile, the connections between these islands are made stretchable [7,8,11].

This segregation is possible because most electronic circuits use highly modular designs, with the interconnections between these modules being less critical—meaning that modifying these connections does not affect the circuit's functionality. As a result of this approach, the inherent capabilities of the circuit, regarding stretchability, are quantifiable by investigating the interconnections themselves, or at least that is the assumption commonly made. Arguments exist against this approach, but comparisons including the islands become difficult without considering specific circuit designs. Because of this reason, and this reason alone, only the interconnects themselves are considered in this manuscript, and for large high-density circuits, the effect of the islands should definitely be included.

The traditional approach to compare stretchable circuit technologies is based on considering the maximum percentage increase in length. This metric works by measuring the maximum increase in length the interconnect can survive, and dividing it by its original starting length—measured between the terminals [12]. This simple approach can be a fair indication of technological abilities when considering planar circuits using similar materials and designs, and is generally called "stretchability".

Closely related is the percentage increase in resistance; this is the increase in resistance divided by the resistance measured between the terminals before stretching. It is relatively easy to couple these resistance increases to physical effects, for example, in metal meanders, an increase of resistance tends to indicate the presence of mechanical damage (e.g., necking and micro-cracks) [13]. Whereas in the case of conductive filler particles in a polymer matrix, it can indicate the average proximity of these conductive particles to each other [14]. Upon unloading the stretchable interconnect, which is to say the elongation is reduced to zero, the latter will recover, assuming no defects form in the polymer matrix and no conductive fillers migrate. So, a permanent increase in resistance here would mean the polymer material did not return entirely to its original length, or significant defects formed. The metal meander, on the other hand, will permanently maintain this damage, even though it might not be directly visible because of the metal being pressed together again, which creates an unreliable electrical connection [5,13].

Sadly, both these metrics can introduce a significant amount of bias towards a particular technology, just because of the way these values are measured. What may initially appear to be an insignificant measurement error can quickly add up and lead to incorrect results. To demonstrate the flaws of the above "stretchability" definition, we design, fabricate, and measure a circuit intended to exploit these flaws, and then define a series of alternative metrics to compensate for them. Additionally, the more subtle points of interconnect resistance measurements for stretchable electronics are highlighted.

2. Materials and Methods

The circuit was fabricated using a 246 mm by 207 mm piece of UBE Upisel-N SR-1220 (UBE EXSYMO Co., Ltd., Tokyo, Japan) polyimide flexible copper clad laminate (FCCL) with 18 μm of rolled-annealed copper on 50 μm of polyimide. Using standard printed circuit board processing techniques [4], photolithography and wet-etching, the FCCL was structured into a flexible circuit board (FCB), as illustrated in Figure 1b. This FCB was attached to a 1.6 mm FR-4 carrier board (Hitachi Chemical Company, Ltd., Tokyo, Japan) covered with Taconic FH20LB Tacsil tape (TACONIC, Seongnam-si, Republic of Korea)—a pressure sensitive adhesive for reflow assembly of flexible circuit boards—using a hand roller. The purpose of this adhesive is to form a tacky surface that will hold onto the FCB to prevent entanglement as it is structured into a long meandering (also sometimes referred to as serpentine [15]) interconnect. Once the FCB is attached, the carrier board is placed underneath a 10 kHz-5 W pulsed Nd/YAG laser cutter (OPTEC, Frameries, Belgium). The FCCL is cut at a rate of 5 mm/s without damaging the TacSil tape and carrier board underneath. The material surrounding the desired circuit was peeled away using tweezers and discarded, as shown in Figure 1b, leaving behind the circuit design shown in Figures 1a and 2.

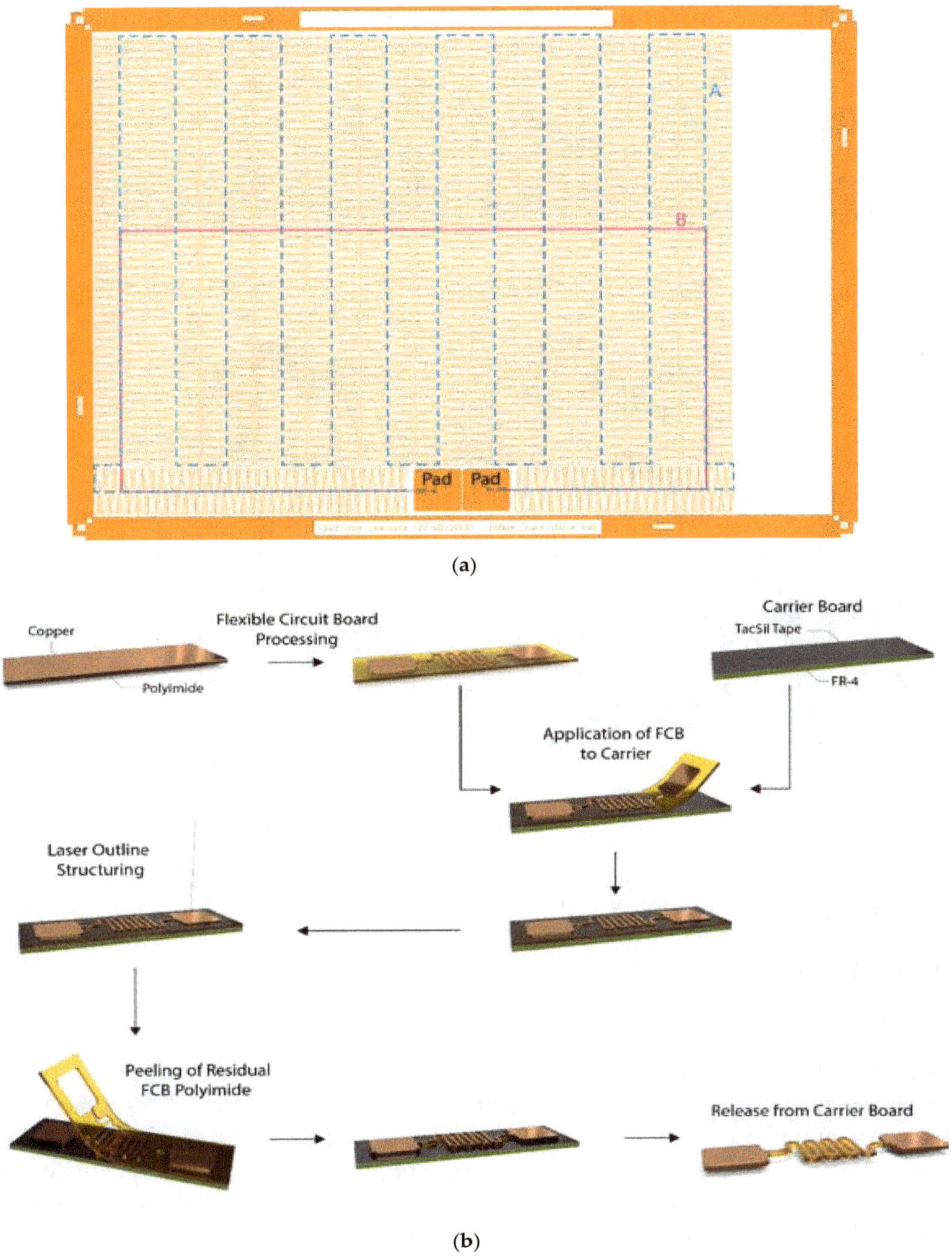

Figure 1. Manufacturing of free-standing stretchable circuits. (**a**) Copper artwork of the manufactured stretchable circuit. (**b**) Process flow used to manufacture the stretchable circuits. First, a flexible copper clad laminate is patterned using photolithography and wet etching. The resulting flexible circuit board (FCB) is then applied to an FR-4 carrier board covered with a pressure-sensitive TacSil Tape adhesive for mechanical support. The outline of the meanders and connection pads is then defined by cutting the flexible circuit board material using a laser cutter. The excess material surrounding the circuit is removed by hand, after which the circuit can be released from the carrier board by carefully peeling it off.

The design of the stretchable interconnect, of which the design file is available as supplementary material, has the goal of optimizing the stretchability. The easiest way to improve the potential increase in length is having this length of conductor available, though this does not necessarily have to be in a practical location. Additionally, placing the pads next to each other increases the achieved stretchability without leading to any practical benefit. A third factor to consider is the reliability, because this is not included in the definition of stretchability, the design can use sharp corners with a small radius, which will hamper reliability. The first step is easily done by placing the 18 mm by 18 mm contact pads against each other. A critical value in this entire operation is the outline spacing, the distance between the edge of the copper artwork and the edge of the polyimide, as indicated in Figure 2. The minimum spacing for the outline of the flexible circuit board in regard to the copper is 100 μm; however, for yield reasons, the outline spacing was increased to 200 μm between the pads, resulting in a distance of 400 μm.

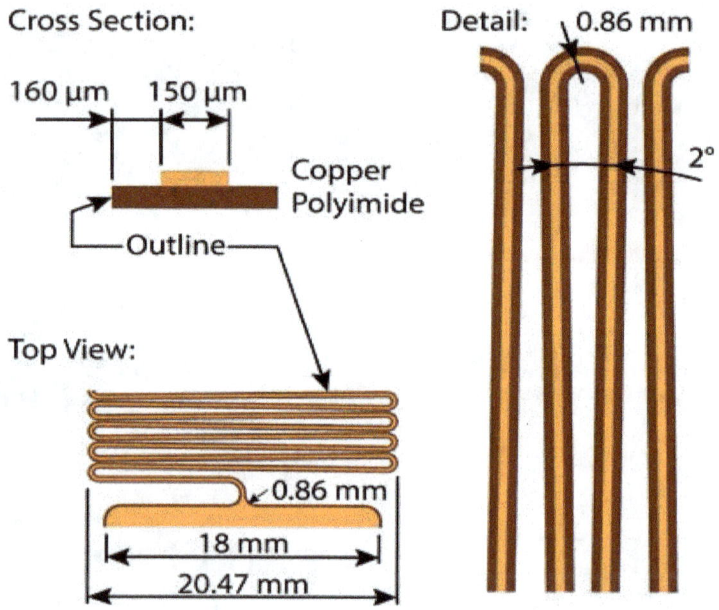

Figure 2. Design details of the stretchable circuit. The side of the polyimide track is the circuit outline. The meanders fold back on themselves at an angle of 2°, optimizing the amount of track in a given surface area without introducing extremely sharp corners.

From these pads, a 150 μm wide copper trace leaves with an outline offset of 160 μm on both sides—as indicated on Figure 2. After the first 90° bend, a U-shaped meander was started, which is folded back on itself—meaning the straight sections connecting the arcs are no longer parallel. This meander has a pitch of 1.5 mm and a height of approximately 20.5 mm and was looped in an accordion pattern above the contact pads. Excluding the length of the contact pads, this meander spans a distance of 66,637.5 mm along its center line. Two additional center lines were defined in Figure 1a, the purpose of which will become clear later on. The first centerline (A), which closely follows the zig-zag pattern, has a length of 2711.7 mm; the second one, however, does not follow this pattern and has a length of only 639.1 mm. The total surface area used by this circuit, including the contact pads, is 50,905.0 mm^2—a number that will become important later on.

Conventional tools, such as a tensile tester, are unable to deal with meanders of these lengths, meaning standard test methods were impossible to use. Instead, the carrier board with the circuit was attached to rigid surface perpendicular to the floor in a long corridor and markings were put on the floor at one-meter intervals using a tape measure, as shown in Figure 3b,c. To determine the resistance of the meander, a Keithley 2400 source meter (Tektronix, Inc., Beaverton, OR, USA) was set up to perform a four-wire measurement with a test current of 1 mA. To extend the test leads, a conventional power cable reel, visible in Figure 3a, was used. To ensure mechanical stability during extension of the meander, one of its pads was attached to the cable reel, as seen in Figure 3a. The reel was then moved backwards at a rate of approximately 10 cm/s to extend the circuit to the desired lengths.

Before measurement, the resistance was allowed to stabilize for at least 30 s to eliminate possible measurement errors due to mechanical vibrations.

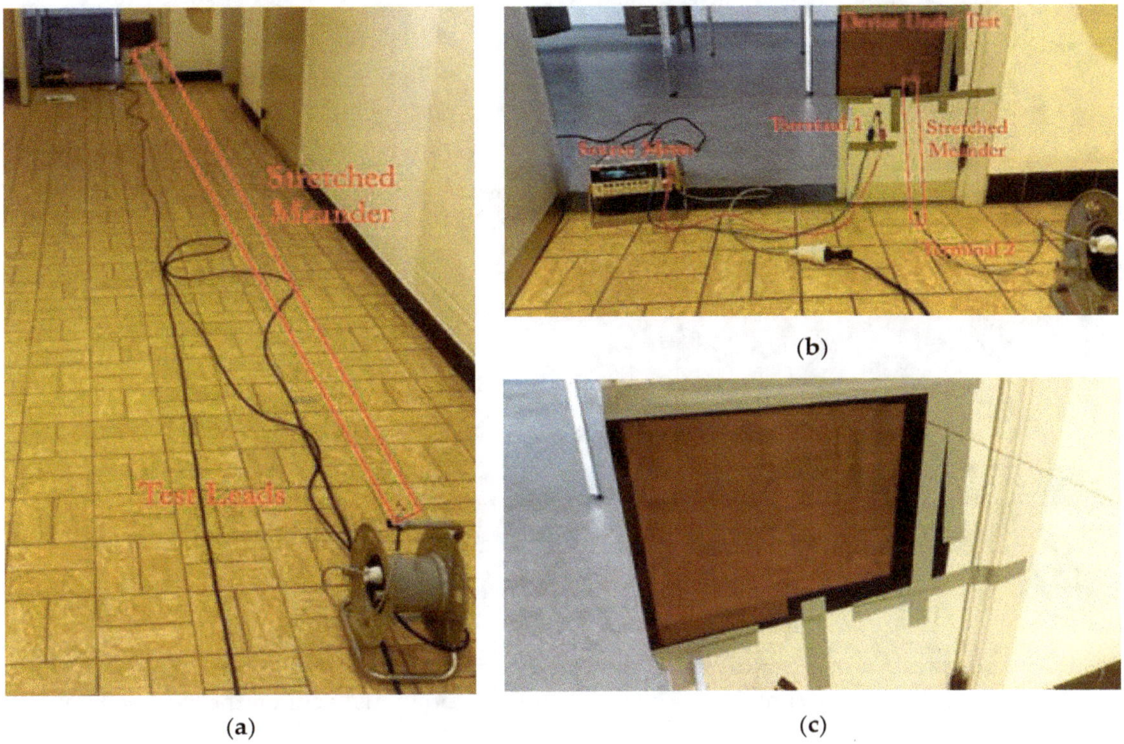

Figure 3. Measurement setup used to determine the meander resistance as a function of the extension: (**a**) measurement setup at the 10 m extension mark; (**b**) measurement setup at the start of the measurement before any extension. The cable reel visible at the bottom right of the image was used to extend the test leads to the desired length; (**c**) detailed view of the carrier board and circuit at the 10 m extension mark.

A second experiment, demonstrating the design dependence of stretchability—even within a given technology—uses a variety of stretchable interconnect designs. For this purpose, 18 designs based on flexible circuit board technology were manufactured and tested until failure using a tensile tester (Instron 5543, Instron, Norwood, MA, USA). As before, the flexible copper clad laminate (Shengyi SF305 101820SR—25 µm of polyimide with 18 µm of adhesively bonded copper) was patterned using industry standard practices. Afterwards, a coverlay film (Shengyi SF305C 1025) was bonded to the resulting flexible circuit board using the manufacturer's recommended press program in a vacuum press (Lauffer RLKV25). The FCB was placed on an identical carrier board and laser structured using a picosecond pulse-length Nd/YAG laser system (3D Micromac microSTRUCT™ vario) to define the meander outline—as illustrated in Figure 1b. The spacing between the copper track, which is 100 µm wide, and the laser path was set to a rather hefty 300 µm—representing the minimum tolerances achievable in an industrial setting. Some of the meanders were encapsulated between two layers of thermoplastic polyurethane (TPU)—100 µm Covestro Platilon U4201 AU—using a vacuum press. To ensure even pressure distribution during the lamination of the TPU, a 1/8-inch silicone–rubber–foam press pad (Rogers Corporation BISCO Foam HT-870) was used between the 25-µm polytetrafluoroethylene (PTFE) release foil and separator plate. The press applies 10 N/cm^2 at 175 °C for 20 min in a <2 mbar vacuum atmosphere to bond the TPU layers to the polyimide and together. The final samples, shown in Figure 4, were then cut using a guillotine cutter. To enable electrical measurements, the TPU was opened up using a soldering iron, and flexible wires were attached to the samples using leaded solder ($Sn_{63}Pb_{37}$ alloy).

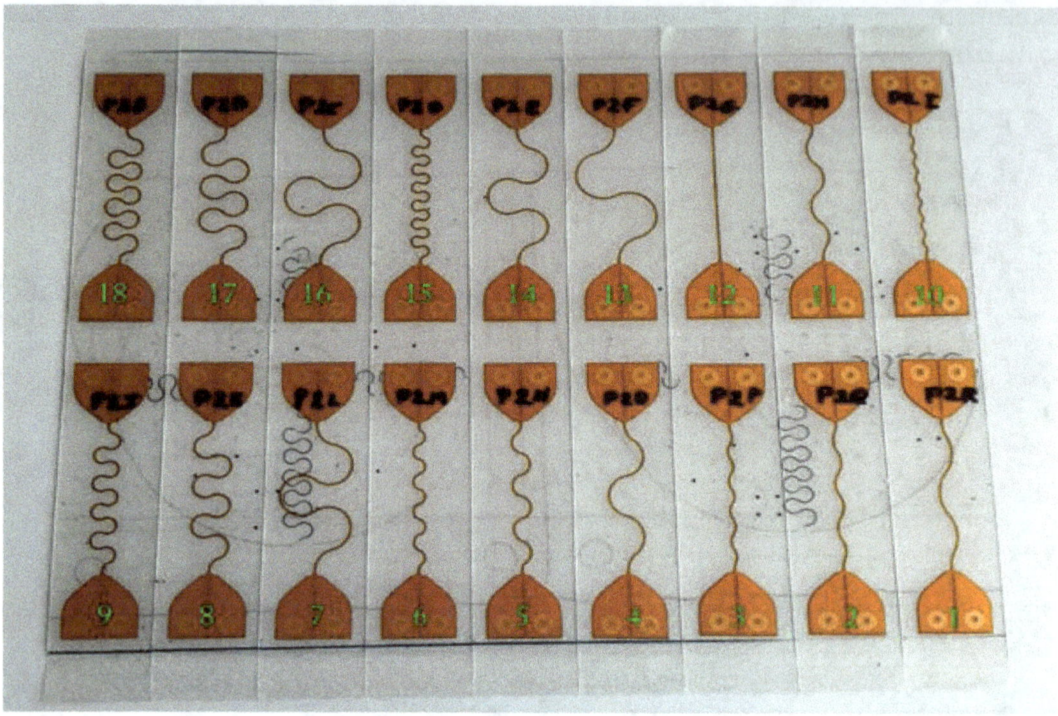

Figure 4. Meander test samples encapsulated in thermoplastic polyurethane using a vacuum press. The black debris visible on the sample originates from FCBs previously cut on the same carrier board.

Each meander has an effective length of 50 mm between the contact pads, of which 3 mm is taken up by the fillets to transition from the connection pads to the meander, as shown in Figure 5. These connections pads at either end of the meander are necessary for both clamping and electrical connectivity—to perform a four-point measurement while the meander is being stretched in a tensile tester. The contact pads were split up and only meet near the start of the meander in an attempt to eliminate their influence on the measurement. The design parameters of these meanders are listed in Table 1, all except design #12 are of the horseshoe-shaped meander type. Design #12 is a reference with a straight track in between the connection pads to determine the inherent capability of the material to meet these deformations. A schematic representation of such a test sample with relevant measurements is visible in Figure 5. Alternatively, the original design file is available as Figure S2.

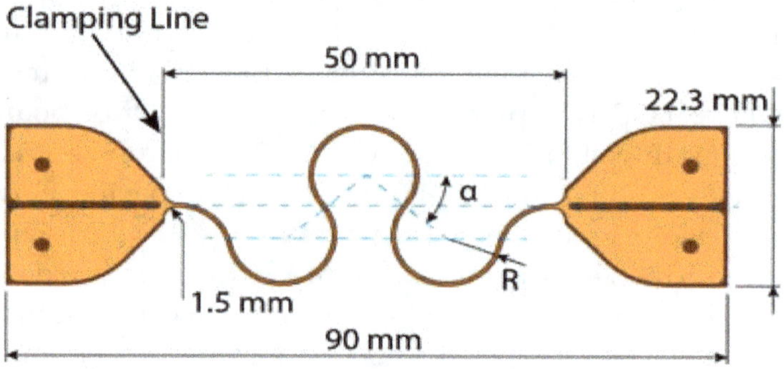

Figure 5. Design of the test vehicle with horseshoe-shaped meanders used for the tensile tests. R is the radius of the horseshoe-shaped meander segment, while α is the opening angle. These two variables combined completely define a horseshoe-shaped meander. To provide a smooth transition to the contact pads, a small 1.5-mm fillet was introduced at both ends. The contact pads at either side of the meander were split up to enable a four-point measurement with separate drive and sense lines.

Table 1. Design parameters for the tested interconnects.

Design	Length [mm]	Radius [mm]	Opening Angle α [°]	Area [mm^2]	Segments	Theoretical Stretchability [%]
1	50.605	10.000	−47.330	281.747	3	8
2	49.516	6.547	−56.166	137.118	6	5
3	49.506	4.555	−56.892	102.417	9	5
4	62.037	6.491	−8.238	555.625	3	32
5	62.046	3.649	−13.373	296.565	6	32
6	62.044	2.532	−14.995	209.313	9	32
7	97.147	6.678	39.037	1056.007	3	107
8	97.063	3.729	29.092	553.856	6	107
9	97.192	2.584	26.249	383.209	9	107
10	49.517	2.652	−57.394	72.190	16	5
11	52.029	6.999	−41.517	254.722	4	11
12	47.000	N/A	N/A	32.900	1	0
13	78.669	8.372	−90.000	1201.706	2	67
14	86.447	6.382	28.966	923.338	3	84
15	104.171	1.891	29.001	296.834	13	122
16	104.732	6.927	45.000	1144.398	3	123
17	124.688	4.262	45.000	716.767	6	165
18	133.551	3.078	45.000	526.740	9	184

The encapsulated samples were clamped in the tensile tester—shown in Figure 6—and elongated at a rate of 0.5 mm/s until failure. At the same time, the DC resistance of the samples was measured using a benchtop multimeter (Keithley 2001), which was triggered by the tensile tester software using a network connection. For this, it was set to auto-ranging with an upper limit of 200 Ω and averaging over 10 measurements—with one measurement per cycle of the 50 Hz power grid—because only the estimated point of failure was of interest, not the precise resistance increase, to determine the point of failure.

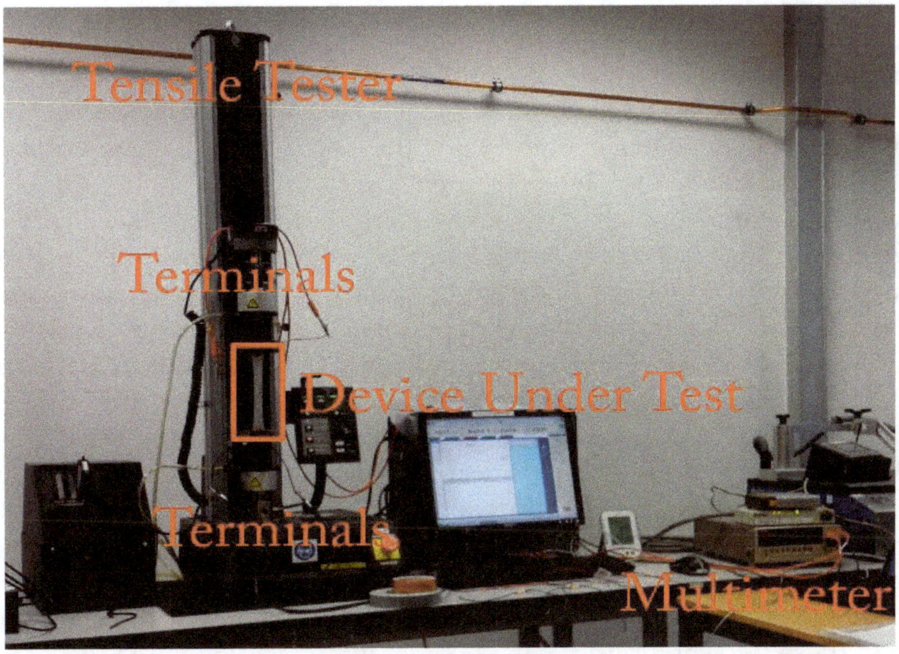

Figure 6. Tensile tester (Instron 5543) with the benchtop multimeter (Keithley 2001) performing the electrical measurements at the right-hand side. The multimeter was triggered over a network connection, and data were recorded using a secondary computer.

3. Results

The meander started out with a resistance of approximately 537 Ω, while it was flat on the carrier board, verified using a zeroed Keysight U1253B multimeter (Keysight Technologies, Santa Rosa, CA, USA). After mounting in the measurement setup, as shown in Figure 3b, this resistance increases slightly to 537.50 Ω, as measured with the Keithley 2400 using a four-wire measurement. The observed resistance as the meander was extended is shown in Figure 7.

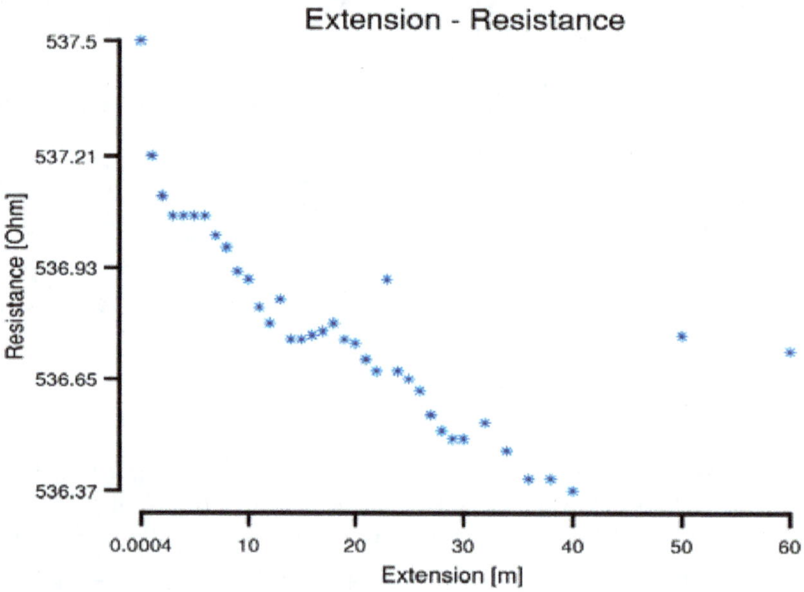

Figure 7. The resistance of the 66.6 m long meander plotted as a function of its extension. A noticeable, but procentual small drop in resistance was observed as the meander was extended, most likely due to the copper folding back on itself at each arc, which would slightly reduce the electric path length when stretching the meander.

A slight drop in resistance from 537.5 Ω to 536.37 Ω was observed during the experiment, contrary to what might be expected from a meander under significant strain. However, this can most likely be explained by the arcs at the end of the U-shaped meanders twisting and folding back on themselves as the meander extends, something slightly visible in Figure 3c. Extension of the meander was stopped at sixty meters because of length limitations of the corridor, limiting the achieved stretchability to 15,000,000%. When attempting to roll the meander on a spool for storage after the experiment, it broke in several places at the points where it twisted during the elongation. More importantly, this demonstrates the design dependence of the term stretchability.

The chosen distance between the start and end point is poorly defined and allows for parlor tricks as demonstrated above. Possible values for this circuit are the width of the cut between the pads (70 μm), the distance between the nearest copper features (400 μm), the distance between the centre of the pads 18.4 mm, the distance between the meander starting points (36.4 mm), or any other ill-devised definition. With each distance leading to its respective extension: 85,714,200%, 14,999,900%, 326,100%, and 164,800%.

The two center lines, A and B, defined in Figure 1a, provide a more realistic length, leading to 2113% and 9288%, respectively. While in this case, center line A provides a more accurate number, the choice between these two center lines can become quite tricky in some instances. For example, using center line B might be justifiable if the radius of 0.86 mm (Figure 2) is increased to 10 mm. In short, defining the initial length against which stretchability is measured has a considerable influence on the result. However, making the *"correct"* choice might not be trivial in some cases.

Needless to say, unless the same definition and measurement guidelines were used, any comparison between these numbers is meaningless. Additionally, this meander will fail whenever the tension on it is released, making it useless for most applications. Clearly, a better definition of stretchability is required to remove ambiguity.

3.1. Alternative Stretchability Metrics

Defeating stretchability as a metric of a stretchable interconnect's capabilities is trivial. As demonstrated above, it only requires creating a long wire using the technology at hand. A useful metric is unbiased and considers the reliability of the technology at hand. Solving this for the case of stretchable electronics seems trivial at first.

For a planar substrate, dividing the maximum achieved elongation by the occupied surface area before elongation would prevent looping the wire around the contact pads, as was done here. In fact, this planar stretchability (PS) should provide a fair assessment of a technology's capabilities at first glance. The maximum PS would depend solely on the minimum feature size and not the design; some small gains could be made by shrinking the contact pads; however, these would be minimal.

$$\text{Planar Stretchability} = \text{Maximum Elongation/Area} \tag{1}$$

This, of course, does not account for technologies that use out-of-plane features to create a stretchable interconnect [10,16]. These could create a tower of wire that might take up a relatively large volume while providing no cyclic endurance, like the above circuit—a single elongation might break the interconnect. An easy method to alleviate this is to include the thickness of the circuit by dividing the maximum elongation by the volume of the bounding box around the stretchable interconnect, leading to the volumetric stretchability (VS):

$$\text{Volumetric Stretchability} = \text{Maximum Elongation/Volume} \tag{2}$$

To eliminate the effect of the units, each value is normalized to millimeters; alternatively, the unit used could be mentioned next to the variable as a subscript (e.g., PS_{mm}). At first glance, this might result in a constant number for a given technology. However, the design of the interconnect still depends on electrical parameters, such as impedance and maximum current carrying capability, and the mechanics of the encapsulating material.

Using these metrics and looking back at the above example, consider the 60 m extension, 50,905.0 mm^2 surface area, and 3461.5 mm^3 volume (68 μm thickness). The planar stretchability then becomes 1.178 and the volumetric stretchability becomes 17.334. These values are significantly more useful; the required substrate surface area or volume is directly related to the cost to fabricate the device. Additionally, merely dividing the desired final length by the achieved PS or VS will indicate the required area or volume to achieve a given wire length. However, this still would not help determine the actual stretchability a designer or engineer might expect from a circuit—the above circuit is still frightfully unreliable, even though it might score well in a comparison.

Defining reliability as a single number is a troublesome prospect. However, for meanders, a few assumptions are possible. For example, it is possible to postulate the following: "Each directional change between segments of a conductor in a stretchable electrical interconnect where the angle between individual segments exceeds 30 degrees is prone to failure over time". This statement is most certainly wrong in an absolute sense, but provides a straight-forward method to quantify the chance of failure by cutting the stretchable interconnect into segments.

Providing a generic definition for a segment is troublesome; however, using the above definition, a few cases can be defined. First, any sharp corner (<90°) between two straight lines would most definitely be a transition at risk of failure. A second case is an interconnect that consists out of multiple identical elements that are repeated to form an interconnect. The third case, which is close to the above definition, would be a generic catch-all case; consider the tangent along the interconnect its center line. If this tangent changes more than 30° versus the tangent at an earlier point on the line this, indicates a transition, and hence a segment. Under ideal circumstances, the first two definitions are used, but these are troublesome to apply to special cases (e.g., fractal meanders).

Consider a segment with a known length, and n segments are required to span this distance. Neglecting the start and end points, the chance of failure for each segment and each transition between individual segments ris $P_1[i]$ and $P_2[i]$, respectively, where i is the cycle number during the cyclic test. Assuming the segments are independent of each other, with an equal chance of failure, and the applied strain is irrelevant, the chance a meander segment will survive a certain stretch cycle becomes the following:

$$P_{Survival}[i] = (1 - P_1[i])(1 - P_2[i]) \quad (3)$$

Hence the chance of failure becomes the following:

$$P_{Failure}[i] = 1 - (1 - P_1[i])(1 - P_2[i]) \quad (4)$$

Taking a few statistical liberties, like assuming the stretch cycles are completely independent of each other, the chance a meander segment breaks after i cycles becomes the following:

$$P_F[i] = 1 - \Pi^{i}_{j=0} (1 - P_1[j])(1 - P_2[j]) \quad (5)$$

While this might appear extreme at first glance, the survival chances P_1 and P_2 are close to 1 in most technologies, resulting in a rather small P_F. Assuming each meander segment is independent of its neighbors, the chance of failure in cycle i for n segments then becomes the following:

$$P_F[i, n] = 1 - (\Pi^{i}_{j=0} (1 - P_1[j])(1 - P_2[j]))^n \quad (6)$$

Clearly, an increase in n will increase the chance of failure. However, this rudimentary approach fails to take into account the strain the meander might experience; but it does demonstrate the effect of the number of segments or transitions n on meander reliability if these are considered weak points. Introducing this into the stretchability metric is trivial; simply dividing the stretchability by n ought to penalize a technology with many transitions. This leads to the definition of compensated planar stretchability (CPS) and compensated volumetric stretchability (CVS):

$$\text{Compensated Planar Stretchability} = \text{Maximum Elongation}/(n \times \text{Area}) \quad (7)$$

$$\text{Compensated Volumetric Stretchability} = \text{Maximum Elongation}/(n \times \text{Volume}) \quad (8)$$

In both cases, the area and volume are those for the complete meander. The above circuit has a staggering 4033 transitions, which meets the above requirement, slicing the CPS and CVS down to 0.003 and 0.004. For the case of a polymer matrix filled with conductive particles, setting n equal to one is an acceptable choice, unless it is patterned as well, because the likeliness of failure will depend on the material's inherent properties instead of the interconnect design.

An even more generic approach would be to consider the chance of failure occurring per length unit after i cycles $P_{FPL}[i]$. If such a number were available, it could easily be included by substituting n by $(1 - P_{FPL}[i])$. However, this would only provide a momentary comparison point during cycle i. A more generic approach would distill the reliability function $P_{FPL}[i]$ into a single number based on the number of life-cycles a device should survive.

Let $i_{expected}$ and k be the number of stretch-cycles the device should survive during regular use and the percentage of acceptable failures within this period, respectively. While setting k to zero might be an attractive prospect, no economical process achieves a 100% yield. Consider the function f[i] that returns the percentage of failed devices after i cycles; multiplying it by a weighing function and calculating the sum over the entire range of i would return a single number that signifies the reliability.

This weighing function should heavily punish crib deaths (early failures), while not significantly penalizing for failures beyond $i_{expected}$. Based on this, we can propose the weighing function w[i]:

$$w[i] = a \times \exp(-b \times i) \wedge \forall i \in \mathbb{N}: w[i] > 0 \wedge a, b \in \mathbb{R}^{+}{}_{0} \wedge \sum^{+\infty}{}_{i = iexpected}\ w[i] = 1. \quad (9)$$

The latter condition in Equation (9) ensures failures after the expect lifetime do not significantly count towards the (un)reliability metric. However, a more reliable technology should still achieve a better result.

Next, factoring in the acceptable percentage of failures is done by the following:

$$a = 1 - k \quad (10)$$

Once a is known, determining b is trivial by solving the following equation numerically:

$$\exp(b) - 1 = (1 - k) \times \exp(b - b \times i_{expected}) \quad (11)$$

The reliability metric $R[i_{expected}, k]$ can then be calculated by multiplying the weighing function point-wise with the cumulative failure percentage given by f[i]:

$$R[i_{expected}, k] = \sum^{+\infty}{}_{i = 1}\ (1 - k) \times \exp(-b \times i) \times f[i] \quad (12)$$

Using this metric, a higher number will indicate a less reliable technology, meaning it can substitute the compensation factor in Equations (7) and (8).

The exact method used to compare stretchable interconnect technologies and designs should be selected based on the application, available data, and desired outcome. For example, a smart health monitoring patch would only be expected to last a day, while a consumer device in the European Union would have to last for over two years. Sadly, calculating these metrics is infeasible at this time because of insufficient data and will most likely only happen at the point of large-scale industrialization.

3.2. Example Case

Analysing the data from the second test using the above methodology—planar stretchability—demonstrates the usefulness of these modified metrics and their potential pitfalls. Per design, four samples were tested, two with and two without TPU encapsulation. Tables 2 and 3 list the mechanical and electrical measurements, respectively. As mechanical failure, the first point at which the material ruptures was chosen, while electrically, a ten-fold resistance increase versus the starting resistance was used. Because of time uncertainty between the trigger signal being sent and the start of the measurement, one millimeter is deducted from the extension before electrical failure.

The observed failure (Run 2—Design #13) in the mechanical measurement occurred because of a loss of air pressure to the pneumatic grips during the test, releasing the sample. The more common failures during the electrical measurement were caused by a variety of problems between synchronizing the mechanical and electrical measurements. In both cases, the actual achievable elongation values are expected to be less, the reason for this is a combination between the sample slipping and cantilevering in the claws. Measuring the exact length after failure was impossible because of delamination and curling of the material, as shown in Figure 8.

Averaging the values for both the free-standing and encapsulated cases, and separating them into mechanical and electrical failure, the stretchability and planar stretchability were calculated for each value, resulting in Table 4. The considered area when calculating the planar stretchability is the actual width the interconnect takes up on the sample multiplied by 47 mm; the values used are listed in Table 1. The starting fillets were neglected, as these are identical in all cases. However, when comparing technologies, or if they differ between designs, these fillets or other transition structures should be included.

Table 2. Mechanical measurements performed on meanders using a tensile tester.

	Free Standing				Encapsulated			
	Run 1		Run 2		Run 3		Run 4	
Design	Force [N]	Ext. [mm]	Force [N]	Ext. [mm]	Force [N]	Ext. [mm]	Force [N]	Ext. [mm]
1	7.46	17.3	7.42	17.5	31.65	17.1	37.05	19.1
2	7.66	18.7	7.85	17.9	33.93	18.1	34.02	15.6
3	7.49	16.8	7.51	16.7	34.41	18.0	34.96	16.7
4	6.69	24.7	7.01	29.8	34.71	30.2	35.81	21.1
5	7.41	31.7	7.54	32.2	36.40	25.7	37.20	23.6
6	7.42	31.7	7.29	29.5	33.58	21.6	37.38	24.5
7	8.06	80.2	8.06	79.9	35.85	28.7	35.99	34.3
8	7.03	67.3	7.62	73.8	35.05	34.7	35.52	41.4
9	6.84	64.8	4.67	55.8	36.85	41.2	39.50	45.7
10	6.72	12.7	6.80	13.0	32.00	14.8	28.55	14.4
11	7.82	22.4	7.98	25.1	34.99	21.4	34.47	20.3
12	**7.56**	**14.7**	**7.58**	**14.1**	**34.30**	**16.2**	**31.22**	**14.1**
13	7.60	52.7	Failure	Failure	36.01	27.2	33.58	26.6
14	7.91	70.0	7.69	52.9	38.88	35.7	35.56	29.2
15	5.39	64.4	5.20	64.0	39.08	48.8	37.83	44.5
16	7.94	89.1	7.97	84.8	35.70	31.8	35.62	28.2
17	7.57	107.7	7.57	101.9	39.61	47.3	36.71	38.4
18	6.98	106.0	7.03	108.4	30.01	48.4	37.24	49.5

Table 3. Electrical measurements performed on the meanders during the tensile test.

	Free Standing				Encapsulated			
	Run 1		Run 2		Run 3		Run 4	
Design	R_{Start} [Ω]	Ext. [mm]	R_{Start} [Ω]	Ext. [mm]	R_{Start} [Ω]	Ext. [mm]	R_{Start} [Ω]	Ext. [mm]
1	0.527	14.5	0.515	16.5	0.521	14.1	0.502	12.2
2	0.509	15.4	0.498	14.3	0.508	11.5	0.496	9.1
3	Failure	Failure	0.470	15.1	0.490	14.0	0.491	13.3
4	0.631	24.1	0.598	23.5	0.613	17.1	0.611	19.7
5	0.618	26.2	0.601	28.4	0.606	19.5	0.582	19.4
6	0.620	25.8	Failure	Failure	0.616	19.8	0.582	21.5
7	0.959	67.4	Failure	Failure	0.879	21.4	0.867	26.9
8	Failure	Failure	0.949	72.8	0.933	26.9	Failure	Failure
9	0.970	64.3	0.979	54.9	0.976	39.5	0.923	45.1
10	0.531	12.2	0.535	12.7	0.531	11.4	0.533	9.7
11	0.515	14.5	Failure	Failure	0.528	15.5	0.532	14.6
12	**0.487**	**7.6**	**0.471**	**11.1**	**0.496**	**9.9**	**0.483**	**12.3**
13	0.820	48.3	Failure	Failure	0.819	21.7	0.779	24.2
14	0.872	63.3	Failure	Failure	0.835	22.4	0.862	20.7
15	1.076	63.9	1.069	63.3	Failure	Failure	1.026	43.9
16	1.046	79.9	1.057	74.0	1.049	28.3	1.044	21.8
17	1.238	104.4	1.206	95.2	1.274	34.1	1.258	27.7
18	1.330	105.4	1.339	107.9	1.413	70.4	1.352	31.5

The straight track reference (Design #12) appears to have a stretchability of 20% to 30%. This high stretchability is not practical because it originates from plastic deformation of the material, combined with the above clamping problems. Additionally, it is non-reversible, limiting its use to one-time deformations. As a result, this should only be considered a baseline measurement.

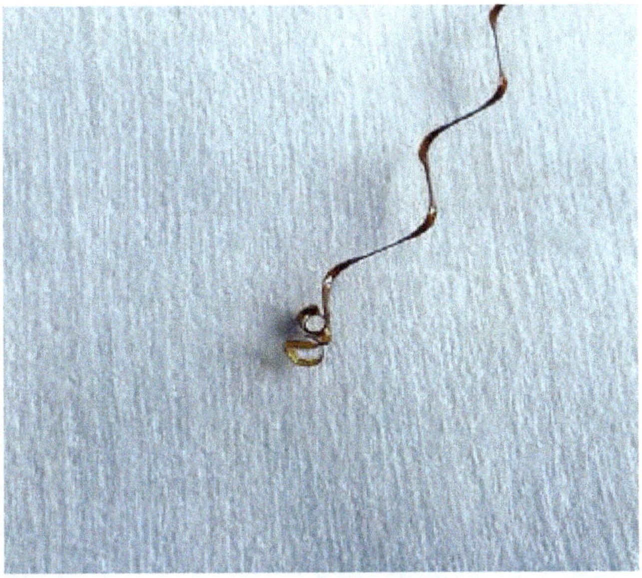

Figure 8. Delamination of the coverlay and curling of the FCB materials after tensile tests performed on the interconnects.

Table 4. Average stretchability and planar stretchability (PS) of the tested interconnects.

			Mechanical				Electrically			
	Theoretical		Free Standing		Encapsulated		Free Standing		Encapsulated	
Design	s [%]	PS [×1000]	s [%]	PS [×1000]	s [%]	PS [×1000]	s [%]	PS [×1000]	s [%]	PS [×1000]
1	8	13	37	62	39	64	33	55	28	47
2	5	18	39	134	36	123	32	108	22	75
3	5	24	36	163	37	169	32	147	29	133
4	32	27	58	49	55	46	51	43	39	33
5	32	51	68	108	52	83	58	92	41	66
6	32	72	65	146	49	110	55	123	44	99
7	107	47	170	76	67	30	143	64	51	23
8	107	90	150	127	81	69	155	131	57	49
9	107	131	128	157	92	113	127	156	90	110
10	5	35	27	178	31	202	26	172	22	146
11	11	20	51	93	44	82	31	57	32	59
12	**0**	**0**	**31**	**438**	**32**	**459**	**20**	**284**	**24**	**337**
13	67	26	112	44	57	22	103	40	49	19
14	84	43	131	67	69	35	135	69	46	23
15	122	193	137	216	99	157	135	214	93	148
16	123	50	185	76	64	26	164	67	53	22
17	165	108	223	146	91	60	212	139	66	43
18	184	164	228	203	104	93	227	203	108	97

Considering the possible deviation caused by plastic deformation and the test (20% to 30%), the experimental stretchability values are closely in line with the theoretical values calculated by considering the meander's centerline—as expected. However, the planar stretchability (PS) values—multiplied times a thousand for the sake of convenience—tell a different story. The clearest example is design #4; while it achieves the same stretchability as designs #5 and #6, it requires significantly more surface area to do so, as illustrated in Figure 9a, making it less attractive from a manufacturing cost perspective.

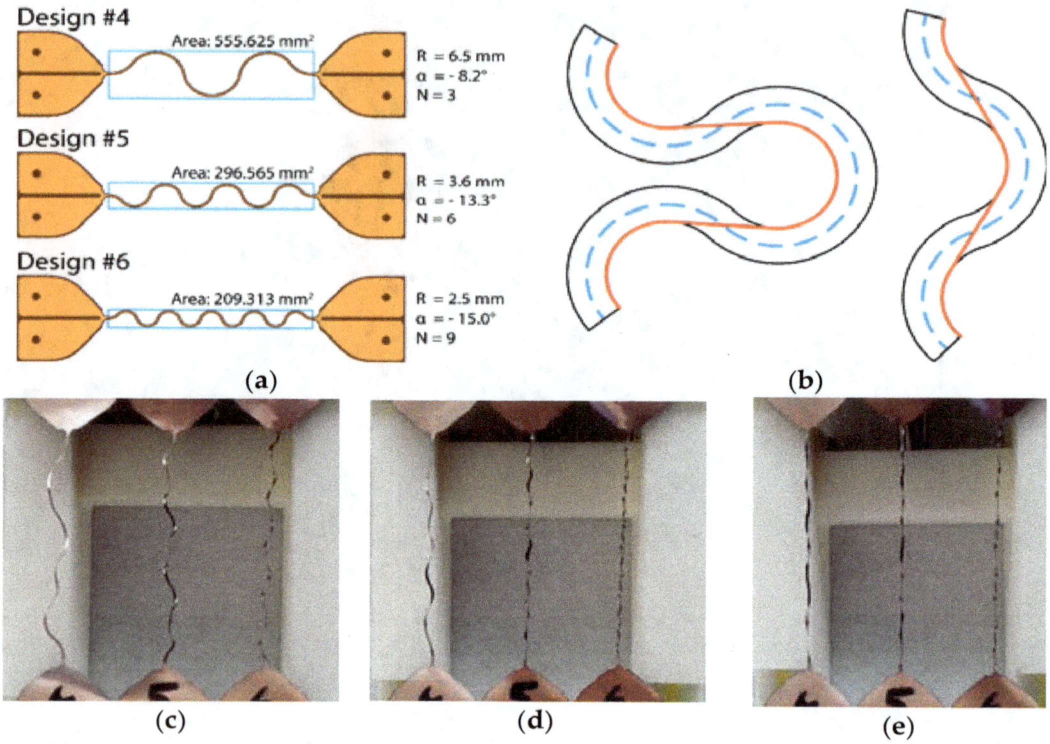

Figure 9. Detailed view of some of the tested samples. (**a**) Parameters and comparison of Designs #4, #5, and #6, illustrating the clear distinctions between designs with similar path lengths. (**b**) The length along the centerline (dashed line) is not the actual length a meander can traverse without undergoing plastic deformation. Instead, the shortest possible distance that traverses the meander track provides a more accurate value. (**c**) Three dummy samples of Design #4, #5, and #6 with a 5 mm extension applied. (**d**) Three dummy samples of Design #4, #5, and #6 with a 10 mm extension applied. (**e**) Three dummy samples of Design #4, #5, and #6 with a 15 mm extension applied.

At the same time, the fallacy of the planar stretchability is magnified by the reference design, which scores a staggering 284 to 438—three to ten times higher than the meanders. This is not surprising, because a straight line is the most efficient way to connect two points on a flat substrate. However, a wider trace would lead to a lower planar stretchability; luckily, the higher stretchability meanders would experience a similar drawback, because they would have to use even wider copper to achieve the same resistance as the much shorter straight trace. However, this method heavily promotes horseshoe-shaped meanders with small radii and a large number of segments because it packs a lot of conductor length per substrate area, even though smaller bending radii might not necessarily be an advantage from a reliability perspective.

The compensated planar stretchability (CPS), listed in Table 5, changes this figure entirely by introducing the number of transitions or segments—available in Table 1. This is, once again, exceptionally clear when comparing designs with similar stretchability, such as Design #4, #5, and #6. While #5 and #6 were previously closely matched because they take up similar substrate surface areas, #4 and #5 now have the advantage because they have less segments. This stands to reason because the increased space between the individual segments would allow more encapsulation material in-between the segments—meaning it can deform more before developing tears in the encapsulation material. An additional reason that Designs #4 and #5 are preferable over #6 can be seen in Figure 9c; at 5 mm elongation, Design #6 is almost entirely stretched, while #4 and #5 still have some headroom. This situation only worsens as the elongation is increased to 10 mm (Figure 9d) and then to 15 mm (Figure 9e). While this might appear counterintuitive at first glance, the reason for this discrepancy can be found in Figure 9b. The length along the meander centerline, as listed in Table 1, is not the actual length a meander can achieve without plastic deformation. Instead, the smallest radii in combination with a tangent running in between these radii provides an accurate number. As a result, the meander with a small number of segments, and hence a lower opening angle α, can potentially span a far longer distance without undergoing plastic deformation.

Instead, the smallest radii in combination with a tangent running in between these radii provides an accurate number. As a result, the meander with a small number of segments, and hence a lower opening angle α, can potentially span a far longer distance without undergoing plastic deformation. For example, when considering the difference between the centerline and this actual length on a per segment basis, Design #4 only loses 4.56% of its length per segment, while this increases to 7.49% and 10.10% for Designs #5 and #6, respectively, explaining the results seen in Figure 9.

Table 5. Compensated planar stretchability (CPS) $\times$1000.

		Mechanical		Electrical	
	Theoretical	Free Standing	Encapsulated	Free Standing	Encapsulated
Design	CPS [×1000]	CPS [×1000]	CPS [×1000]	CPS [×1000]	CPS [×1000]
1	4	21	21	18	16
2	3	22	21	18	13
3	3	18	19	16	15
4	9	16	15	14	11
5	24	18	14	15	11
6	16	16	12	14	11
7	16	25	10	21	8
8	15	21	11	22	8
9	15	17	13	17	12
10	2	11	13	11	9
11	5	23	20	14	15
12	**0**	**438**	**459**	**284**	**337**
13	13	22	11	20	10
14	14	22	12	23	8
15	15	17	12	16	11
16	17	25	9	22	7
17	18	24	10	23	7
18	18	23	10	23	11

The advantages of compensated planar stretchability are also apparent when comparing Designs #16, #17, and #18. While Designs #18 and #17 should significantly exceed the stretchability of Design #16, both their CPS figures are significantly scaled down by introducing the number of segments as a factor. Given that the minimum spacing between two separate meander segments decreases from 5.038 mm to 2.831 mm, and finally to 1.849 mm, respectively, for these three designs, this is a correct figure; especially when considering a larger radius should decrease the strain experienced by the copper.

3.3. *Resistance Measurements*

DC resistance measurements are the golden standard to electrically verify stretchable electronics. The general idea consists of proving the prowess of the technology or design at its intended application; that is, passing an electrical signal. However, many measurements downplay the actual effects observed or are fundamentally flawed from the start by failing to consider realistic use conditions.

The above circuit was tested using a 1 mA current, a typical test current for resistance measurement to place the returned voltage in the multimeter's 2.1 V measurement range. However, if the function of the device was to carry a large current (e.g., 10 A), the performed measurement would be pointless because the heating might significantly affect the mechanical reliability of the circuit. Carrying large currents will pose different challenges as opposed to, for example, capacitive sensing, the measurement should use a test current close to the one encountered in the intended application whenever possible. This is especially important when considering the failure mode of conductors can change dramatically as the width and thickness increases, as commonly seen with flexible circuit boards, meaning a small dimensional change to improve electrical performance might cause significant issues mechanically.

Under any condition, the first step is correctly measuring the observed resistance. Three circuit topologies, shown in Figure 10, are in everyday use when measuring the resistance of an electrical interconnect. The first topology (Figure 10a) is equivalent to placing multimeter probes on the contact pads and performing a two-point measurement. Here, both the contact and lead resistance come into full effect. The next possible topology is the quasi four-wire approach (Figure 10b), where separate wires are used for driving the measurement current and sensing the voltage, but they are connected before contacting the device-under-test (DUT). As a result, the contact resistance is still present. The final case, a true four-wire measurement (Figure 10c), is the only correct method in most cases [17,18].

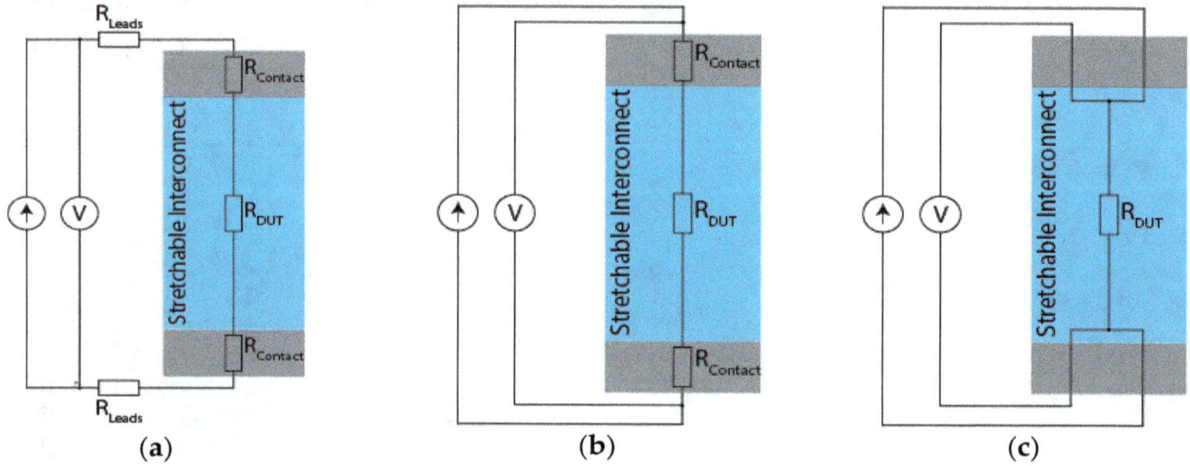

Figure 10. Measurement setups for resistance measurement: (**a**) two-wire setup affected by lead and contact resistance; (**b**) quasi four-wire setup affected by the contact resistance; and (**c**) true four-wire setup that eliminates the effect of lead and contact resistance.

The method in Figure 10a is only suitable for quick process verification, secondary checks, or when dealing with resistances of multiple kilo-ohms. For the tested sixty-meter meander, Figure 10b presents the minimum acceptable measurement. The movement of the test leads, and especially the coiling on the reel, might significantly affect the resistance of the test leads during the measurement. However, during movement, the contact resistance can change dramatically, making the true four-wire measurement the only correct method.

To understand the importance, it is worth looking at ballpark figures. Consider a stretchable interconnect in an undefined technology; this hypothetical copper-based interconnect has a resistance of 300 mΩ. A two-wire measurement (Figure 10a) is performed on the interconnect while it is stretched, hoping to monitor the formation of micro-cracks over time. The test leads have a resistance of 100 mΩ each, while the nickel-plated probes have a contact resistance of 25 mΩ with interconnect. Before stretching the interconnect, the multimeter will see 550 mΩ, which, for example, increases to 700 mΩ when the interconnect is fully extended—or a 27.3% increase. While actually, the resistance increase the meander experienced was 50%, a significantly different result.

The next measurement, performed by a slightly more skilled operator, uses a zeroed multimeter—meaning the probes were placed on the same contact pad to make a reference measurement. This action will indeed subtract 250 mΩ from the measured resistance and provide a 50% figure in theory. However, in practice, the measurement is still flawed and the formation of small micro-cracks will still be lost to the measurement noise and error. The movement during stretching will slightly change the resistance of the test leads, and shifting of the contacts will result in a slight, but noticeable alteration of the contact resistance because it is a function of pressure, surface condition, and location [19–21]. The magnitude of this effect is difficult to estimate as it heavily depends on the surface roughness and type of probe. However, it is safe to say it will create an uncertainty in the milliohm range or larger (e.g., ± 5 mΩ). Additionally, the multimeter might not be in its optimal measurement range; the software will dutifully subtract the 250 mΩ and present a 150 mΩ in ideal circumstances. However, if the measurement ranges were, for example, 500 mΩ and 5 Ω with a resolution of 1 mΩ and 10 mΩ, respectively, the measurement uncertainty would increase tenfold for this scenario—because the multimeter is now operating in its 500 mΩ range. This means that it is crucial to consider the measurement range the multimeter is in, and not blindly base the result on the displayed value, especially when dealing with low-cost handheld units.

Of course, not every effect comes into play in every situation; therefore, it is essential to understand the magnitude of the problem for a specific situation. Micro-cracks are initially virtually impossible to detect electrically; this is easy to prove using simple electrical theory. The resistance of a rectangular conductor at DC depends on its resistivity ρ, length L, thickness t, and width w [22]:

$$R = \rho \times L/(t \times w). \tag{13}$$

Consider a copper trace with a thickness of 18 μm and width of 100 μm at 20 °C, it will have a resistivity of 16.8 nΩm [23,24]. If the slice in which the micro-crack occurs has a length of 1 μm, it will start with a resistance of 9.3 μΩ. To add 1 μΩ of resistance, the crack has to propagate 9.7 μm into the trace, and another 7.95 μm to add another 1 μΩ. In short, detecting micro-cracks early requires exact high-resolution resistance measurements. This also explains why failures might appear suddenly; the trace might have broken most of the way before the actual resistance increase becomes noticeable. Of course, detecting a single micro-crack tends to be impossible under most circumstances, luckily, they tend to occur in large numbers. However, based on this, we can conclude analyzing the mechanical condition of a circuit based on electrical measurements is a precarious method unless the measurement limitations are considered.

What is also interesting to study is the effect of this localized resistance increase when carrying larger currents. Consider, for example, the case of a constant current LED driver supplying 100 mA through the above 18 μm by 100 μm long interconnect. Assuming the interconnect spans a distance of 200 mm, it will have a resistance of 1.867 Ω and will dissipate approximately 18.7 mW over its entire length—a trivial amount not worth mentioning for a highly conductive material like copper. However, at the location of micro-cracks, things can turn for the worse quickly; a 1-μm long slice will dissipate only 93.33 nW, but this goes up to 933.3 nW for a 90-μm wide crack. Considering this slice weighs only 16.13 ng and copper has a specific heat capacity of 384.4 J/(kg K) [24], it requires only 620 nJ to increase its temperature by 100 °C—or less than one second at this current assuming no thermal energy is conducted to the surrounding material. While conduction to the surroundings will dampen the temperature rise, it does point out that thermal degradation of the encapsulating material, which supports the interconnect, can occur at the site of failing conductors. It is not far-fetched that this might significantly aggravate the situation and lead to earlier failures depending on the encapsulation material.

Another effect worth considering when measuring small resistances is the Seebeck effect, meaning any metal-to-metal junction starts behaving as a thermocouple [17]. In most measurement setups, this effect is trivial and readily eliminated by using a parallel measurement path on both sides of the sample, because the entire setup is at the same temperature. Alternatively, specific alloys can be used on the contacts to limit this effect further—usually labelled as "low thermal electromotive force (EMF)" by the manufacturer. However, heating caused by the circuit or mechanical movement can upset this isothermal environment and result in the introduction of large thermal voltages.

Consider a typical nickel-plated probe; the copper–nickel Seebeck coefficient is 10 μV/K for pure copper and nickel [25]. A temperature gradient of 10 °C will then lead to a thermal voltage of 100 μV. Using the initial example once more, a source meter sends a 100 mA measurement current through 300 mΩ, which results in a voltage of 30 mV over the sense leads in ideal conditions. The 100 μV offset by the thermal voltage skews this measurement to 301 mΩ, a relatively small error. The problem stems from the fact that many handheld multimeters are incapable of providing high test currents, unlike source meters or high-end bench meters, which will happily provide higher test currents. For example, the Keysight U1253B uses a current of 1.04 mA, which would result in a 312 μV output, adding 100 μV onto that results in a 400 mΩ value on the display (10 mΩ resolution). An even worse situation exists when considering copper–copper oxide contacts, which can exhibit Seebeck coefficients of 1 mV/K [17]. For this reason, choosing the correct probe for the intended application is of vital importance when measuring small resistances.

A final consideration is the time it takes the multimeter to perform the measurement. This measurement is rarely instant because the sample-and-hold circuit requires time to charge a capacitor [26]. A changing resistance during this interval—called the aperture time—is a potential cause for error, meaning movement of either the probes or the sample can introduce an unintended bias to the measurement [18]. This is exceptionally clear when attempting to eliminate the noise of the power grid by using aperture times that are multiples of 20 ms or 16.7 ms to average one or more power cycles. Ideally, the mechanical measurement stops, triggers the multimeter, and waits for it to complete a measurement before continuing the test.

In short, measuring small resistance changes requires a correctly configured measurement setup. The most substantial errors can be avoided by using a four-wire measurement with appropriate test currents to emulate self-heating. Samples should have clean electrical contacts at the same temperature, and the measurement instrument has to be configured to the correct range with a suitable test current.

4. Discussion

Stretchability is commonly used to describe the performance of a stretchable electronic interconnect. However, as demonstrated above, it is an unreliable metric for the performance of a technology because of its design dependence. A series of alternative metrics taking into account the area, volume, and reliability of the interconnect were proposed to limit the influence of the design. Additionally, the considerations for accurate resistance measurement were highlighted using simple examples, providing a straightforward way to estimate their importance for typical stretchable interconnect measurements.

Author Contributions: Conceptualization, B.P. and J.V.; Investigation, B.P.; Methodology, B.P.; Visualization, B.P.; Writing—original draft, B.P.; Writing—review & editing, B.P., F.B., and J.V.

Acknowledgments: The authors would like to thank Sheila Dunphy, Kristof Dhaenens, and Steven Van Put for their assistance in fabricating the samples.

References

1. Wagner, S.; Bauer, S. Materials for stretchable electronics. *MRS Bull.* **2012**, *37*, 207–213. [CrossRef]
2. Rogers, J.A.; Someya, T.; Huang, Y. Materials and mechanics for stretchable electronics. *Science* **2010**, *327*, 1603–1607. [CrossRef] [PubMed]
3. Isaac, S.; Leah, S. Conductor Reel. U.S. Patent 2,031,434, 18 Feburary 1936.
4. Coombs, C.F.; Holden, H. *Printed Circuits Handbook*, 7th ed.; McGraw-Hill Education: New York, NY, USA, 2016.

5. Someya, T. *Stretchable Electronics*; Wiley: Hoboken, NJ, USA, 2013.
6. Gray, D.S.; Tien, J.; Chen, C.S. High-conductivity elastomeric electronics. *Adv. Mater.* **2004**, *16*, 393–397. [CrossRef]
7. Plovie, B.; Yang, Y.; Joren, G.; Sheila, D.; Kristof, D.; Steven, V.P.; Björn, V.; Thomas, V.; Frederick, B.; Jan, V. Arbitrarily shaped 2.5d circuits using stretchable interconnects embedded in thermoplastic polymers. *Adv. Eng. Mater.* **2017**, *19*, 1700032. [CrossRef]
8. Vanfleteren, J.; Löher, T.; Gonzalez, M.; Bossuyt, F.; Vervust, T.; Wolf, I.D.; Jablonski, M. Scb and smi: Two stretchable circuit technologies, based on standard printed circuit board processes. *Circuit World* **2012**, *38*, 232–242. [CrossRef]
9. Yang, Y.; Thomas, V.; Sheila, D.; Steven, P.; Bjorn, V.; Kristof, D.; Lieven, D.; Lothar, M.; Linde, V.; Tom, M.; et al. 3d multifunctional composites based on large-area stretchable circuit with thermoforming technology. *Adv. Electron. Mater.* **2018**. [CrossRef]
10. Sun, Y.; Choi, W.M.; Jiang, H.; Huang, Y.Y.; Rogers, J.A. Controlled buckling of semiconductor nanoribbons for stretchable electronics. *Nat. Nanotechnol.* **2006**, *1*, 201. [CrossRef] [PubMed]
11. Carta, R.; Jourand, P.; Hermans, B.; Thoné, J.; Brosteaux, D.; Vervust, T.; Bossuyt, F.; Axisa, F.; Vanfleteren, J.; Puers, R. Design and implementation of advanced systems in a flexible-stretchable technology for biomedical applications. *Sens. Actuators A Phys.* **2009**, *156*, 79–87. [CrossRef]
12. Kubo, M.; Li, X.; Kim, C.; Hashimoto, M.; Wiley, B.J.; Ham, D.; Whitesides, G.M. Stretchable microfluidic radiofrequency antennas. *Adv. Mater.* **2010**, *22*, 2749–2752. [CrossRef] [PubMed]
13. Hsu, Y.-Y.; Gonzalez, M.; Bossuyt, F.; Axisa, F.; Vanfleteren, J.; De Wolf, I. The effects of encapsulation on deformation behavior and failure mechanisms of stretchable interconnects. *Thin Solid Films* **2011**, *519*, 2225–2234. [CrossRef]
14. Kwon, S.; Cho, H.W.; Gwon, G.; Kim, H.; Sung, B.J. Effects of shape and flexibility of conductive fillers in nanocomposites on percolating network formation and electrical conductivity. *Phys. Rev. E* **2016**, *93*, 032501. [CrossRef] [PubMed]
15. Tao, C.; Yizhou, Z.; Wen-Yong, L.; Wei, H. Stretchable thin-film electrodes for flexible electronics with high deformability and stretchability. *Adv. Mater.* **2015**, *27*, 3349–3376.
16. Kim, D.H.; Xiao, J.; Song, J.; Huang, Y.; Rogers, J.A. Stretchable, curvilinear electronics based on inorganic materials. *Adv. Mater.* **2010**, *22*, 2108–2124. [CrossRef] [PubMed]
17. Keithley. *Low Level Measurements Handbook*, 7th ed.; Tektronix: Beaverton, OR, USA, 2013; p. 244.
18. Horowitz, P.; Hill, W. *The Art of Electronics*; Cambridge University Press: Cambridge, UK, 2015.
19. Kogut, L.; Komvopoulos, K. Electrical contact resistance theory for conductive rough surfaces. *J. Appl. Phys.* **2003**, *94*, 3153–3162. [CrossRef]
20. Hyun, C.; Park, H.; Hahm, S. *An Analysis of Probing Cres in Gold Bumping Pad Using Automatic Test Equipment*; IEEE SW Test Workshop: San Diego, CA, USA, 2012.
21. Timsit, S. Electrical Contact Resistance: Properties of Stationary Interfaces, Electrical Contacts–1998. In Proceedings of the Forty-Fourth IEEE Holm Conference on Electrical Contacts (Cat. No.98CB36238), Arlington, VA, USA, 26–28 October 1998; pp. 1–19.
22. Kumar, N. *Comprehensive Physics Xii*; Laxmi Publications: New Delhi, India, 2004.
23. Matula, R.A. Electrical resistivity of copper, gold, palladium, and silver. *J. Phys. Chem. Ref. Data* **1979**, *8*, 1147–1298. [CrossRef]
24. Kaye, G.W.C.; Laby, T.H. *Tables of Physical and Chemical Constants*; Longman: Harlow, UK, 1995.
25. Kidd, M. Watch out for those thermoelectric voltages! *Int. J. Metrol.* **2012**, *19*, 18–21.
26. Analog Devices Inc., E; Zumbahlen, H. *Linear Circuit Design Handbook*; Elsevier Science: New York, NY, USA, 2011.

13

One-Dimensional Thermal Analysis of the Flexible Electronic Devices Integrated with Human Skin

Yun Cui [1], Yuhang Li [1,2,3,*], Yufeng Xing [1], Tianzhi Yang [4] and Jizhou Song [2,5,*]

1 Institute of Solid Mechanics, Beihang University (BUAA), Beijing 100191, China; kyalex@sina.com (Y.C.); xingyf@buaa.edu.cn (Y.X.)
2 Key Laboratory of Soft Machines and Smart Devices of Zhejiang Province, Zhejiang University, Hangzhou 310027, China
3 State Key Laboratory of Digital Manufacturing Equipment and Technology, Huazhong University of Science and Technology, Wuhan 430074, China
4 School of Aerospace Engineering, Shenyang Aerospace University, Shenyang 110136, China; yangtz@sau.edu.cn
5 Department of Engineering Mechanics and Soft Matter Research Center, Zhejiang University, Hangzhou 310027, China
* Correspondence: liyuhang@buaa.edu.cn (Y.L.); jzsong@zju.edu.cn (J.S.)

Academic Editors: Seung Hwan Ko, Daeho Lee and Zhigang Wu

Abstract: A one-dimensional analytic thermal model for the flexible electronic devices integrated with human skin under a constant and pulsed power is developed. The Fourier heat conduction equation is adopted for the flexible electronics devices while the Pennes bio-heat transfer equation is adopted for the skin tissue. Finite element analysis is performed to validate the analytic model through the comparison of temperature distributions in the system. The influences of geometric and loading parameters on the temperature increase under a pulsed power are investigated. It is shown that a small duty cycle can reduce the temperature increase of the system effectively. A thin substrate can reduce the device temperature but increase the skin surface temperature. The results presented may be helpful to optimize the design of flexible electronic devices to reduce the adverse thermal influences in bio-integrated applications.

Keywords: flexible electronics; thermal analysis; human skin

1. Introduction

Recent advances in flexible electronics enable the development of epidermal electronics [1,2], which could be mounted onto the skin and retain conformal contact with the skin under compression and tension. Knowledge on heat transfer within skin tissue due to the physical coupling of epidermal electronics to the skin is critical for the use of epidermal electronics since even only a few degrees in temperature increase may induce uncomfortable feelings.

Many researchers have performed the thermal analysis of flexible electronic devices by adopting the Fourier heat conduction equation. Kim et al. [3], Lu et al. [4], and Cui et al. [5] studied the heat conduction of flexible micro-scale inorganic light-emitting diodes under a constant power experimentally and analytically. Relevant experimental results are presented as surface temperature contours, which can be captured by QFI Infra-Scope Micro-Thermal Imager [3]. Results of experiment agree well with analytic ones. Kim et al. [6] and Li et al. [7] investigated the thermal properties of flexible devices on various substrates under a pulsed power. They developed an analytic axi-symmetric model to explore the dependence of device temperature on the geometric dimensions, material properties and loading parameters. There exist several reviews of thermal and mechanical analysis of flexible electronics [8,9].

The heat transfer in skin tissue is much more complicated than that in flexible electronic devices due to the influences of blood perfusion and metabolism and the complex multilayered structure consisting of four layers: stratum corneum, epidermis, dermis and fat as shown in Figure 1a. The Pennes bioheat equation has been widely used to investigate the thermal response of skin tissue under surface heating. Jiang et al. [10] studied the skin burn process resulting from a high temperature heat source to the skin surface by using finite difference method to solve the Pennes bio-heat equation. Ozen et al. [11] presented a one-dimensional multi-layer model to characterize the temperature rise resulting from skin exposure to microwaves. Im et al. [12] carried out a numerical study on the temperature profiles as a result of local heating of human skin. An extensive review on heat transfer of skin tissue was given by Xu et al. [13].

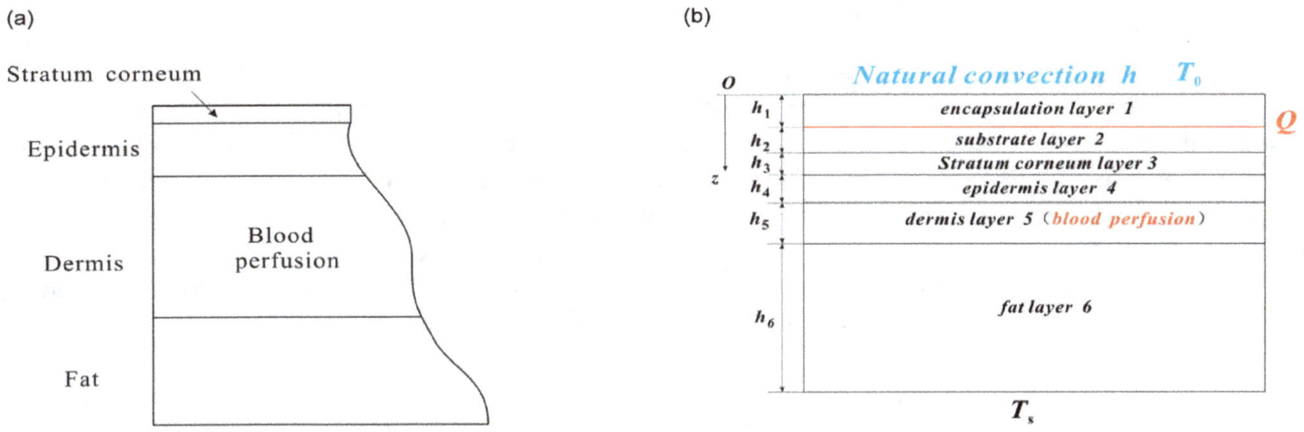

Figure 1. (**a**) Schematic illustration of the skin tissue structure; and (**b**) schematic illustration of one-dimensional geometry of the analytic modeled device-skin system.

The above heat transfer analyses are either for flexible electronic devices or human skin. There are few studies on the system of flexible electronic devices integrated with the human skin. For better understanding of the thermal properties of device and establishing design guidelines for flexible electronic devices to minimize the adverse thermal effect, we aims to develop an analytical model for the flexible electronic devices integrated with human skin under both constant and pulsed power. Basing on the Kim's experimental results and analytical model for the device on top of a metal trunk [3,4], this paper aims at the device integrated with human skin. For flexible electronic devices with the in-plane dimension on the order of a few millimeters, which is much larger than the device thickness, the heat mainly transfers along the thickness direction. It is reasonable to perform a one-dimensional thermal analysis. Such treatment simplifies the theoretical model significantly with enough accuracy. Moreover, the finite element models are also established to validate the analytic solutions.

2. Thermal Analysis under a Constant Power

The human skin is modeled as a multi-layer structure consisting of four layers: stratum corneum, epidermis, dermis and fat. The flexible electronic device consists of the functional component on a flexible substrate (e.g., polydimethylsiloxane (PDMS)) encapsulated by an encapsulation layer (e.g., SU8). The functional component could be modeled as a planar heat source with heat generation power density Q (W/m^2) since its thermal conductivity (~100 W/m/K) is much larger than that of substrate or encapsulation layer. Figure 1b shows the one-dimensional geometry of the analytic model with the flexible device on the human skin. The origin of the coordinate (z) is established on the top surface of encapsulation layer with the positive direction pointing from the flexible device to the skin tissue. The total thickness of the system is denoted by H. h_i denotes the thickness for each layer with the substrate i (i = 1, 2, 3, 4, 5 and 6) for the encapsulation, substrate, stratum corneum, epidermis, dermis and fat layer, respectively. The top surface of encapsulation layer has the natural convection boundary with h as the coefficient of heat convection. The bottom surface of fat layer has constant core body temperature T_s.

The temperature in the device satisfies the Fourier heat conduction equation

$$k_i \frac{d^2 T_i}{dz^2} = 0 \, (i = 1, 2) \tag{1}$$

while the temperature in the skin tissue satisfies the Pennes bio-heat equation [14]

$$\begin{cases} k_i \dfrac{d^2 T_i}{dz^2} + q_{\text{met}} = 0 \, (i = 3, 4, 6) \\ k_5 \dfrac{d^2 T_5}{dz^2} - \omega_b \rho_b c_b \left(T_5 - T_s \right) + q_{\text{met}} = 0 \end{cases} \tag{2}$$

where k_i represents the thermal conductivity of corresponding layers; ρ_b and c_b are the density and specific heat of blood, respectively; ω_b is the blood perfusion rate; and q_{met} is the metabolic heat generation. Here, we assumed that the blood temperature is just the same as the core temperature [14]. The main difference between two heat conduction models mentioned above is confined to blood perfusion effect and metabolism. The blood perfusion is effected by ω_b, which varies from 0–1 mL/(mL·s) [15]. While the metabolic heat generation is almost constant for healthy person. It should be noted that the blood perfusion only exists inside the dermis layer [13].

Let $\Delta T = T(z) - T_s$ denote the temperature increase from the core temperature. Equations (1) and (2) then become

$$\begin{cases} k_i \dfrac{d^2 \Delta T_i}{dz^2} = 0 \, (i = 1, 2) \\ k_i \dfrac{d^2 \Delta T_i}{dz^2} + q_{\text{met}} = 0 \, (i = 3, 4, 6) \\ k_5 \dfrac{d^2 \Delta T_5}{dz^2} - \omega_b \rho_b c_b \Delta T_5 + q_{\text{met}} = 0 \end{cases} \tag{3}$$

The temperature increase in each layer can be given

$$\begin{cases} \Delta T_1 = A_1 z + B_1 \\ \Delta T_2 = A_2 z + B_2 \\ \Delta T_3 = -q_{\text{met}} z^2 / 2k_3 + B_3 z + C_3 \\ \Delta T_4 = -q_{\text{met}} z^2 / 2k_4 + B_4 z + C_4 \\ \Delta T_5 = A_5 \exp\left(z\sqrt{\eta / k_5} \right) + B_5 \exp\left(-z\sqrt{\eta / k_5} \right) + q_{\text{met}} / \eta \\ \Delta T_6 = -q_{\text{met}} z^2 / 2k_6 + B_6 z + C_6 \end{cases} \tag{4}$$

where $\eta = \omega_b \rho_b c_b$, and the coefficients A_i, B_i and C_i are to be determined by the boundary conditions and the interfacial continuity condition. At the top surface of encapsulation layer (z = 0), the natural convection condition gives

$$-k_1 \left. \frac{d\Delta T}{dz} \right|_{z=0} = -h \cdot \left(\Delta T - \Delta T_0 \right) \big|_{z=0} \tag{5}$$

where $\Delta T_0 = T_0 - T_s$, T_0 denotes the ambient temperature, k_1 is the thermal conductivity of encapsulation layer and h is the coefficient of heat convection. At the encapsulation/substrate interface

($z = h_1$), the temperature is continuous while the heat flux satisfies the surface heat source condition, which gives,

$$\Delta T|_{z=h_1^+} = \Delta T|_{z=h_1^-} \text{ and } k_1 \left.\frac{d\Delta T}{dz}\right|_{z=h_1^-} - k_2 \left.\frac{d\Delta T}{dz}\right|_{z=h_1^+} = Q \tag{6}$$

where k_2 is the thermal conductivity of substrate layer. At the interface between any other two layers, both the temperature and heat flux are continuous, which give

$$\Delta T|_{z=(\sum_{i=1}^{n} h_i)^+} = \Delta T|_{z=(\sum_{i=1}^{n} h_i)^-} \text{ and } k_{(n+1)} \left.\frac{d\Delta T}{dz}\right|_{z=(\sum_{i=1}^{n} h_i)^+} = k_n \left.\frac{d\Delta T}{dz}\right|_{z=(\sum_{i=1}^{n} h_i)^-} \tag{7}$$

where n = 2, 3, 4, and k_n is the thermal conductivity of layer n with n = 2 for substrate n = 3 for stratum corneum, n = 4 for epidermis, and n = 5 for dermis.

The ambient temperature at the bottom surface ($z = \sum_{i=1}^{6} h_i$) of fat layer gives

$$\Delta T|_{z=\sum_{i=1}^{6} h_i} = 0 \tag{8}$$

It should be noted that the stratum corneum usually has the same thermal property as that of epidermis, i.e., $k_3 = k_4$. The coefficients in Equation (4) are then determined by the boundary and continuity conditions in Equations (5)–(8) as

$$\begin{Bmatrix} A_1 \\ B_1 \\ A_2 \\ B_2 \\ B_3 \\ C_3 \\ B_4 \\ C_4 \\ A_5 \\ B_5 \\ B_6 \\ C_6 \end{Bmatrix} = \begin{Bmatrix} \dfrac{y\,[-E\sinh(y) + F\cosh(y)] - E\cosh(y) + F\sinh(y) - M}{y\,[C\sinh(y) + D\cosh(y)] + C\cosh(y) + D\sinh(y)} \\ \dfrac{k_1}{h}A_1 + \Delta T_0 \\ (k_1 A_1 - Q)/k_2 \\ h_1 A_1 + B_1 - h_1 A_2 \\ [k_1 A_1 - Q + q_{\text{met}}(h_1 + h_2)]/k_3 \\ A_2(h_1 + h_2) + h_1 A_1 + B_1 - h_1 A_2 + q_{\text{met}}(h_1 + h_2)^2/(2k_3) - B_3(h_1 + h_2) \\ [k_1 A_1 - Q + q_{\text{met}}(h_1 + h_2)]/k_3 \\ A_2(h_1 + h_2) + h_1 A_1 + B_1 - h_1 A_2 + q_{\text{met}}(h_1 + h_2)^2/(2k_3) - B_3(h_1 + h_2) \\ [(C + D)A_1 + E - F]\,e^{-(h_1+h_2+h_3+h_4)\sqrt{\eta/k_5}}/2 \\ [(C - D)A_1 + E + F]\,e^{(h_1+h_2+h_3+h_4)\sqrt{\eta/k_5}}/2 \\ \begin{bmatrix} y(CA_1 + E)\sinh(y)/h_6 + y(DA_1 - F)\cosh(y)/h_6 \\ +q_{\text{met}}(h_1 + h_2 + h_3 + h_4 + h_5)/k_6 \end{bmatrix} \\ \begin{bmatrix} (CA_1 + E)\cosh(y) + (DA_1 - F)\sinh(y) + q_{\text{met}}/\eta \\ +q_{\text{met}}(h_1 + h_2 + h_3 + h_4 + h_5)^2/2k_6 - B_6(h_1 + h_2 + h_3 + h_4 + h_5) \end{bmatrix} \end{Bmatrix} \tag{9}$$

where,

$$\begin{aligned} C &= h_1 + k_1/h + k_1 h_2/k_2 + k_1(h_3 + h_4)/k_3 \\ D &= k_1/\sqrt{k_5\eta} \\ E &= -(h_2/k_2 + (h_3 + h_4)/k_3)Q - q_{\text{met}}(h_3 + h_4)^2/(2k_3) - q_{\text{met}}/\eta + \Delta T_0 \\ F &= (Q + q_{\text{met}}(h_3 + h_4))/\sqrt{k_5\eta} \\ M &= q_{\text{met}}/\eta - q_{\text{met}} h_6^2/(2k_6) \\ y &= h_5\sqrt{\eta/k_5} \end{aligned}$$

In order to validate the analytical solutions in Equation (9), finite element analysis (FEA) is performed using ABAQUS software (6.13-1, Dassault Simulia, Waltham, MA, USA) to study the thermal properties of this system with continuum element DC3D8. The thicknesses of encapsulation, substrate, stratum corneum, epidermis, dermis, and fat are taken as 7 μm, 2 mm, 0.02 mm, 0.08 mm, 1.5 mm and 4.4 mm, respectively [15]. The thermal conductivities of encapsulation, substrate, stratum corneum, epidermis, dermis, and fat are 0.2 W/(m·K), 0.15 W/(m·K), 0.21 W/(m·K), 0.21 W/(m·K), 0.37 W/(m·K) and 0.16 W/(m·K) [4,15]. The top surface of encapsulation layer has a natural convection boundary with the coefficient of heat convection $h = 25$ W/(m^2·K) [2,16]. The ambient temperature is set as 25 °C. The core body temperature at the bottom surface of substrate is 37 °C. The metabolic heat generation in the skin tissue is 368 W/m^3 [15]. The product of mass density and specific heat capacity of the blood is 4.218×10^6 J/(m^3·K) [17]. The blood perfusion rate inside the dermis layer is 0.03 mL/(mL·s) [15]. Figure 2 shows the comparison of temperature increase along the thickness direction between the analytic predictions and FEA under the input power density 2500 W/m^2. The good agreement validates the analytic model. With the distance to the top surface of encapsulation layer increasing, the temperature increase first increases inside the encapsulation layer then decreases in other layers. The temperature increase of the heat source can reach about 28 °C while the temperature increase at the device/skin interface is about 9 °C. Compared to the temperature increase of skin away from electronic device, which is about −1.5 °C, the constant input power causes a remarkable temperature increase.

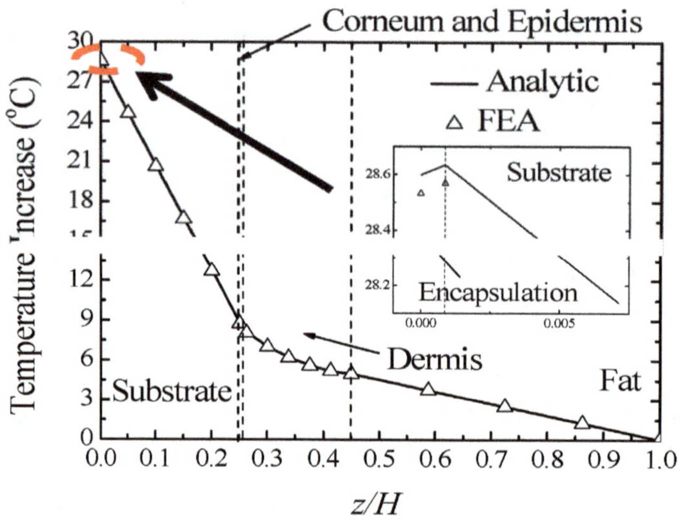

Figure 2. The comparison of temperature increase along the thickness direction between the analytic prediction and finite element analysis with the total thickness $H = 8.007$ mm.

3. Thermal Analysis under a Pulsed Power

As shown in Figure 2, the temperature increase of the device/skin interface is much more than that the human can stand. It may induce uncomfortable feelings or even tissue lesion in bio-integrated applications. Clearly, the temperature increase can be reduced by decreasing the input power density. However, the devices can hardly operate properly under small input power. In order to solve this dilemma, Kim et al. [6] and Li et al. [7] changed the input loading method from constant power to pulsed protocol, which has proven pretty effective on modern devices. The operation of flexible electronic devices (e.g., flexible light-emitting diodes) in a pulsed mode could significantly reduce the temperature increase to reach the goal in thermal management. In this section, an analytic model is developed to investigate the thermal response of flexible electronic devices on human skin.

Under a pulsed power, the temperature in the system increases to saturation in a fluctuation way [12]. We are interested in the saturated temperature since it gives the maximum temperature that could reach in the system. The pulsed power $Q(t)$ is defined in Figure 3 with Q_0 as the peak value and τ as the heating duration in one period t_0. The duty cycle D is given by $D = \tau/t_0$. Under a pulsed power density $Q(t)$, the temperature increase from the core body temperature in the system satisfies,

$$\begin{cases} k_i \dfrac{\partial^2 \Delta T_i}{\partial z^2} = \rho_i c_i \frac{\partial \Delta T_i}{\partial t} \; (i = 1, 2) \\ k_i \dfrac{\partial^2 \Delta T_i}{\partial z^2} + q_{\text{met}} = \rho_i c_i \frac{\partial \Delta T_i}{\partial t} \; (i = 3, 4, 6) \\ k_5 \dfrac{\partial^2 \Delta T_5}{\partial z^2} - \omega_b \rho_b c_b \Delta T_5 + q_{\text{met}} = \rho_5 c_5 \frac{\partial \Delta T_5}{\partial t} \end{cases} \tag{10}$$

where ρ_i and c_i are the mass density and the specific heat capacity for each layer. The boundary and continuity conditions are the same as those for the case under a constant power in Equations (5)–(8) except the constant power density Q in Equation (6) should be changed to $Q(t)$, i.e.,

$$\Delta T|_{z=h_1^+} = \Delta T|_{z=h_1^-} \text{ and } k_1 \left.\frac{\partial \Delta T}{\partial z}\right|_{z=h_1^-} - k_2 \left.\frac{\partial \Delta T}{\partial z}\right|_{z=h_1^+} = Q(t) \tag{11}$$

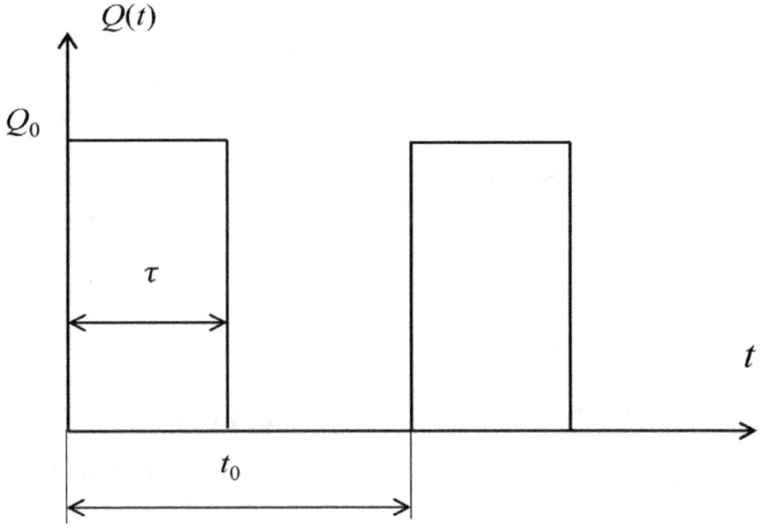

Figure 3. The pulsed power density $Q(t)$ with Q_0 as the peak power density, t as the during time and t_0 as the period.

The solution for Equation (10) is the summation of the specific solution, which is given in Section 2 without the input power, and the homogeneous solution, which is to be determined below. The homogenous solution for Equation (10) satisfies

$$\begin{cases} \dfrac{\partial \Delta T_i}{\partial t} - \lambda_i \dfrac{\partial^2 \Delta T_i}{\partial z^2} = 0 \; (i = 1, 2, 3, 4, 6) \\ \dfrac{\partial \Delta T_5}{\partial t} - \lambda_5 \dfrac{\partial^2 \Delta T_5}{\partial z^2} + \omega_b \Delta T_5 = 0 \end{cases} \tag{12}$$

where $\lambda_i = k_i/(c_i\rho_i)$ is the thermal diffusivity with c, ρ and k as the specific heat capacity, mass density and thermal conductivity, respectively.

The method of superposition is adopted to obtain the temperature increase after saturation (i.e., homogeneous solution) under a pulsed power density, which could be written via Fourier series by

$$Q(t) = Q_0 \begin{cases} 1 & 0 < t \le \tau \\ 0 & \tau < t \le t_0 \end{cases} = Q_0 \left[a_0 + \sum_{n=1}^{\infty} (a_n \cos n\omega t + b_n \sin n\omega t) \right] \tag{13}$$

where $\omega = 2\pi/t_0$, $a_0 = D = \tau/t_0$, $a_n = \sin(2n\pi D)/(n\pi)$, and $b_n = \frac{[1-\cos(2n\pi D)]}{n\pi}$. The temperature increase after saturation due to each sinusoidal power density $Q_0\cos n\omega t$ (or $Q_0\sin n\omega t$) in Equation (13) corresponds to the real (or imaginary part) of the solution due to a power of $Q_0 e^{n\omega t \cdot i}$. The power of $Q_0 e^{n\omega t \cdot i}$ yields the temperature increase as $\theta(z; n\omega) e^{n\omega t \cdot i}$, which gives the real and imaginary part as $|\theta(z; n\omega)| \cos(n\omega t + \beta_n)$ and $|\theta(z; n\omega)| \sin(n\omega t + \beta_n)$, respectively. Here $\beta_n(n\omega)$ is the phase angle of $\theta(z; n\omega)$. Therefore, the temperature increase due to the pulsed power can be obtained by

$$\Delta T(z,t) = D\theta(z;0) + \sum_{n=1}^{\infty} |\theta(z;n\omega)| \cdot \left[\begin{array}{l} \dfrac{\sin(2n\pi D)}{n\pi}\cos(n\omega t + \beta_n) \\ + \dfrac{1-\cos(2n\pi D)}{n\pi}\sin(n\omega t + \beta_n) \end{array} \right] \tag{14}$$

where $\theta(z; n\omega)$ is to be determined from Equation (11) and boundary conditions. The substitution of $\theta(z; n\omega) e^{n\omega t \cdot i}$ into Equation (11) gives the governing equation of $\theta(z; n\omega)$ as

$$\left\{ \begin{array}{l} \dfrac{d^2\theta_i}{dz^2} - q_i^2\theta_i = 0 \; (i = 1,2,3,4,6) \\ \dfrac{d^2\theta_5}{dz^2} - \left(q_5^2 + \eta/k_5\right)\theta_5 = 0 \end{array} \right. \tag{15}$$

where the $q^2 = n\omega i/\lambda$. $\theta(z; n\omega)$ satisfies the following boundary and continuity conditions: $-k_1 \partial\theta/\partial z|_{z=0} = -h \cdot \theta|_{z=0}$, $\theta|_{z=h_1^+} = \theta|_{z=h_1^-}$, $k_1\, \partial\theta/\partial z|_{z=h_1^-} - k_2\, \partial\theta/\partial z|_{z=h_1^+} = Q_0$, $\theta|_{z=(\sum_{i=1}^{n} h_i)^+} = \theta|_{z=(\sum_{i=1}^{n} h_i)^-}$, $k_{(n+1)}\, \partial\theta/\partial z|_{z=(\sum_{i=1}^{n} h_i)^+} = k_n\, \partial\theta/\partial z|_{z=(\sum_{i=1}^{n} h_i)^-}$, $\theta|_{z=\sum_{i=1}^{6} h_i} = 0$.

The solution of Equation (15) takes the form of

$$\left\{ \begin{array}{l} \theta_i = A_i \exp\left(z\sqrt{q_i^2}\right) + B_i \exp\left(-z\sqrt{q_i^2}\right) \; (i = 1,2,3,4,6) \\ \theta_5 = A_5 \exp\left(z\sqrt{q_5^2 + \eta/k_5}\right) + B_5 \exp\left(-z\sqrt{q_5^2 + \eta/k_5}\right) \end{array} \right. \tag{16}$$

where the unknown coefficients in Equation (16) are to be determined by the boundary and continuity conditions as

$$\left\{ \begin{array}{c} A_1' \\ B_1' \\ A_2' \\ B_2' \\ A_3' \\ B_3' \\ A_4' \\ B_4' \\ A_5' \\ B_5' \\ A_6' \\ B_6' \end{array} \right\} = \left\{ \begin{array}{l} \dfrac{T_1 R\cosh\left(h_6\sqrt{q_6^2}\right) + m_5 R T_2 \sinh\left(h_6\sqrt{q_6^2}\right)}{\left(P_1 S_1 + m_1 P_2 T_1\right)\cosh\left(h_6\sqrt{q_6^2}\right) + m_5\left(P_1 S_2 + m_1 P_2 T_2\right)\sinh\left(h_6\sqrt{q_6^2}\right)} \\ \left(k_1\sqrt{q_1^2} - h\right) A_1 / \left(k_1\sqrt{q_1^2} + h\right) \\ \left[\left(P_1 + m_1 P_2\right) A_1 - R\right] e^{-h_1\sqrt{q_2^2}} \\ \left[\left(P_1 - m_1 P_2\right) A_1 + R\right] e^{h_1\sqrt{q_2^2}} \\ (G+H)\, e^{-(h_1+h_2)\sqrt{q_3^2}} \\ (G-H)\, e^{(h_1+h_2)\sqrt{q_3^2}} \\ (G+H)\, e^{-(h_1+h_2)\sqrt{q_3^2}} \\ (G-H)\, e^{(h_1+h_2)\sqrt{q_3^2}} \\ (I+J)\, e^{-(h_1+h_2+h_3+h_4)\sqrt{q_5^2+\eta/k_5}} \\ (I-J)\, e^{(h_1+h_2+h_3+h_4)\sqrt{q_5^2+\eta/k_5}} \\ (K+L)\, e^{-(h_1+h_2+h_3+h_4+h_5)\sqrt{q_6^2}} \\ (K-L)\, e^{(h_1+h_2+h_3+h_4+h_5)\sqrt{q_6^2}} \end{array} \right\} \tag{17}$$

where

$$P_1 = \frac{k_1\sqrt{q_1^2}\cosh\left(h_1\sqrt{q_1^2}\right) + h\sinh\left(h_1\sqrt{q_1^2}\right)}{k_1\sqrt{q_1^2} + h}$$

$$P_2 = \frac{k_1\sqrt{q_1^2}\sinh\left(h_1\sqrt{q_1^2}\right) + h\cosh\left(h_1\sqrt{q_1^2}\right)}{k_1\sqrt{q_1^2} + h}$$

$$R = \frac{Q_0}{2k_2\sqrt{q_2^2}}$$

$$S_1 = \left[\begin{array}{l} \cosh\left(h_2\sqrt{q_2^2}\right)\cosh\left((h_3+h_4)\sqrt{q_3^2}\right)\cosh\left(h_5\sqrt{q_5^2+\eta/k_5}\right) \\ +m_2\sinh\left(h_2\sqrt{q_2^2}\right)\sinh\left((h_3+h_4)\sqrt{q_3^2}\right)\cosh\left(h_5\sqrt{q_5^2+\eta/k_5}\right) \\ +m_4\cosh\left(h_2\sqrt{q_2^2}\right)\sinh\left((h_3+h_4)\sqrt{q_3^2}\right)\sinh\left(h_5\sqrt{q_5^2+\eta/k_5}\right) \\ +m_2m_4\sinh\left(h_2\sqrt{q_2^2}\right)\cosh\left((h_3+h_4)\sqrt{q_3^2}\right)\sinh\left(h_5\sqrt{q_5^2+\eta/k_5}\right) \end{array}\right]$$

$$T_1 = \left[\begin{array}{l} \sinh\left(h_2\sqrt{q_2^2}\right)\cosh\left((h_3+h_4)\sqrt{q_3^2}\right)\cosh\left(h_5\sqrt{q_5^2+\eta/k_5}\right) \\ +m_2\cosh\left(h_2\sqrt{q_2^2}\right)\sinh\left((h_3+h_4)\sqrt{q_3^2}\right)\cosh\left(h_5\sqrt{q_5^2+\eta/k_5}\right) \\ +m_4\sinh\left(h_2\sqrt{q_2^2}\right)\sinh\left((h_3+h_4)\sqrt{q_3^2}\right)\sinh\left(h_5\sqrt{q_5^2+\eta/k_5}\right) \\ +m_2m_4\cosh\left(h_2\sqrt{q_2^2}\right)\cosh\left((h_3+h_4)\sqrt{q_3^2}\right)\sinh\left(h_5\sqrt{q_5^2+\eta/k_5}\right) \end{array}\right]$$

$$S_2 = \left[\begin{array}{l} \cosh\left(h_2\sqrt{q_2^2}\right)\cosh\left((h_3+h_4)\sqrt{q_3^2}\right)\sinh\left(h_5\sqrt{q_5^2+\eta/k_5}\right) \\ +m_2\sinh\left(h_2\sqrt{q_2^2}\right)\sinh\left((h_3+h_4)\sqrt{q_3^2}\right)\sinh\left(h_5\sqrt{q_5^2+\eta/k_5}\right) \\ +m_4\cosh\left(h_2\sqrt{q_2^2}\right)\sinh\left((h_3+h_4)\sqrt{q_3^2}\right)\cosh\left(h_5\sqrt{q_5^2+\eta/k_5}\right) \\ +m_2m_4\sinh\left(h_2\sqrt{q_2^2}\right)\cosh\left((h_3+h_4)\sqrt{q_3^2}\right)\cosh\left(h_5\sqrt{q_5^2+\eta/k_5}\right) \end{array}\right]$$

$$T_2 = \left[\begin{array}{l} \sinh\left(h_2\sqrt{q_2^2}\right)\cosh\left((h_3+h_4)\sqrt{q_3^2}\right)\sinh\left(h_5\sqrt{q_5^2+\eta/k_5}\right) \\ +m_2\cosh\left(h_2\sqrt{q_2^2}\right)\sinh\left((h_3+h_4)\sqrt{q_3^2}\right)\sinh\left(h_5\sqrt{q_5^2+\eta/k_5}\right) \\ +m_4\sinh\left(h_2\sqrt{q_2^2}\right)\sinh\left((h_3+h_4)\sqrt{q_3^2}\right)\cosh\left(h_5\sqrt{q_5^2+\eta/k_5}\right) \\ +m_2m_4\cosh\left(h_2\sqrt{q_2^2}\right)\cosh\left((h_3+h_4)\sqrt{q_3^2}\right)\cosh\left(h_5\sqrt{q_5^2+\eta/k_5}\right) \end{array}\right]$$

$$m_i = \frac{k_i\sqrt{q_i^2}}{k_{i+1}\sqrt{q_{i+1}^2}}\ (i = 1,2,3)$$

$$m_4 = \frac{k_4\sqrt{q_4^2}}{k_5\sqrt{q_5^2+\eta/k_5}}$$

$$m_5 = \frac{k_5\sqrt{q_5^2+\eta/k_5}}{k_6\sqrt{q_6^2}}$$

$$G = \left[P_1\cosh\left(h_2\sqrt{q_2^2}\right) + m_1P_2\sinh\left(h_2\sqrt{q_2^2}\right)\right]A_1' - R\sinh\left(h_2\sqrt{q_2^2}\right)$$

$$H = m_2\left\{\left[P_1\sinh\left(h_2\sqrt{q_2^2}\right) + m_1P_2\cosh\left(h_2\sqrt{q_2^2}\right)\right]A_1' - R\cosh\left(h_2\sqrt{q_2^2}\right)\right\}$$

$$I = G\cosh\left[(h_3+h_4)\sqrt{q_3^2}\right] + H\sinh\left[(h_3+h_4)\sqrt{q_3^2}\right]$$

$$J = m_4 \left\{ G\sinh\left[(h_3 + h_4)\sqrt{q_3^2}\right] + H\cosh\left[(h_3 + h_4)\sqrt{q_3^2}\right]\right\}$$

$$K = I\cosh\left(h_5\sqrt{q_5^2 + \eta/k_5}\right) + J\sinh\left(h_5\sqrt{q_5^2 + \eta/k_5}\right)$$

$$L = m_5\left[I\sinh\left(h_5\sqrt{q_5^2 + \eta/k_5}\right) + J\cosh\left(h_5\sqrt{q_5^2 + \eta/k_5}\right)\right]$$

Thus, the temperature increase of heat source and device/skin interface can be given as follows,

$$\begin{aligned}\Delta T_{\text{device}}(t;\omega) = \; & \Delta T_{\text{initial}}(h_1) + D\theta_1(h_1;0) \\ & + \sum_{n=1}^{\infty} |\theta_1(h_1;n\omega)| \cdot \left[\begin{array}{l} \frac{\sin(2n\pi D)}{n\pi}\cos\left(n\omega t + \beta_n^{\text{surface}}\right) \\ + \frac{1-\cos(2n\pi D)}{n\pi}\sin\left(n\omega t + \beta_n^{\text{surface}}\right)\end{array}\right]\end{aligned} \tag{18}$$

$$\begin{aligned}\Delta T_{\text{interface}}(t;\omega) = \; & \Delta T_{\text{initial}}(h_1 + h_2) + D\theta_2(h_1 + h_2;0) \\ & + \sum_{n=1}^{\infty} |\theta_2(h_1 + h_2;n\omega)| \cdot \left[\begin{array}{l} \frac{\sin(2n\pi D)}{n\pi}\cos\left(n\omega t + \beta_n^{\text{surface}}\right) \\ + \frac{1-\cos(2n\pi D)}{n\pi}\sin\left(n\omega t + \beta_n^{\text{surface}}\right)\end{array}\right]\end{aligned} \tag{19}$$

where $\Delta T_{\text{initial}}$ denotes the specific solution [initial temperature increase field in Equation (4) when there is no input power, given by

$$\Delta T_{\text{initial}}(h_1) = \left(\frac{k_1}{h} + h_1\right) A_1 + \Delta T_0 \tag{20}$$

and

$$\Delta T_{\text{initial}}(h_1 + h_2) = \left(\frac{k_1}{h} + h_1 + \frac{k_1 h_2}{k_2}\right) A_1 + \Delta T_0 \tag{21}$$

Here $\theta_1(h_1;n\omega)$ and $\theta_2(h_1 + h_2;n\omega)$ can be obtained from Equations (16) and (17) as

$$\theta_1(h_1;n\omega) = 2\left[k_1\sqrt{q_1^2}\cosh\left(h_1\sqrt{q_1^2}\right) + h\sinh\left(h_1\sqrt{q_1^2}\right)\right] A_1' / \left(k_1\sqrt{q_1^2} + h\right) \tag{22}$$

and

$$\theta_2(h_1 + h_2;n\omega) = 2\left[P_1 A_1'\cosh\left(-h_2\sqrt{q_2^2}\right) + (m_1 P_2 - R)\sinh\left(-h_2\sqrt{q_2^2}\right)\right] \tag{23}$$

To validate the analytic model under a pulsed power, we also performed FEA to obtain the temperature increase after saturation due to the pulsed power. The continuum element DC3D8 in ABAQUS is used to discretize the geometry. The material and geometry parameters such as thicknesses and thermal conductivity of each layer are the same as those in Section 2. The thermal diffusivity of encapsulation, substrate, stratum corneum, epidermis, dermis and fat are 1.4×10^{-7} m²/s, 1.1×10^{-7} m²/s, 6.6×10^{-8} m²/s, 6.6×10^{-8} m²/s, 1.3×10^{-7} m²/s, 8.1×10^{-8} m²/s, respectively [6,14]. The metabolic heat generation is set as 368 W/m³ [14]. The natural convection boundary with the coefficient of heat convection 25 W/(m²·K) is applied on the top surface of encapsulation layer. The ambient temperature is set as 25 °C. The core body temperature at the bottom surface of substrate is 37 °C. The thermal properties of blood and skin tissue can be found in Section 2 or in reference [14].

Figure 4 shows the influence of duty cycle on the maximum and minimum heat source temperature increase under the pulsed peak power density of 2500 W/m² (here we fix this peak power density to ensure that devices operate properly) with period as 500 ms. The analytic prediction agrees very well with FEA. Both the maximum temperature increase and minimum temperature increase of the heat source decrease as the duty cycle decreases. This trend can qualitatively be understood since the average emitted heat from the electronic device is reduced. For the case of constant power density corresponding to a duty cycle of 100%, the maximum temperature increase of heat source is 28.6 °C. As the duty cycle decreases to a smaller duty cycle of 10%, the maximum temperature increase of the heat source drops rapidly to about 0 °C. When the duty cycle is smaller than 10%, the temperature increase may become negative. This is because the temperature increase is defined from the core body temperature instead of the ambient temperature. A negative temperature increase means that the temperature is lower than the core body temperature.

It is necessary to mention that time period also has an influence on the results in a pulsed protocol. According to a transient heat transfer FEA, we obtained that the characteristic time of the temperature profile to reach a steady state is about 280 s. When the time periods used in the pulsed protocol are much smaller than the characteristic time, just as 500 ms we set above, the changes in time periods have only minor effects on the maximal heat source temperature. However, if the duration (τ) of the pulse is much longer than the characteristic time, the maximal heat source temperature is expected to be the same as if the electronic device is always turned on (constant protocol).

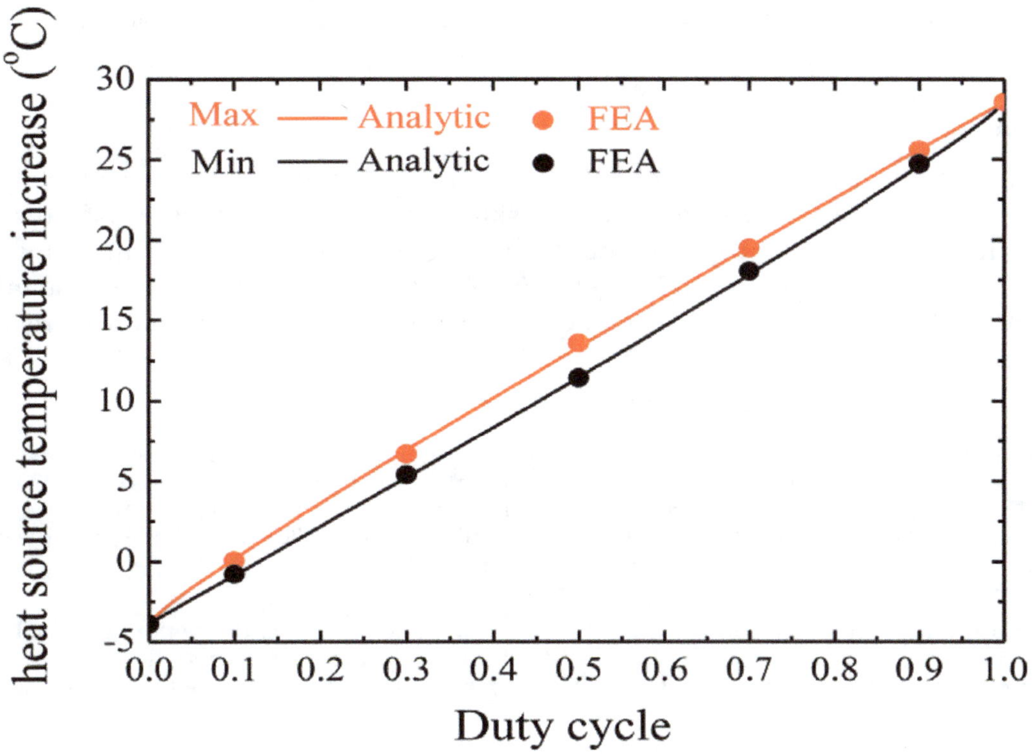

Figure 4. The maximum and minimum heat source temperature increase after saturation versus duty cycle.

Figure 5a compares the temperature increase of heat source after saturation from the analytic model in Equation (18) and FEA under the pulsed peak power density of 2500 W/m^2 with duty cycle D as 50% and period as 500 ms. The good agreement between the analytical prediction and FEA validates the analytical model. The temperature increase of heat source decreases with the increase of substrate thickness. Figure 5b compares the temperature increase of skin surface after saturation versus time from the analytic model in Equation (19) and FEA under the same condition in Figure 5a. With the change of substrate thickness, the temperature increase of skin surface shows the opposite trend to Figure 5a, i.e., the temperature increase of skin surface increases with the increase of the substrate thickness. These results clearly show that the pulsed operation is very effective at reducing the temperature increase in the system. A thin substrate can reduce the device temperature increase due to the heat sink effect of skin while increase the temperature increase of skin surface due to the influence of heat source.

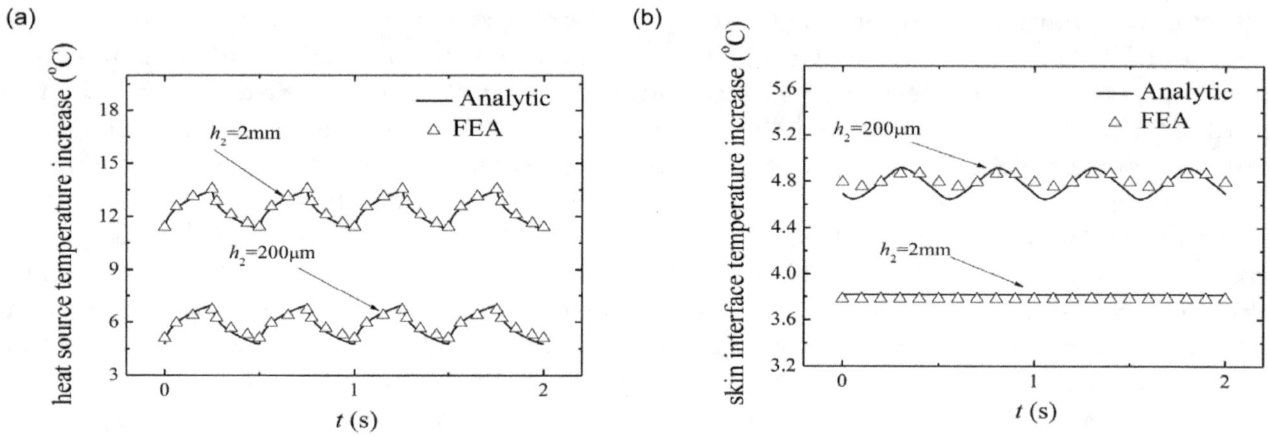

Figure 5. Temperature increase comparison between the analytic prediction and finite element analysis for the pulsed peak power density 2500 W/m^2 with 50% duty cycle and period 500 ms: (**a**) heat source; and (**b**) skin surface.

4. Conclusions

A one-dimensional analytic thermal model, as validated by the finite element analysis, for the flexible electronic devices integrated with human skin under a constant power and pulsed power is presented in this paper. This model combines the Fourier heat conduction equation for the flexible electronic devices and the Pennes bio-heat transfer equation for the skin tissue. The influences of geometric and loading parameters on the temperature increase under a pulsed power are investigated. The results could provide design guidelines for flexible electronic devices to minimize the adverse thermal effect in bio-integrated applications.

Acknowledgments: This work is supported by the National Basic Research Program of China (Grant No. 2015CB351901), the National Natural Science Foundation of China (Grant Nos. 11502009, 11172028 and 11372021), the Fundamental Research Funds for the Central Universities (Grant No. 2016XZZX001-05), the Research Fund for the Doctoral Program of Higher Education of China (20131102110039) and the open fund of Key Laboratory of Soft Machines and Smart Devices of Zhejiang Province.

Author Contributions: Yuhang Li and Jizhou Song conceived and designed the paper construction; Yun Cui performed the FEA simulations and derived the analytical solutions; Yufeng Xing and Tianzhi Yang provided the discussion about the analysis results; Yun Cui and Yuhang Li and Jizhou Song wrote the paper.

References

1. Kim, D.H.; Lu, N.; Ma, R.; Kim, Y.S.; Kim, R.H.; Wang, S.; Wu, J.; Won, S.M.; Tao, H.; Islam, A.; et al. Epidermal electronics. *Science* **2011**, *333*, 838–843. [CrossRef] [PubMed]
2. Webb, R.C.; Bonifas, A.P.; Behnaz, A.; Zhang, Y.; Yu, K.J.; Cheng, H.; Shi, M.; Bian, Z.; Liu, Z.; Kim, Y.S.; et al. Ultrathin conformal devices for precise and continuous thermal characterization of human skin. *Nat. Mater.* **2013**, *12*, 938–944. [CrossRef] [PubMed]
3. Kim, H.S.; Brueckner, E.; Song, J.Z.; Li, Y.H.; Kim, S.; Lu, C.F.; Sulkin, J.; Choquette, K.; Huang, Y.G.; Nuzzo, R.G.; et al. Unusual strategies for using indium gallium nitride grown on silicon (111) for solid-state lighting. *Proc. Natl. Acad. Sci. USA* **2011**, *108*, 10072–10077. [CrossRef] [PubMed]
4. Lu, C.F.; Li, Y.H.; Song, J.Z.; Kim, H.S.; Brueckner, E.; Fang, B.; Hwang, K.C.; Huang, Y.G.; Nuzzo, R.G.; Rogers, J.A. A thermal analysis of the operation of microscale, inorganic light-emitting diodes. *Proc. R. Soc. A Math. Phys. Eng. Sci.* **2012**, *468*, 3215–3223. [CrossRef]
5. Cui, Y.; Xing, Y.F.; Li, Y.H.; Song, J.Z. Thermal management for microscale inorganic light-emitting diodes. *Sci. Sin. Phys. Mech. Astron.* **2016**, *46*, 044612. (In Chinese) [CrossRef]

6. Kim, T.I.; Jung, Y.H.; Song, J.Z.; Kim, D.; Li, Y.H.; Kim, H.S.; Song, I.S.; Wierer, J.J.; Pao, H.A.; Huang, Y.G.; et al. High-efficiency, microscale GaN Light-emitting diodes and their thermal properties on unusual substrates. *Small* **2012**, *8*, 1643–1649. [CrossRef] [PubMed]
7. Li, Y.H.; Shi, Y.; Song, J.Z.; Lu, C.F.; Kim, T.I.; Rogers, J.A.; Huang, Y.G. Thermal properties of microscale inorganic light-emitting diodes in a pulsed operation. *J. Appl. Phys.* **2013**, *113*, 144505. [CrossRef]
8. Song, J.; Feng, X.; Huang, Y. Mechanics and thermal management of stretchable inorganic electronics. *Nat. Sci. Rev.* **2016**, *3*, 128–143. [CrossRef] [PubMed]
9. Li, Y.; Gao, Y.; Song, J. Recent advances on thermal analysis of stretchable electronics. *Theor. Appl. Mech. Lett.* **2016**, *6*, 32–37. [CrossRef]
10. Jiang, S.C.; Ma, N.; Li, H.J.; Zhang, X.X. Effects of thermal properties and geometrical dimensions on skin burn injuries. *Burns* **2002**, *28*, 713–717. [CrossRef]
11. Ozen, S.; Helhel, S.; Bilgin, S. Temperature and burn injury prediction of human skin exposed to microwaves: A model analysis. *Radiat. Environ. Biophys.* **2011**, *50*, 484–489. [CrossRef] [PubMed]
12. Im, I.; Youn, S.B.; Kim, K. Numerical study on the temperature profiles and degree of burns in human skin tissue during combined thermal therapy. *Numer. Heat Transf. A Appl.* **2015**, *67*, 921–933. [CrossRef]
13. Xu, F.; Lu, T.J.; Seffen, K.A.; Ng, E.Y.K. Mathematical modeling of skin bioheat transfer. *Appl. Mech. Rev.* **2009**, *62*, 050801. [CrossRef]
14. Pennes, H.H. Analysis of tissue and arterial blood temperatures in the resting human forearm. *J. Appl. Physiol.* **1948**, *1*, 93–122. [PubMed]
15. Xu, F.; Lu, T.J.; Seffen, K.A. Biothermomechanics of skin tissues. *J. Mech. Phys. Solids* **2008**, *56*, 1852–1884. [CrossRef]
16. Incropera, F.P.; Dewitt, D.P. *Fundamental of Heat and Mass Transfer*; Wiley: Hoboken, NJ, USA, 1996.
17. Zhao, G.; Liu, Z.F.; Zhang, A.L.; Zhang, H.F.; Cheng, S.X. Theoretical analyses of thermal stress of blood vessel during cryopreservation. *Cryo Lett.* **2005**, *26*, 239–250.

14

Flexible Thermo-Optic Variable Attenuator based on Long-Range Surface Plasmon-Polariton Waveguides

Jie Tang [1,2], Yi-Ran Liu [2,3], Li-Jiang Zhang [3], Xing-Chang Fu [3], Xiao-Mei Xue [1,2], Guang Qian [1], Ning Zhao [2,3] and Tong Zhang [1,2,3,*]

1 Key Laboratory of Micro-Inertial Instrument and Advanced Navigation Technology, Ministry of Education, and School of Instrument Science and Engineering, Southeast University, Nanjing 210096, China; tangjieck@126.com (J.T.); xiaomei_xue2011@163.com (X.-M.X.); chinaqgll@163.com (G.Q.)

2 Suzhou Key Laboratory of Metal Nano-Optoelectronic Technology, Suzhou Research Institute of Southeast University, Suzhou 215123, China; duduranla@163.com (Y.-R.L.); njzhao88@163.com (N.Z.)

3 Joint International Research Laboratory of Information Display and Visualization, School of Electronic Science and Engineering, Southeast University, Nanjing 210096, China; ljzhv@163.com (L.-J.Z.); pasf365@163.com (X.-C.F.)

* Correspondence: tzhang@seu.edu.cn

Abstract: A flexible thermo-optic variable attenuator based on long-range surface plasmon-polariton (LRSPP) waveguide for microwave photonic application was investigated. Low-loss polymer materials and high-quality silver strip were served as cladding layers and core layer of the LRSPP waveguide, respectively. By using finite element method (FEM), the thermal distribution and the optical field distribution have been carefully optimized. The fabricated device was characterized by end-fire excitation with a 1550 nm laser. The transmission performance of high-speed data and microwave modulated optical signal was measured while using a broadband microwave photonics link. The results indicated that the propagation loss of the LRSPP waveguide was about 1.92 dB/cm. The maximum attenuation of optical signal was about 28 dB at a driving voltage of 4.17 V, and the variable attenuation of microwave signals was obviously observed by applying different driving voltage to the heater. This flexible plasmonic variable attenuator is promising for chip-scale interconnection in high-density photonic integrated circuits and data transmission and amplitude control in microwave photonic systems.

Keywords: variable optical attenuator (VOA); surface plasmon-polariton (SPP); microwave photonics

---

1. Introduction

The flexible electronics and optical devices are promising for smart wearable devices, large capacity data communications, integrated optical sensing elements, and integrated optoelectronics [1–3]. Recently, the demands for high speed and large capacity interconnections in the optical communications and chip-to-chip signal transmissions increase rapidly. Flexible optical interconnection technology is an attractive solution for these applications, owing to the large bandwidth and the flexibility of the optical signals [4,5].

The variable optical attenuator (VOA) is an important optical component that is widely used in optical communication systems [6,7] and optical signal processing in microwave photonics [8,9]. Especially in microwave photonic systems, microwave signals are transmitted and processed in optical domain and the VOAs are always used to adjust and equalize the power of the optical signals in different parallel channels. With the development of integrated optics, many chip-scale integrated microwave photonic circuits that are used for signal processing and data transceiving have been realized [10,11]. In the future, large-scale multifunctional integrated optical systems will be implemented by using chip-to-chip hybrid interconnections. Hence, the VOAs with wide attenuation range, high-density integration, and flexible bending are necessary for parallel multi-channel optical interconnections.

In recent years, many kinds of VOAs with different structure and mechanism have been reported, such as radiation loss tuning in the S-bend waveguide [12], phase adjusting in Mach–Zehnder interferometers (MZIs) [13,14], optical evanescent field absorption in the waveguide [15], and light mode tuning in a multimode interference device (MMI) [16–20], etc. Although these works above present excellent performance in different aspects, such as low-power, low insertion loss, large attenuation range, but still not adequate to meet the requirements of sub-wavelength optical manipulation as well as transmission in higher integrated circuits in the future for the optical diffraction limit [12,21].

The surface plasmon-polariton (SPP), which is a transverse magnetic (TM) polarized plasmon mode that exists in the interface between a metal and a dielectric, has been shown to transmit optical signals beyond the diffraction limit [22–24]. It is very advantageous for high-density integrated optical systems [21]. With a symmetrical dielectric cladding and thin film core, the surface plasmon modes that are associated with the upper and lower of the metal-dielectric interfaces couple and form a low-loss symmetric mode, known as the long-range SPP (LRSPP) [25]. In comparison with the asymmetric short-range SPP (SRSPP) mode, the LRSPP mode has a large propagation length due to its low propagation loss and it can be applied to chip-to-chip interconnections. But, the SRSPP mode generally has a small mode size and it is suitable for compact integrated optic devices and hybrid integrated chips [26,27]. The SPP waveguides have been employed in various optical devices, such as VOAs [28], filters [29], switches [30], and modulators [31–33] for the refractive indices of the claddings can be adjusted by thermo-optic or electro-optic effect. The LRSPP waveguide consisting of dielectric claddings with similar refractive indices and the metal thin film is extremely sensitive to the symmetry of refractive indices. The mode cut-off of LRSPP in the metal strip can be acquired with an asymmetry in the refractive indices of the dielectric layers above and below the thin metal strip. For a slight asymmetry, the LRSPP mode is still supported, while if the asymmetry is too large the LRSPP mode becomes cut-off and no low loss modes are supported, resulting in asymmetric SRSPP mode with high propagation loss. Light input into the structure radiates away from the metal strip core. Besides, the mode size of the LRSPP waveguide can be adjusted by changing the thickness and width of the stripe core to obtain a low coupling loss with other kinds of waveguide devices or fibers.

Based on the sensitivity to interface property of the LRSPP waveguide mentioned above, several VOAs have been proposed and investigated. For example, a LRSPP VOA that is based on electro-optic controllable liquid crystals and polymers is investigated and it exhibits a good performance of low power, high extinction ratio, and low insertion loss [34]. For both polarization use, a plasmonic nanowire-based thermo-optic VOA with a cross-section of 250 nm $\times$ 250 nm shows a low polarization dependent loss of $\pm$2.5 dB with a compact footprint of 1 mm [35]. Another thermo-optic LRSPP modulator with a length of 1 cm is investigated at telecom wavelengths and it shows low driving power and high extinction ratio [36]. However, most of these VOAs use metal strip core as the heater, which will limit the physical lifetime of the device because the thin metal strip core may be wrinkled and damaged at a high temperature.

In this paper, we design a flexible VOA with a structure consisting of LRSPP waveguide and a thermal heater. By using a simple spin-coating, photolithography, and wet etching process, the device has been fabricated and then characterized at a wavelength of 1.55 μm. The performance of optical attenuation and high speed data transmission are measured.

2. Device Design and Simulation

The schematic diagram of the VOA based on LRSPP waveguide is shown in Figure 1. The bottom left inset is the cross-section of the device. It consists of a gold (Au) layer, lower polymer cladding, thin metal strip core, upper polymer cladding and a heater aligned on top of the upper cladding.

The thicknesses of the upper and lower polymer cladding layers are both 15 μm. The silver (Ag) strip core of the LRSPP waveguide is carefully designed with an optimized parameter and the width and thickness is 4 μm and 12 nm, respectively. The heater is made of aluminum (Al). Its cross-section dimension is 10 μm wide and 150 nm thick and the length is 10 mm. The refractive indices of the polymer claddings are both 1.45 and that of the Ag strip core is 0.15 + 11.38i at 1550 nm. The calculated optical field distribution by the finite element method (FEM) is shown in the top right inset of Figure 1. It indicates that the optical field is symmetrically distributed in the interface between the claddings and the strip core, and most of the energy is distributed in the claddings next to the interface, which ensures a long-range propagation mode.

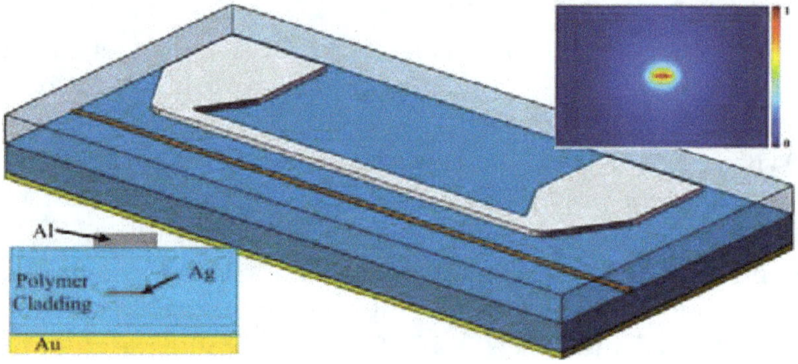

Figure 1. The schematic diagram of the variable optical attenuator (VOA). The bottom left inset is the cross-section and the top right inset is the optical field distribution of the long-range surface plasmon-polariton (LRSPP) waveguide.

The principle of the VOA is based on changing the radiation loss of the LRSPP waveguide resulting from the refractive index difference between the upper and lower claddings. When a voltage is applied to the heater, thermal energy being produced by resistive heating is transferred to the upper cladding, Ag strip, and lower cladding in turn. Then, a temperature gradient is established by the thermal energy and it increases with the applied voltage, as shown in Figure 2. The temperature gradient subsequently results in the refractive index difference between the upper and lower cladding due to the thermo-optic effect. Finally, the asymmetry of refractive indices results in the radiation loss of the LRSPP mode. Therefore, the optical output power of the device can be controlled by changing the temperature gradient by applying different voltages to the heater. The thermo-optic coefficient of the polymer is $-1.758 \times 10^{-4}/°C$. The negative thermo-optic coefficient results in the refractive index decrease with the increase in temperature.

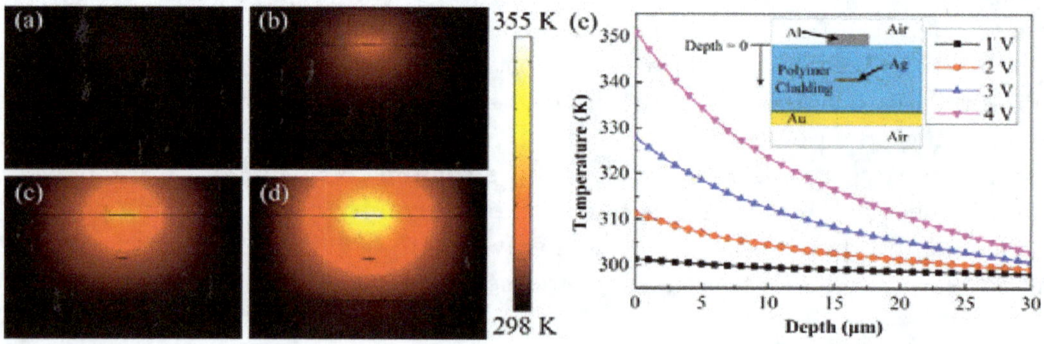

Figure 2. The calculated thermal distribution with applied voltage of (**a**) 1 V, (**b**) 2 V, (**c**) 3 V, (**d**) 4 V to the heater. (**e**) Temperature gradient in the depth of the device at different applied voltage.

According to the dimension of the heater, we can calculate resistance of the heater. By using FEM, we calculate the thermal distribution in the device while applying different voltages to the heater, as shown in Figure 2a–d. The corresponding temperature distribution is shown in Figure 2e. It shows a clear asymmetry distribution of the temperature in the upper and lower claddings when the voltage is applied. Temperature gradient also increases rapidly with the increasing applied voltage (see Figure 2e), which will result in a larger refractive index gradient and eventually enlarge the radiation loss of the LRSPP mode. Therefore, the propagation loss of the device can be adjusted by changing the applied voltage to the heater.

3. Experiments

Figure 3 presents the fabrication process of the flexible thermo-optic VOA based on LRSPP waveguide. Firstly, a layer of 40 nm thick Au was thermally evaporated on the silicon wafer as an adhesion layer, which made it easy to peel off the flexible device from the wafer in the last step. Then, a layer of 15 μm thick ultraviolet (UV) cured epoxy resin was spin-coated on the Au layer as the lower polymer cladding. After a subsequent bake at 160 °C for 30 min, a layer of 2.5 μm thick negative photoresist was spin-coated and baked at 110 °C for 3 min. With a UV photolithography and development, the photoresist was patterned to fabricate a 4 μm wide strip groove. The electron beam evaporation method was used to deposit high-quality Ag thin metal film of 12 nm thick onto the negative photoresist layer. In the evaporation process, a high-purity (>99.999%) solid Ag and a very low deposit speed of 0.02 nm/s in a high vacuum of 1×10^{-4} Pa were used to ensure a high-quality Ag film. After that, the patterned negative photoresist layer was lift-off and a 4 μm wide Ag strip was left over to form the core layer of LRSPP waveguide. Then, another layer of 15 μm thick UV-cured epoxy resin was spin-coated and cured as upper cladding. To fabricate the heater, a layer of 150 nm Al film was thermally evaporated. Then, a layer of positive photoresist was spin-coated and patterned by UV photolithography. After that, wet etching was used to fabricate the Al heater with sodium hydroxide (NaOH) solution. Finally, the flexible device was lift-off from the silicon wafer. After the fabrication process above, the two end faces of the device were polished to minimize the coupling loss between the fiber arrays and the waveguide.

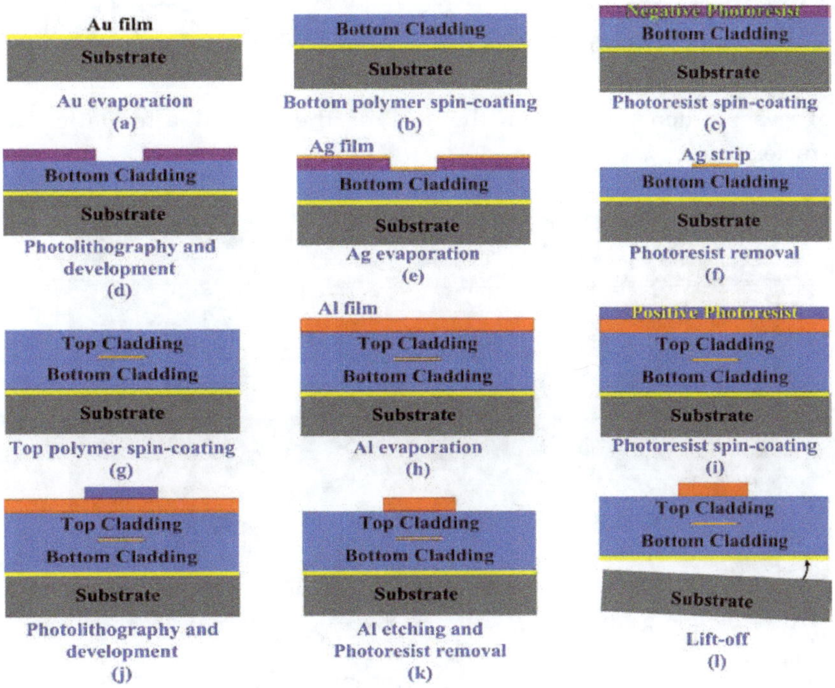

Figure 3. Fabrication process of the flexible LRSPP based VOA.

The schematic diagram of the measurement setup for the thermo-optic VOA is shown in Figure 4. To study the transmission characteristics of high frequency microwave signals, we constructed a broadband microwave photonics link with a bandwidth over 25 GHz. Vector network analyzer (VNA) with a bandwidth of 50 GHz was used to measure the transmission characteristics of the high frequency microwave signals. The light at 1550 nm from a distributed feedback (DFB) laser source was modulated by high frequency microwave signals via a lithium niobate ($LiNbO_3$) electro-optic modulator (EOM). The modulated light was perpendicularly polarized to the waveguide plane by passing through a polarization controller (PC) and then coupled into the waveguide to excite LRSPP mode through a standard single mode fiber (SMF) array. To measure the optical attenuation characteristics of the VOA, a direct-current (DC) power supply was used to apply different voltages to the heater through two probes. In the output port of the device, a standard SMF array was used in order to couple the output light signal into an optical power meter to measure the propagation loss of the LRSPP waveguide and the optical attenuation characteristics of the VOA. Alternatively, the output light was also coupled into a broadband photodetector to demodulate the high frequency microwave signals. Herein, we used an Erbium-doped fiber amplifier (EDFA) with a gain of 20 dB to compensate for the insertion loss of the LRSPP waveguide-based VOA. Meanwhile, a microwave power amplifier (PA) with a gain of 19 dB was employed to compensate the total loss of the microwave photonic link. The fabricated thermo-optic VOA under test is shown in the top left inset of Figure 5.

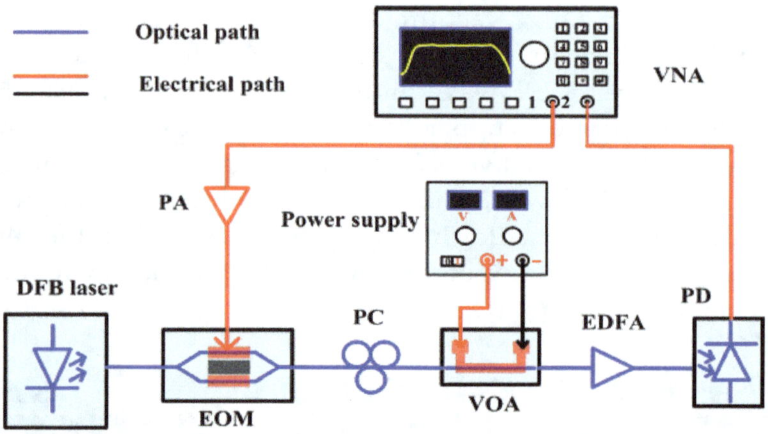

Figure 4. The measurement setup for the test of the optical attenuation and microwave transmission characteristics.

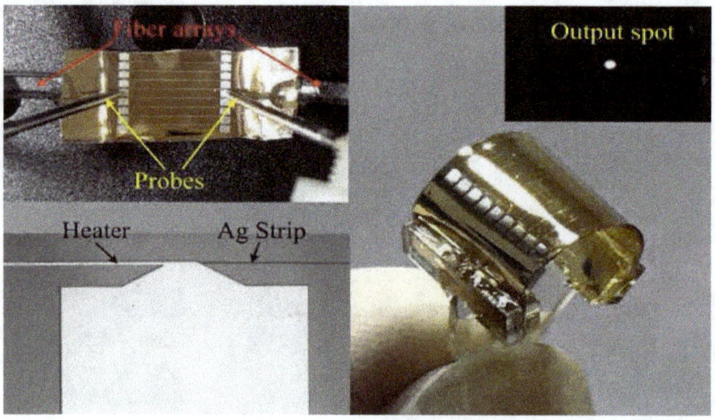

Figure 5. The photograph of the curved flexible thermo-optic VOA. Top left inset is the device under test and the driving voltage is applied through two probes. Top right inset is the near-field output spot of the LRSPP waveguide. Bottom left inset is the photograph taken by an optical microscope.

4. Results and Discussion

The output light of LRSPP waveguide was observed before packing with the fiber array by using an infrared (IR) charge-coupled device (CCD). The near-field output spot of the device is shown in the top right inset of Figure 5. The bright spot shows that the fabricated LRSPP waveguide has a good performance in terms of light propagation. By coupling the output light into the optical power meter, the propagation loss of the LRSPP waveguide is measured at about 1.92 dB/cm by cut-back method, as shown in Figure 6. The results show that the propagation loss is at least two times lower than other previously reported flexible LRSPP waveguide [37,38]. It also indicates that the quality of the fabricated LRSPP waveguide is very high. In order to investigate the performance of the optical attenuation, the voltage is applied to the heater. As the applied voltage increases, the optical attenuation can be observed clearly, as shown in Figure 7. The measured results show that a maximum optical attenuation of over 28 dB is acquired at a driving voltage of 4.17 V. At the same time, we can see that the optical attenuation is inconspicuous when the applied voltage is lower than about 2.5 V. While the applied voltage is larger than 2.5 V, the output optical power decreases rapidly. The reason of this phenomenon can be attributed to the fact that the LRSPP mode still exists in spite of a slight asymmetry of the refractive index that is produced by a relatively low driving voltage. When a larger voltage is applied, the asymmetry is too large and the LRSPP mode becomes cut-off. In other words, the SPP mode changes from nonradiative mode to radiative mode in the case of a large asymmetry of the refractive index.

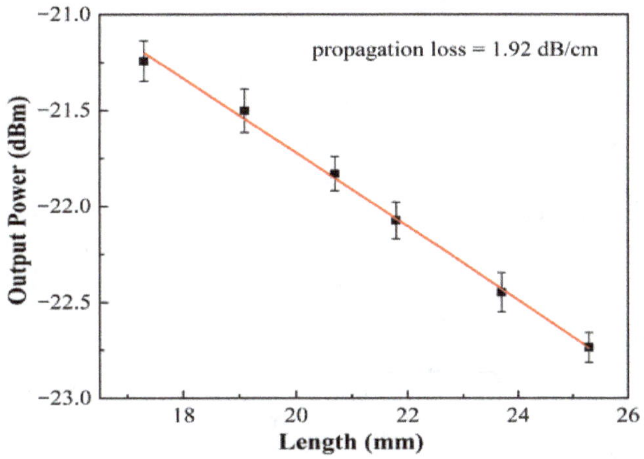

Figure 6. Measured propagation loss of the LRSPP waveguide by cut-back method.

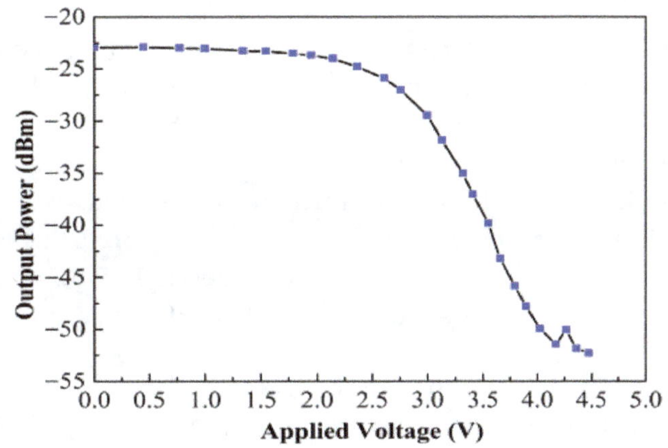

Figure 7. Measured optical output power of the thermo-optic VOA as a function of applied voltage to the heater.

By using the microwave photonics link and the VNA, we can observe the transmission and attenuation characteristics of high frequency microwave signals. The microwave attenuation can be attributed to the decrease of the optical light coupled into the photodetector. According to the measured results of the optical attenuation above, the maximum optical attenuation of 28 dB can result in a maximum microwave attenuation of 56 dB for the square relation between optical loss and microwave loss. The measured microwave transmission and attenuation characteristics under different applied voltage are shown in Figure 8. In the experiment, we carefully adjust the voltage in order to acquire each microwave attenuation interval of 5 dB. Similar to the optical attenuation characteristics, the microwave attenuation increases with the increase of applied voltage. The microwave attenuation is also inconspicuous when the applied voltage is lower than about 2.5 V. Only 5 dB attenuation is acquired when the applied voltage increases from 0 V to 2.51 V. While the applied voltage is larger than 2.5 V, the microwave attenuation increases rapidly. Over 50 dB attenuation is acquired as the voltage increases from 2.51 V to 4.05 V. As the applied voltage increases, more noise appears in the transmission curve. The reason is that the optical power reached in the photodetector is too low to demodulate the high frequency microwave signals. In spite of the noise in the curve, a maximum attenuation of about 55 dB is acquired.

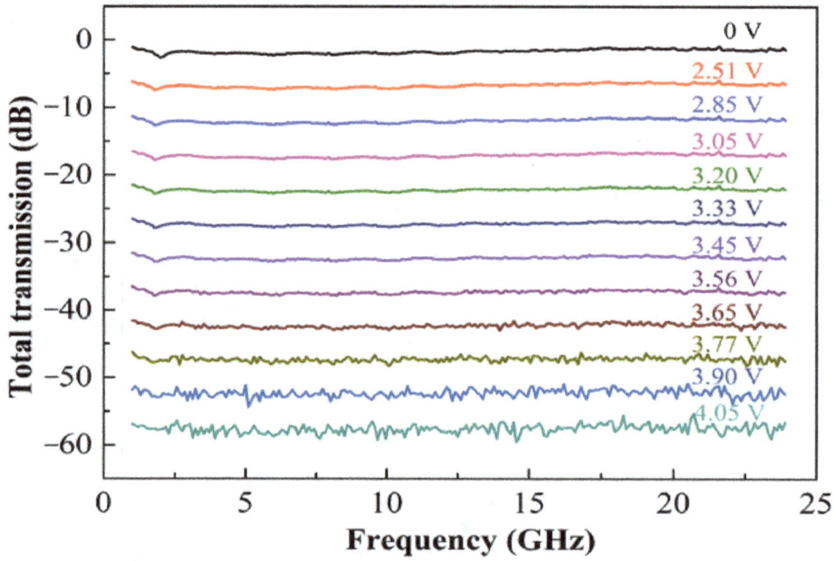

Figure 8. The total microwave transmission characteristics of the microwave photonics link at different applied voltage to the heater.

After the characterization of optical propagation loss and microwave transmission performance, we also carried out a data transmission experiment with the flexible LRSPP waveguide by using a bit-error-rate (BER) test system (N4903B, Agilent Technology, Santa Clara, CA, USA). Figure 9 shows the measured eye diagrams of the total microwave photonics link. A modulated light with 3 Gb/s, 6 Gb/s, 10 Gb/s, and 12.5 Gb/s pseudorandom-binary sequence (PRBS) ($2^{31} - 1$) generated by a pulse-pattern-generator (PPG) was transmitted through the flexible LRSPP waveguide. As shown in Figure 9, the eye is well open at a different transmission speed. The BER at 12.5 Gb/s was measured about 1.13×10^{-10}. The results indicate that the flexible LRSPP waveguide is suitable for high speed optical interconnections.

Evidently, the fabricated devices still suffer from too high insertion loss to meet practical applications. In comparison with the other low power VOAs [39–41], the power consumption is a little larger as well. It is mainly due to the thick polymer cladding that extends the distance of thermal diffusion. However, a good performance of the optical and microwave attenuation has been acquired. Furthermore, the transmission of a light signal modulated by high frequency microwave signals through the flexible LRSPP waveguide-based VOA has been tested in spite of obstacles of large insertion loss, wave vector difference, and dispersion of the LRSPP mode. It is a potential solution to realize the flexible chip-scale interconnections and power equalization in the optical systems. The problem of large insertion loss can be solved by further optimizing the coupling loss between fiber and waveguide or using gain media doped polymer to reduce the propagation loss in the future.

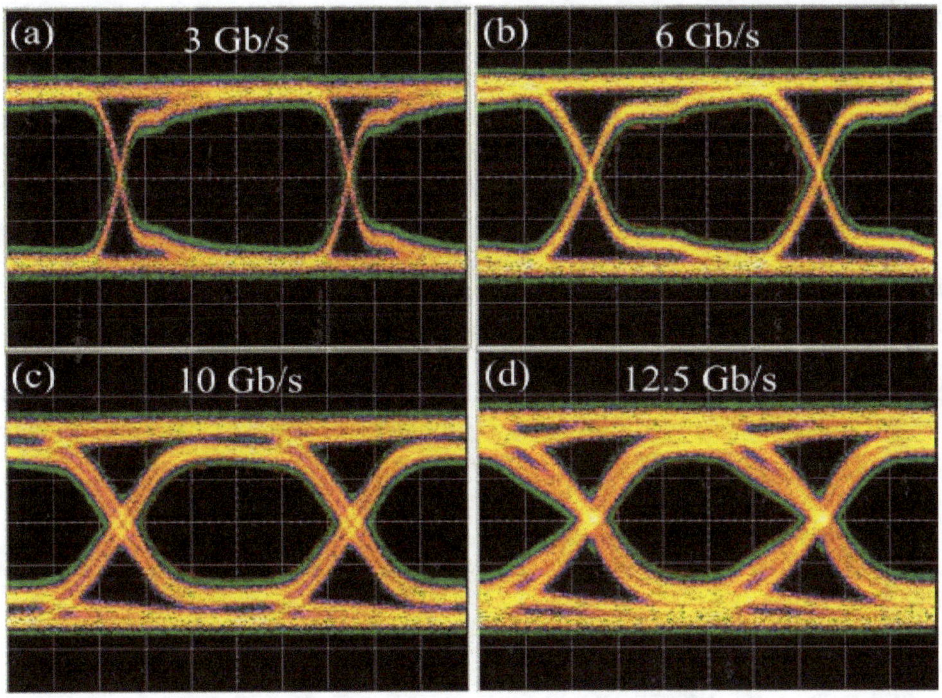

Figure 9. Measured eye diagrams of the total microwave photonics link at (**a**) 3 Gb/s, (**b**) 6 Gb/s, (**c**) 10 Gb/s, and (**d**) 12.5 Gb/s.

5. Conclusions

We demonstrated a flexible thermo-optic variable attenuator based on LRSPP waveguide that can be used to transmit and equalize high frequency microwave signals in microwave photonic systems. The controllable optical attenuation has been realized by applied different driving voltage to the thermo-optic VOA. The experimental results show that the maximum optical attenuation of the VOA is over 28 dB at a driving voltage of 4.17 V. The maximum attenuation of microwave signals is about 55 dB. The flexible thermo-optic VOA using LRSPP waveguide is promising for chip-scale interconnection in high-density photonic integrated circuits and data transmission, amplitude control, and equalization in microwave photonic systems.

Author Contributions: J.T. and T.Z. conceived and designed the experiments; Y.-R.L. and L.-J.Z. performed the experiments; X.-C.F. and X.-M.X. analyzed the data; G.Q. and N.Z. contributed reagents and materials; J.T. wrote the paper.

Acknowledgments: We appreciate Sami Iqbal for the manuscript revision. We also appreciate Wei Li at School of Information Science and Engineering, Southeast University for granting us access to their BER test system.

References

1. Rogers, J.A.; Someya, T.; Huang, Y. Materials and mechanics for stretchable electronics. *Science* **2010**, *327*, 1603–1607. [CrossRef] [PubMed]
2. Khan, Y.; Ostfeld, A.E.; Lochner, C.M.; Pierre, A.; Arias, A.C. Monitoring of vital signs with flexible and wearable medical devices. *Adv. Mater.* **2016**, *28*, 4373–4395. [CrossRef] [PubMed]

3. Fan, F.R.; Tang, W.; Wang, Z.L. Flexible nanogenerators for energy harvesting and self-powered electronics. *Adv. Mater.* **2016**, *28*, 4283–4305. [CrossRef] [PubMed]
4. Kim, J.T.; Ju, J.J.; Park, S.; Kim, M.S.; Park, S.K.; Lee, M.H. Chip-to-chip optical interconnect using gold long-range surface plasmon polariton waveguides. *Opt. Express* **2008**, *16*, 13133–13138. [CrossRef] [PubMed]
5. Prajzler, V.; Neruda, M.; Nekvindova, P. Flexible multimode polydimethyl-diphenylsiloxane optical planar waveguides. *J. Mater. Sci. Mater. Electron.* **2018**, *29*, 5878–5884. [CrossRef]
6. Tsuchizawa, T.; Yamada, K.; Watanabe, T.; Park, S.; Nishi, H.; Kou, R.; Shinojima, H.; Itabashi, S.-I. Monolithic integration of silicon-, germanium-, and silica-based optical devices for telecommunications applications. *IEEE J. Sel. Top. Quantum Electron.* **2011**, *17*, 516–525. [CrossRef]
7. Dong, P. Silicon photonic integrated circuits for wavelength-division multiplexing applications. *IEEE J. Sel. Top. Quantum Electron.* **2016**, *22*, 370–378. [CrossRef]
8. Capmany, J.; Mora, J.; Ortega, B.; Pastor, D. Microwave photonic filters using low-cost sources featuring tunability, reconfigurability and negative coefficients. *Opt. Express* **2005**, *13*, 1412–1417. [CrossRef] [PubMed]
9. Li, X.; Dong, J.; Yu, Y.; Zhang, X. A tunable microwave photonic filter based on an all-optical differentiator. *IEEE Photonics Technol. Lett.* **2011**, *23*, 308–310. [CrossRef]
10. Marpaung, D.; Roeloffzen, C.; Heideman, R.; Leinse, A.; Sales, S.; Capmany, J. Integrated microwave photonics. *Laser Photonics Rev.* **2013**, *7*, 506–538. [CrossRef]
11. Wu, J.Y.; Xu, X.Y.; Nguyen, T.G.; Chu, S.T.; Little, B.E.; Morandotti, R.; Mitchell, A.; Moss, D.J. Rf photonics: An optical microcombs' perspective. *IEEE J. Sel. Top. Quantum Electron.* **2018**, *24*, 1–20. [CrossRef]
12. Wang, L.F.; Song, Q.Q.; Wu, J.Y.; Chen, K.X. Low-power variable optical attenuator based on a hybrid sion-polymer s-bend waveguide. *Appl. Opt.* **2016**, *55*, 969–973. [CrossRef] [PubMed]
13. Maese-Novo, A.; Zhang, Z.Y.; Irmscher, G.; Polatynski, A.; Mueller, T.; de Felipe, D.; Kleinert, M.; Brinker, W.; Zawadzki, C.; Keil, N. Thermally optimized variable optical attenuators on a polymer platform. *Appl. Opt.* **2015**, *54*, 569–575. [CrossRef]
14. Chen, S.T.; Shi, Y.C.; He, S.L.; Dai, D.X. Variable optical attenuator based on a reflective mach-zehnder interferometer. *Opt. Commun.* **2016**, *361*, 55–58. [CrossRef]
15. Mohsin, M.; Schall, D.; Otto, M.; Chmielak, B.; Porschatis, C.; Bolten, J.; Neumaier, D. Graphene based on-chip variable optical attenuator operating at 855 nm wavelength. *Opt. Express* **2017**, *25*, 31660–31669. [CrossRef] [PubMed]
16. Yu, Y.Y.; Sun, X.Q.; Ji, L.T.; He, G.B.; Wang, X.B.; Yi, Y.J.; Chen, C.M.; Wang, F.; Zhang, D.M. The 650-nm variable optical attenuator based on polymer/silica hybrid waveguide. *Chin. Phys. B* **2016**, *25*, 054101. [CrossRef]
17. Sun, X.; Xie, Y.; Liu, T.; Chen, C.; Wang, F.; Zhang, D. Variable optical attenuator based on long-range surface plasmon polariton multimode interference coupler. *J. Nanomater.* **2014**, *2014*, 1. [CrossRef]
18. Gu, Y.L.; Chen, C.M.; Zheng, Y.; Shi, Z.S.; Wang, X.B.; Yi, Y.J.; Jiang, T.C.; Sun, X.Q.; Wang, F.; Cui, Z.C.; et al. Heat-induced multimode interference variable optical attenuator based on novel organic-inorganic hybrid materials. *J. Opt.* **2015**, *17*, 085802. [CrossRef]
19. Oh, M.-C.; Chu, W.-S.; Shin, J.-S.; Kim, J.-W.; Kim, K.-J.; Seo, J.-K.; Lee, H.-K.; Noh, Y.-O.; Lee, H.-J. Polymeric optical waveguide devices exploiting special properties of polymer materials. *Opt. Commun.* **2016**, *362*, 3–12. [CrossRef]
20. Noh, Y.O.; Lee, C.H.; Kim, J.M.; Hwang, W.Y.; Won, Y.H.; Lee, H.J.; Han, S.G.; Oh, M.C. Polymer waveguide variable optical attenuator and its reliability. *Opt. Commun.* **2004**, *242*, 533–540. [CrossRef]
21. Sorger, V.J.; Oulton, R.F.; Ma, R.M.; Zhang, X. Toward integrated plasmonic circuits. *MRS Bull.* **2012**, *37*, 728–738. [CrossRef]
22. Fang, Y.; Sun, M. Nanoplasmonic waveguides: Towards applications in integrated nanophotonic circuits. *Light Sci. Appl.* **2015**, *4*, e294. [CrossRef]
23. Zhang, X.Y.; Hu, A.; Wen, J.Z.; Zhang, T.; Xue, X.J.; Zhou, Y.; Duley, W.W. Numerical analysis of deep sub-wavelength integrated plasmonic devices based on semiconductor-insulator-metal strip waveguides. *Opt. Express* **2010**, *18*, 18945–18959. [CrossRef] [PubMed]
24. Gramotnev, D.K.; Bozhevolnyi, S.I. Plasmonics beyond the diffraction limit. *Nat. Photonics* **2010**, *4*, 83–91. [CrossRef]
25. Berini, P. Long-range surface plasmon polaritons. *Adv. Opt. Photonics* **2009**, *1*, 484–588. [CrossRef]

26. Dabos, G.; Manolis, A.; Papaioannou, S.; Tsiokos, D.; Markey, L.; Weeber, J.C.; Dereux, A.; Giesecke, A.L.; Porschatis, C.; Chmielak, B.; et al. Cmos plasmonics in wdm data transmission: 200 Gb/s (8 × 25 Gb/s) transmission over aluminum plasmonic waveguides. *Opt. Express* **2018**, *26*, 12469–12478. [CrossRef] [PubMed]
27. Dabos, G.; Ketzaki, D.; Manolis, A.; Markey, L.; Weeber, J.C.; Dereux, A.; Giesecke, A.L.; Porschatis, C.; Chmielak, B.; Tsiokos, D.; et al. Plasmonic stripes in aqueous environment co-integrated with si3n4 photonics. *IEEE Photonics J.* **2018**, *10*, 1–8. [CrossRef]
28. Gagnon, G.; Lahoud, N.; Mattiussi, G.A.; Berini, P. Thermally activated variable attenuation of long-range surface plasmon-polariton waves. *J. Lightwave Technol.* **2006**, *24*, 4391–4402. [CrossRef]
29. Lee, J.; Lu, F.; Belkin, M.A. Broadly wavelength tunable bandpass filters based on long-range surface plasmon polaritons. *Opt. Lett.* **2011**, *36*, 3744–3746. [CrossRef] [PubMed]
30. Papaioannou, S.; Kalavrouziotis, D.; Vyrsokinos, K.; Weeber, J.-C.; Hassan, K.; Markey, L.; Dereux, A.; Kumar, A.; Bozhevolnyi, S.I.; Baus, M.; et al. Active plasmonics in wdm traffic switching applications. *Sci. Rep.* **2012**, *2*, 652. [CrossRef] [PubMed]
31. Kim, J.T.; Chung, K.H.; Choi, C.-G. Thermo-optic mode extinction modulator based on graphene plasmonic waveguide. *Opt. Express* **2013**, *21*, 15280–15286. [CrossRef] [PubMed]
32. Haffner, C.; Heni, W.; Fedoryshyn, Y.; Niegemann, J.; Melikyan, A.; Elder, D.L.; Baeuerle, B.; Salamin, Y.; Josten, A.; Koch, U.; et al. All-plasmonic mach-zehnder modulator enabling optical high-speed communication at the microscale. *Nat. Photonics* **2015**, *9*, 525–529. [CrossRef]
33. Fan, H.; Berini, P. Thermo-optic characterization of long-range surface-plasmon devices in cytop. *Appl. Opt.* **2013**, *52*, 162–170. [CrossRef] [PubMed]
34. Zografopoulos, D.C.; Beccherelli, R. Plasmonic variable optical attenuator based on liquid-crystal tunable stripe waveguides. *Plasmonics* **2013**, *8*, 599–604. [CrossRef]
35. Leosson, K.; Rosenzveig, T.; Hermannsson, P.G.; Boltasseva, A. Compact plasmonic variable optical attenuator. *Opt. Express* **2008**, *16*, 15546–15552. [CrossRef] [PubMed]
36. Nikolajsen, T.; Leosson, K.; Bozhevolnyi, S.I. In-line extinction modulator based on long-range surface plasmon polaritons. *Opt. Commun.* **2005**, *244*, 455–459. [CrossRef]
37. Vernoux, C.; Chen, Y.T.; Markey, L.; Sparchez, C.; Arocas, J.; Felder, T.; Neitz, M.; Brusberg, L.; Weeber, J.C.; Bozhevolnyi, S.I.; et al. Flexible long-range surface plasmon polariton single-mode waveguide for optical interconnects. *Opt. Mater. Express* **2018**, *8*, 469–484. [CrossRef]
38. Nikolajsen, T.; Leosson, K.; Salakhutdinov, I.; Bozhevolnyi, S.I. Polymer-based surface-plasmon-polariton stripe waveguides at telecommunication wavelengths. *Appl. Phys. Lett.* **2003**, *82*, 668–670. [CrossRef]
39. Wu, X.P.; Liu, W.; Yuan, Z.L.; Liang, X.R.; Chen, H.; Xu, X.H.; Tang, F.F. Low power consumption voa array with air trenches and curved waveguide. *IEEE Photonics J.* **2018**, *10*, 1–8. [CrossRef]
40. Sun, S.Q.; Niu, D.H.; Sun, Y.; Wang, X.B.; Yang, M.; Yi, Y.J.; Sun, X.Q.; Wang, F.; Zhang, D.M. Design and fabrication of all-polymer thermo-optic variable optical attenuator with low power consumption. *Appl. Phys. Mater. Sci. Process.* **2017**, *123*, 646. [CrossRef]
41. Ren, M.Z.; Zhang, J.S.; An, J.M.; Wang, Y.; Wang, L.L.; Li, J.G.; Wu, Y.D.; Yin, X.J.; Hu, X.W. Low power consumption 4-channel variable optical attenuator array based on planar lightwave circuit technique. *Chin. Phys. B* **2017**, *26*, 074221. [CrossRef]

15

Materials, Mechanics and Patterning Techniques for Elastomer-Based Stretchable Conductors

Xiaowei Yu, Bikram K. Mahajan, Wan Shou and Heng Pan *

Department of Mechanical and Aerospace Engineering, Missouri University of Science and Technology, Rolla, MO 65401, USA; xy5w8@mst.edu (X.Y.); bkm6g7@mst.edu (B.K.M.); ws9n5@mst.edu (W.S.)
* Correspondence: hp5c7@mst.edu

Academic Editors: Seung Hwan Ko, Daeho Lee and Zhigang Wu

Abstract: Stretchable electronics represent a new generation of electronics that utilize soft, deformable elastomers as the substrate or matrix instead of the traditional rigid printed circuit boards. As the most essential component of stretchable electronics, the conductors should meet the requirements for both high conductivity and the capability to maintain conductive under large deformations such as bending, twisting, stretching, and compressing. This review summarizes recent progresses in various aspects of this fascinating and challenging area, including materials for supporting elastomers and electrical conductors, unique designs and stretching mechanics, and the subtractive and additive patterning techniques. The applications are discussed along with functional devices based on these conductors. Finally, the review is concluded with the current limitations, challenges, and future directions of stretchable conductors.

Keywords: stretchable conductors; elastomers; patterning techniques; direct printing; transfer printing

1. Introduction

Since the pioneering work on buckling phenomena in metallic film/elastomer composite [1], researchers started to pay attentions to the field of "stretchable electronics". Stretchable electronics represent a new generation of electronics that utilize soft, deformable elastomers as the substrate or matrix instead of rigid printed circuit boards. Unlike "flexible electronics" which utilize thin plastic substrates to endow the devices with conformity to bending and twisting [2,3], "stretchable electronics" are fascinating and challenging because of their capability to stretch and compress over a large scale, in addition to conformability [4–11].

Materials have been playing an important role in the development of stretchable conductors in the past two decades. Due to their high stretchability, elastomeric materials are commonly chosen as the supporting substrate or matrix for stretchable conductors and integrated device systems. Along with the traditional metallic conductors [1,12–14], several research groups have focused on developing conductive nanomaterials, including silver nanowires (AgNWs) [15], carbon nanotubes (CNTs) [16], etc. A substantial amount of work has been done to combine the rigid, brittle conductors with the flexible and soft elastomers, and overcoming the mismatches between their elastic behaviors. Unique designs and stretching mechanics have been proposed to harmonize the mismatches and integrate materials with widely different properties as one unique system. Patterning the conductors is one of the key techniques for the successful fabrication of stretchable electronic devices. The common patterning techniques for stretchable conductors include lithography [17,18], screen/stencil printing [19–21], direct printing [14,22–31], and transfer printing [32–34]. The patterning methods used in wafer-based electronics have also been adopted in stretchable electronics. Besides that, direct printing is emerging as an alternative to the conventional subtractive patterning method, with the recent development of additive manufacturing.

Despite their short history and limited scalability in manufacturing, it is quite clear that stretchable electronic devices would have a huge impact on future consumer electronics [4]. Because of their soft and conformable nature, stretchable electronics have shown great potential in biomedical engineering, e.g., epidermal electronic devices [18,35] and implantable devices [36,37]. Many studies have demonstrated the application of stretchable conductors in various devices and integrated systems, with functions such as sensing, display, and energy storage and conversion.

In this review, we focus on the basic building block of stretchable electronics, the elastomer-based stretchable conductors. The materials for elastomeric substrates and electrical conductors, and the mechanics of various types of stretchable conductors, are summarized in Sections 2 and 3. Section 4 reviews the patterning techniques used in stretchable conductors, followed by a brief discussion on their applications in Section 5. Finally, in Section 6, current limitations, challenges, and the future directions of stretchable conductors are discussed.

2. Materials

It has been recognized that there are two basic approaches to making stretchable conductors: to exploit structures that are stretchable and materials that are stretchable [4,8,10,38,39]. The first approach involves designing commonly used conductors into particular structures that are stretchable. The latter employs a simple design, but utilizes materials that are stretchable, such as carbon nanotubes, metallic nanowires, liquid metals, and ionic liquids. Some studies combine these two approaches to achieve enhanced stretchability [40–47].

In this section, the materials for elastomeric substrate and electrical conductors are summarized. The preparation, modification, and performance of the stretchable conductive materials are discussed. Along with these materials, the traditional metallic conductors are discussed, with an emphasis on the recently developed additive manufacturing methods.

2.1. Elastomers as the Substrate or Supporting Matrix

Elastomers are the fundamental supporting materials for stretchable conductors. They serve either as the matrix for conductive fillers and networks [48–50], or the substrate for the conductive films, tracks, and functional devices [13,18,51]. Elastomers not only provide the stretchability when the whole system is under strain, but also protect the devices and interconnects from large-scale deformation [12,52–54] and corrosive environments [55]. In epidermal electronics, elastomers facilitate the conformal contact of human skin with the sensing electrodes, and prevent the unnecessary contact with other electrical components [18,35,56]. Besides that, elastomers can also be used as functional dielectric materials [20,57–59].

The most important characteristic of the elastomers is their elasticity. Elastomers can easily sustain repetitive strain (typically larger than 100%), for thousands of stretch/release cycles. Desirable properties also include optical transparency, which facilitates optical applications in optoelectronics, photodetectors, light-emitting devices, solar cells, etc. [8,10]. They also have good biocompatibility, which makes the elastomers suitable for biomedical applications [60,61]. Elastomers typically have a high thermal expansion coefficient and have an inherent tendency of swelling in common solvents [62], which can result in device failures [63,64]. However, thermal expansion and induced deformation can be exploited to make 3D buckling structures [1], which is one of the most common approaches to fabricating stretchable conductors. Swelling can also be cleverly utilized to introduce wrinkling patterns on the substrate surface [65,66].

The most extensively used elastomer is polydimethylsiloxane (PDMS), a commercially available silicone-based elastomer [55]. PDMS is an ideal choice for a stretchable substrate/matrix as it is low-cost, transparent, biocompatible, easy to process, and permeable to air [67]. It is also known to have surface hydrophobicity, which can be modified to hydrophilicity by O_2 plasma [68,69], UV/ozone [24,70,71], and chemical treatment [68]. However, after exposure to air for a certain time, the surface reverts back to its hydrophobic nature. Such property has been utilized to modify the surface stiffness (Young's modulus) of PDMS [66,70,72], and to control the interfacial adhesion of PDMS with conductive materials [24,69]. Besides acting as the substrate/matrix, PDMS has been widely used in soft lithography [73], as an intermediate substrate (stamp) to transfer functional materials from the donor substrate to the target substrate.

Along with PDMS, some other elastomers are also employed as the substrate/matrix for various purposes. For example, Ecoflex [40,74,75] and polyurethane (PU) [49,76,77] are used in systems that require stretchability beyond the range of PDMS. Poly(isobutylene-co-isoprene) (IIR) is suitable for air- or moisture-sensitive applications [55] as it has a good gas diffusion barrier. Some rubbers, like polyolefin (POE) and poly (styrene-block-butadiene-block-styrene) (SBS), can be made into elastic fiber mats, taking advantage of their spinnability. These fiber mats have better stretchability than bulk elastomer blocks due to the network structure [78,79]. SBS fiber mat substrate is shown in Figure 1 as an example. Other reported elastomers include Dragon Skin [80], Solaris [81], acrylic elastomer [82–85], and nitrile butadiene rubber (NBR) [86]. Their elastic properties are summarized in Table 1.

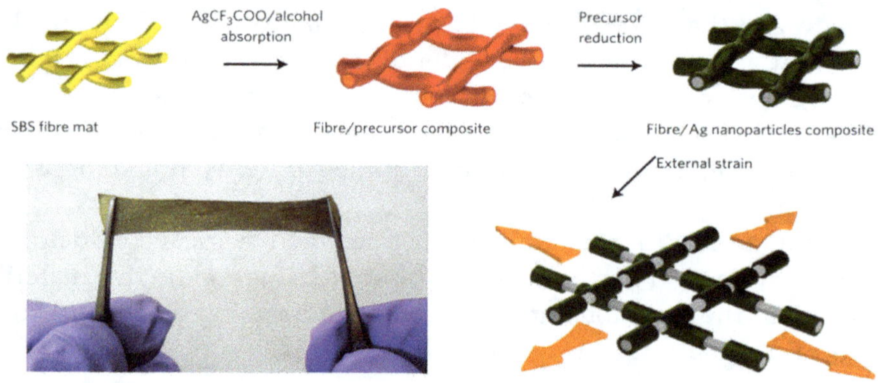

Figure 1. The schematic of fabrication process and the photograph of an SBS (styrene-block-butadiene-block-styrene) fiber mat-based stretchable conductor.

Table 1. Elastomers for stretchable conductors and their elastic properties.

Material	Tensile Strength (MPa)	Maximum Strain (%)	Young's Modulus (MPa)	Reference
PDMS	6.25	120–160	2.05 (strain < 40%)	[52,55]
Ecoflex® 00-30	1.38	900	0.07 (strain = 100%)	[87]
Polyurethane	7.32	760	7.82 (initial)	[76]
IIR	3.51	170	0.41 (strain < 40%)	[55]
POE fiber mat	-	>600	-	[78]
SBS fiber mat	-	>530	0.47 (low strain)	[79]
Dragon Skin® 30	3.45	364	0.59 (strain = 100%)	[88]
Solaris	1.24	290	0.17 (strain = 100%)	[89]
Acrylic elastomer (VHB 4910)	0.69	-	-	[90]
NBR	10	250	-	[91]

2.2. *Electrical Conductors*

The conductive materials for stretchable conductors have more diversity compared to the elastomeric materials. They vary from inorganic to organic materials, solids to liquids, and bulk films to nano-scale percolation systems. In this section, the commonly used electrical conductors are introduced, including bulk metal films, metallic nanowires, carbon-based nanomaterials, conductive polymers, liquid metals, and ionic liquids. Bulk metal films have almost the same performance as the regular metallic conductors, with the exception that they are deformable owing to their ultra-small thickness (typically smaller than 1 μm) [12]. They have been extensively studied [1,12,13,92,93] and incorporated into various stretchable devices [4,35,94–96]. Recently, additive manufacturing has brought the bulk metal films and tracks to the research scope as materials for maskless, printable conductors [97–99]. Comparatively, conductive nanomaterials are relatively new, but their development has opened up a whole new area for stretchable conductors. Conductive nanomaterials include carbon-based nanomaterials and metallic nanowires. They can be incorporated into elastomeric materials by direct blending, or by fine architectures. Compared to metallic materials, polymeric materials have better intrinsic flexibility. The discovery of conductive polymers has brought another promising option for stretchable conductors. Liquid conductors, including liquid metals and ionic liquids, are advantageous in mobility in comparison to solid conductors. They can easily fit into the shape defined by the channel and have repeatable conductivity without forming any crack.

2.2.1. Bulk Metal Films and Tracks

Bulk metal films were studied extensively at the initial stages of development of stretchable conductors. It was found by Whitesides and his colleagues that disordered and ordered buckling (Figure 2a,b) occurred as the thin gold film evaporated on PDMS due to thermal contraction [1,72]. Researchers utilized this phenomenon to make stretchable gold stripes, which can be stretched up to 22%, as seen in Figure 2c [12]. Shortly after that, the lithographic method was incorporated and sinuous patterned gold tracks encapsulated in PDMS were fabricated, with a stretchability of 54% [13]. Gold was initially used because it is the most ductile metal. Later, emerging serpentine geometry designs were proposed to reduce strain in the metal [100–103]. Copper turned out to be the preferred material for serpentine-shape conductors due to lower cost and comparable conductivity.

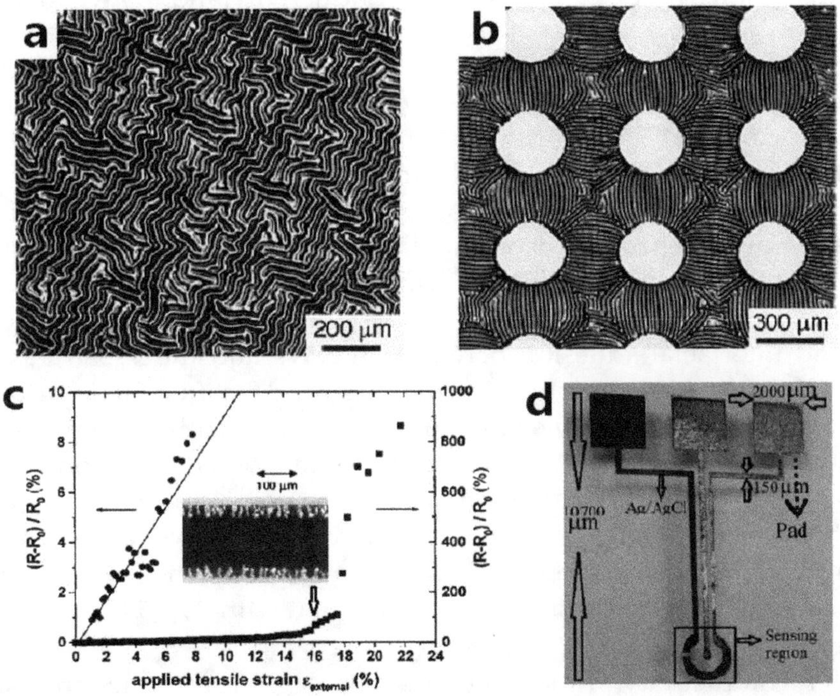

Figure 2. (**a**) Disordered buckling of Au film on flat, unconstraint PDMS (polydimethylsiloxane); (**b**) Ordered buckling of Au film on PDMS with flat circles (150 μm in radius) elevated by 10–20 μm; (**a**,**b**) are reprinted by permission from Macmillan Publishers Ltd.: *Nature* [1]; (**c**) Normalized change in resistance of the Au stripe on PDMS under tensile strain. The left curve represents the linear behavior when the strain is below 8%. The inset is a photograph of the stripe stretched to 16.4%. Adapted from [12], with permission of AIP Publishing; (**d**) Ink-jet-printed Ag three-electrode electrochemical sensor.

With the development of additive manufacturing, direct writing techniques are showing up as an alternative to the traditional subtractive patterning methods. In the direct printing method, nanoparticle (NP) or metal-organic decomposition (MOD) inks are used, which are deposited from a printing head or nozzle to the substrate [98]. NP inks are usually a suspension of metallic nanoparticles dispersed in solvent. Silver NP inks are primarily used due to their high conductivity, good solvent-stability, air-inertness, and accessibility. MOD inks contain metal precursor and need additional steps (i.e., reduction) to generate metallic materials in-situ. In addition, sintering is required for both types of inks to evaporate solvent, decompose organic additives, and solidify the metallic particles. Figure 2d shows an optical image of an electrochemical sensor fabricated by ink-jet printing silver NP ink. The details of printing methods are introduced in Section 4.3.

2.2.2. Metallic Nanowires

Despite the high conductivity of bulk metal films and tracks, their utilization in stretchable electronics has to rely on special designs based on stretching mechanics due to their rigid and brittle nature. On the other hand, metallic nanowires are typically arranged into networks (shown in Figure 3a) where the nanowires with high aspect ratio are connected to each other and form the conductive paths. The intrinsic stretchability of the networks makes metallic nanowires an ideal candidate for stretchable conductors.

Silver nanowires (AgNWs) [20,40,59,82,104], copper nanowires (CuNWs) [85,105], and gold nanowires (AuNWs) [106,107] were reported serving as the conductors in stretchable electronics. AuNWs are costly, while CuNWs are easily oxidized in air [105]. Comparatively, AgNWs turned out to be the most popular choice because of their high conductivity and the well-established synthesis methods [108].

AgNWs were largely studied for transparent conductors [109,110] and were not incorporated into stretchable conductors until about five years ago [15,40,104]. AgNWs networks are usually fabricated by solution processes such as drop casting [20,59,82,104,111], vacuum filtration [40], and spray deposition [112,113]. Because elastomers are at risk to be attacked by the solvent and heat, AgNWs are typically solution-processed on glass or silicon, and then transferred to elastomeric substrate after evaporating the solvent. The performance of the AgNW-based stretchable conductors is related to various factors, such as the aspect ratio of AgNWs [40], the loading density of AgNWs [82,113], and interfacial adhesion between AgNWs and the substrates [82,112]. A short AgNW percolation network fabricated on 50% pre-stretched Ecoflex failed at 60% strain, while a long AgNW percolation network was able to retain conductivity with a strain of more than 80% [40]. Figure 3b shows the conductivity of an AgNW/PDMS conductor as a function of the areal density of AgNW, in which the conductivity increases with the loading density until the percolation threshold is reached [113]. The bonding between polyacrylate substrate and AgNWs was enhanced by introducing acrylic acid (AA) into the polymerization reaction. The sheet resistance of the AgNW/polyacrylate composite without AA was found to be 26% higher than that with AA [112].

Despite the high conductivity of the nanowires, the significant junction resistance makes the actual resistance of the whole nanowire network higher than expected [6]. The junction resistance can be effectively reduced by soldering. Thermal annealing, an effective method to get the junctions fused, was reported to reduce the sheet resistance of an AgNW network by one third [114]. Soldering with secondary functional materials or networks [74,115,116] would further improve the performance of the AgNW-based conductors under large tensile deformation. There are reports where AgNW networks were soldered by silver nanoparticles [116] and graphene oxides [74]. The conductor made by the former approach exhibited no obvious change in electrical conductivity, with strain up to 120%, and that by the latter was stretched to greater than 100% strain without losing electrical conductivity.

2.2.3. Carbon-Based Nanomaterials

Carbon-based nanomaterials are a burning topic in materials science due to their unusual structures, which endow these materials with excellent mechanical and electrical properties. The application of the carbon-based nanomaterials in stretchable conductors is dated back from 1982 [117], where carbon black particles were dispersed in a silicon rubber matrix to make conductive elastomers. With the progress in materials research, carbon nanotubes [77] and graphene [118] were subsequently introduced to stretchable conductors. However, carbon-based nanomaterials offer lower electrical conductivity compared to metallic materials [8]. The simplest way of increasing the conductivity is to increase the loading density of the carbon-based nanomaterials, at the cost of sacrificing some of the mechanical properties and the transparency. High loading density would also make it difficult to obtain a uniform dispersion in solvents. Hence, there is always a trade-off between electrical performance and stretchability, transparency, and ease of processing. However, researchers have found clever ways to bypass this challenge and obtain good conductivity as well as good stretchability.

Dispersing carbon black [50,117,119] or graphite [120] in the elastomer matrix is a simple but effective method to fabricate stretchable conductors. Such composites have relatively low conductivity, which is desirable for micro-heaters [50]. Single-wall carbon nanotubes (SWCNTs) were reported to be made into composite dispersion and aerogel. The elastic conductor made from the dispersion of SWCNT, ionic liquid, and fluorinated copolymer retained good conductivity up to 134% strain [121]. The resistivity of SWCNT aerogels was found to increase only slightly (14%) at a strain of 100% [122]. Multi-wall carbon nanotubes (MWCNTs) can be dispersed in elastomer using a similar method to carbon blacks. The MWCNT/elastomer composites were reported to have percolation concentration ranging from 0.2–5 wt % [48].

Besides direct dispersion, many designs and processing techniques to assemble CNTs into stretchable conductors have been proposed. The vertically grown MWCNTs were embedded into polyurethane, retaining their conductivity when stretched up to 300% strain [76], and into PDMS for a load-bearing antenna [123]. CNT ribbons drawn from vertically grown CNT forest [124] were embedded into PDMS and stretched to 120% strain. Figure 3c,d show the SEM images of the vertically aligned MWCNT forest and the CNT ribbons, respectively. The CNT ribbons were fabricated into "out-of-plane" and "lateral" buckling architectures, which further enhanced the stretchability [42,43]. Cross-stacked super-aligned CNT film possessed a high intrinsic tensile strain of more than 35%, which was further improved by embedding the film into PDMS [125]. In addition, gap-island networks made by the overlapping SWCNT films [41], spring-like CNT ropes [126], hierarchical reticulate SWCNT architecture [44], etc. were demonstrated for stretchable, transparent, and highly conductive carbon-based conductors.

Graphene is a two-dimensional material with hexagonal honeycomb architecture and boasts of very high conductivity, stretchability, and transparency [46,47]. Successful synthesis of graphene in a large scale by chemical vapor deposition facilitated its application in stretchable conductors [118,127]. Graphene film transferred to a pre-stretched PDMS substrate was able to sustain isotopic strain up to 12% without change in resistance, as seen in Figure 3e [118]. Graphene was further patterned into structures such as meshes [47], ribbons [46,128], and serpentine tracks [27,129] for better stretchability by using pre-designed metallic templates or photolithography.

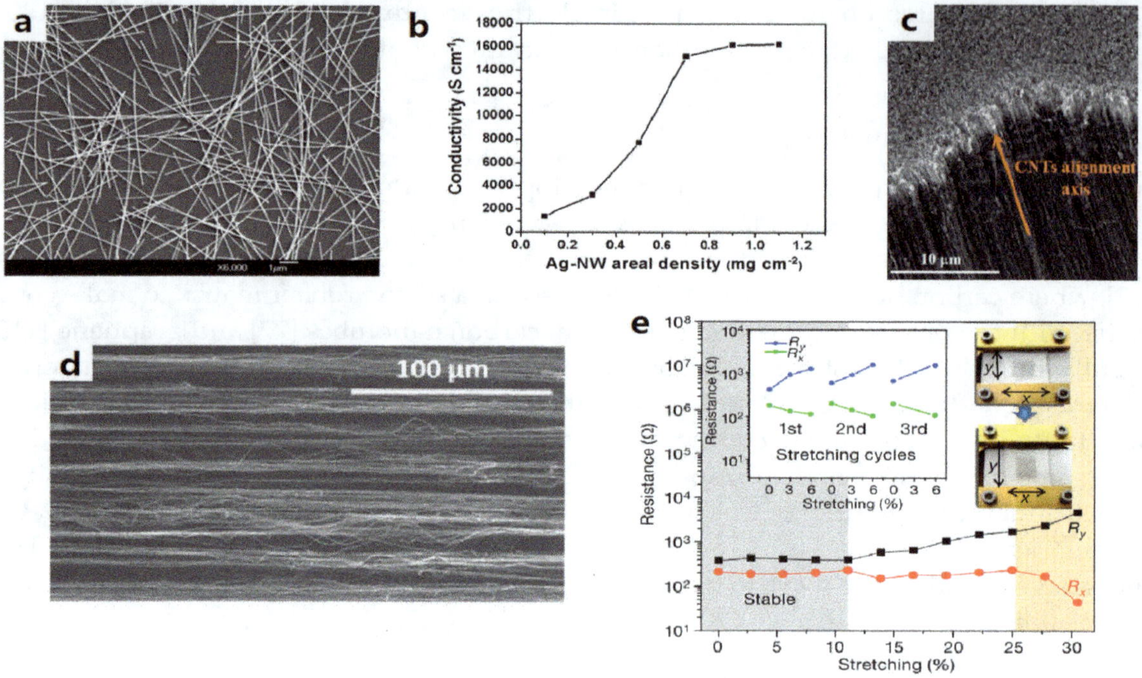

Figure 3. (**a**) SEM image of an AgNW (silver nanowire) network on glass [82]. © IOP Publishing. Reproduced with permission. All rights reserved; (**b**) Conductivity of an AgNW/PDMS conductor as a function of the areal density of AgNW. Reused from [113]; (**c**) SEM image of vertically aligned CNT (carbon nanotube) forests. Reproduced from [130] with permission of The Royal Society of Chemistry; (**d**) SEM image of CNT ribbons drawn from a vertically aligned CNT forest. Reproduced from [131] with permission of The Royal Society of Chemistry; (**e**) Resistance of a graphene film transferred to a pre-stretched PDMS substrate isotopically stretched by 12%. The left inset shows the case in which the graphene film is transferred to an unstretched PDMS substrate. The right insets are photographs of the graphene film under vertical and horizontal strain.

2.2.4. Conductive Polymers

Compared to metals, polymers have better intrinsic flexibility. The discovery of conductive polymers has brought another promising option for stretchable conductors. Conductive polymers like polyaniline (PAni) [132], polypyrrole [133,134], poly(3,4-ethylenediox-ythiophene) (PEDOT) [49], and poly(3,4-ethylenediox-ythiophene):poly(styrene sulfonate) (PEDOT:PSS) [27,51,71] have been used in stretchable conductors despite some limitations such as environmental instability [135] and low stretchability [49,136]. PEDOT:PSS receives most of the attention among the conductive polymers because of its good conductivity and ease of processing. However, the breaking strain of a pristine PEDOT:PSS thin film is below 10% [136,137]. The stretchability of PEDOT:PSS film was improved by adding dimethylsulfoxide (DMSO) and Zonyl fluorosurfactant with the maximum strain of more than 20% [137]. The Zonyl modified PEDOT:PSS film on PDMS was found to form a 3D wavy buckling structure which promotes stretchability. The resistance of the film increased by a factor of only two at 50% strain [71]. Recently, a strain-insensitive PEDOT:PSS/acrylamide organogel was reported, which was stretched to more than 350% strain while it retained invariant resistance up to 50% strain [138]. Besides that, elastomeric conducting polyaniline networks with maximum elongation up to 90% $\pm$ 10% were fabricated by synthesizing polyaniline via molecular templates [139].

2.2.5. Liquid Metals and Ionic Liquids

The most common liquid metal used in stretchable conductors is eutectic gallium indium alloy (EGaIn, with 75.5% Ga and 24.5% In). The advantages of EGaIn include mobility, self-healability [140], and processability at room temperature [141]. Typically, it is injected into microchannels fabricated by molding elastomers [53]. Some other methods, such as microcontact printing, stencil printing [142], and photolithography [143,144], are also used to fabricate stretchable conductors based on liquid metals.

Ionic liquids are molten salts that have conductivity comparable to many organic electrolyte solutions at room-temperature and are stable up to 300–400 °C [145]. Ma et al. reported the integration of ionic liquids with various supporting materials, such as cotton fabric, rubber film, rubber band, and sponge, which exhibited exceptional performance with high conductivity at strain greater than 600% [78].

3. Mechanics of the Stretchable Conductive Structure

As introduced in last section, elastomers serve as the stretchable substrate or matrix for stretchable conductors, while conductive materials provide the electrical functionalities. However, they have huge differences in elastic properties. Substantial effort has been made to combine the rigid, brittle conductors with the flexible and soft elastomers. From the aspect of the mechanics, many fancy structures and clever designs were proposed to harmonize the mismatches and integrate materials with widely different properties as a whole system.

3.1. Blending

Blending is a pure mechanical process which is simple and cost-effective. Several material, like carbon blacks [50,117,119], silver particles/flakes [19,50,86,146,147], MWCNTs [48,77], and PEDOTs [49] were reported as blending into elastomers. Such composites can be analyzed by the percolation theory, where the electrical conductivity is determined by the concentration of the fillers [86]:

$$\sigma = \sigma_0 (V_f - V_c)^s \tag{1}$$

where σ is the electrical conductivity, σ_0 is the conductivity of the conductive filler, V_f is the volumetric fraction of the filler, V_c is the volumetric fraction at the percolation threshold, and s is the fitting exponent. The percolation threshold is the key to control the electrical conductivity of composites. Table 2 gives the percolation thresholds of the conductive fillers in elastomer matrix (in weight percent) for comparison. As seen in the table, the percolation threshold has a strong dependence on the size and shape of the materials. Compared to the particle fillers, MWCNT (as a 1D nanomaterial), has lower percolation threshold due to its high aspect ratio, and thus is the ideal candidate for transparent stretchable conductors. The electrical properties of these conductive composites are also dependent on dispersion techniques [48] and properties of matrix [77,119].

Table 2. Conductive fillers for stretchable conductors based on blending.

Material	Size	Percolation Threshold (wt %)	Reference
Silver particles	1–2 μm	83	[50]
Carbon black	40–100 nm	10	[50]
MWCNT	Diameter: 60–100 nm Length: 5–15 μm	2	[48]
PEDOT	-	23	[49]

The blended conductive composites can be further patterned or used as conductive inks. Photo-patternable conductive PDMS was fabricated by blending in both conductive fillers and photosensitizers [146]. A composite ink of soluble silver salt and adhesive rubber was directly filled into ballpoint pens to write on different substrate to form adhesive conductive tracks [147]. Ag/PDMS composite ink was stencil printed and screen printed onto PDMS substrate to fabricate soft printed circuit boards (PCBs) [19].

3.2. Conductors in Microchannels

Filling the PDMS microchannels with liquid metals is another straightforward method of fabricating stretchable conductors. The PDMS microchannels are typically fabricated by molding top and bottom layers from the SU-8 master and adhering these two layers by oxygen plasma treat, heat, and pressure [144,148]. In the initial stages, solder was melted and filled into PDMS channels to fabricate stretchable conductors [149]. This process needed heating (up to 180 °C) and cooling steps, which was hence incompatible with heat-sensitive materials [141]. Filling liquid metals into the microchannels circumvents the heating during fabrication. The liquid nature endows them with excellent stretchability and self-healability [140]. Conductors fabricated by filling diamond-shape PDMS microchannels with EGaIn were stretched biaxially to 100% strain [144] while retaining their characteristics. Wetting of PDMS with the liquid metal was an issue which was solved effectively by putting on a metal layer [144] or applying silane adhesion promoter [149]. This strategy was also exploited to strengthen the brittle areas, such as the sharp corners of the diamond-shaped metallic conductive tracks [150].

3.3. Stretchable Network on/Embedded in Elastomers

As discussed in Section 2, the nanomaterials with high aspect ratio are typically arranged into networks possessing stretchability. Such network films are ready to be fabricated onto the elastomeric substrate without applying any other mechanics. Some nanomaterials can be directly solution- processed onto the modified elastomeric substrate [112]. For example, the self-assembly of functionalized SWCNTs onto PDMS substrate was reported [151]. Besides the direct fabrication of network films onto elastomer, transferring a solid film to the elastomeric substrate by conformal contact (due to the good adhesion between PDMS substrate and the conductive network film), or using the transfer media [27,41], provides alternative routes to obtain conductive network films on the elastomer surface [42,43]. In addition to these methods, vacuum suction is also an effective method to transfer conductive network film from membrane filter to elastomeric substrate [40,75,122].

The nanomaterial-based conductive networks can be embedded into the elastomer as well. The conductive networks are fabricated on a donor substrate. The conductive composites are obtained by coating the liquid elastomer precursor onto the donor substrate (thus the precursor infiltrates into the networks), curing and peeling off [82,104]. This method is also suitable for fabricating conductive networks with relatively large thickness, such as the vertically aligned CNT forests [45,76,123]. Figure 4a gives a schematic of such fabrication process of an AgNW/PDMS stretchable conductor.

3.4. Geometry Design

Designing the geometry of conductive tracks to reduce the strain in the metals, is an intelligent strategy in the fabrication of metallic film-based stretchable conductors. The realization of the geometry largely relies upon the patterning techniques, which are discussed in Section 4.

Here, we categorize those geometries which endow stretchability to the conductors into three types: polygonous, serpentine, and fractal designs. The polygonous designs make metallic tracks into connected polygonous networks, such as diamond [150,152] and hexagon (honeycomb) [153]. Serpentine designs include zigzag [100], half circle (U-shape) [101,154], sinuous [6,13], and horseshoe [100,101,155]. Among them, the horseshoe design provided the best stretchability [101]. Parameters such as the linewidth of metal, thickness of metal, and height/spacing ratio of geometry in the serpentine design play an important role in the performance of these conductive tracks. It was proven experimentally [13] and with the finite element method (FEM) analyses [101] that subdividing a wide conductive line into several thinner lines largely improved the stretchability and alleviated the stretching-induced stress. Analysis of postbuckling of the serpentine interconnects revealed that the elastic stretchability of the serpentine interconnects increased with decreased thickness [154,156].

The height/spacing ratio of geometry refers to the ratio of height (perpendicular to the stretching direction) to spacing width (parallel to the stretching direction). Increased height/spacing aspect ratio benefits the stretchcability when it is under a certain range [102,154]. In Figure 4b, the simulated uniaxial stretchability of the horseshoe patterns, with varied height/spacing ratio (arc length), is given [102]. As shown in the figure, the stretchability keeps increasing until the arc angle reaches 235°. Fractal designs can be perceived as the assembly of serpentine designs, which enable unusual mechanics with implications in stretchable device design [102]. In Figure 4c, the stretchability of a freestanding U-shape serpentine copper interconnect in symmetric buckling mode is demonstrated both experimentally and with FEM analyses. Some other representative geometry designs are shown in Figure 4d–g.

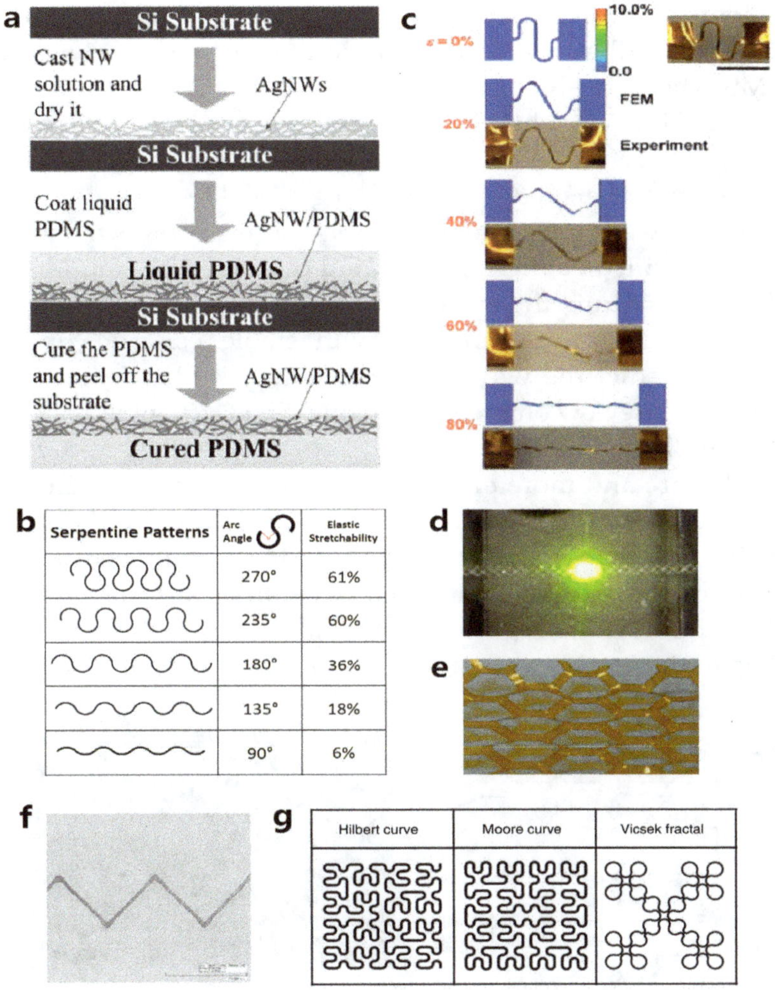

Figure 4. (**a**) Schematic of the fabrication process of an AgNW/PDMS stretchable conductor. Reprinted with permission from [21]. (**b**) Simulated uniaxial elastic stretchability for horseshoe patterns as a function of height/spacing ratio (arc angle). Reprinted by permission from Macmillan Publishers Ltd.: *Nat. Commun.* [102]; (**c**) Experiments and FEM analyses on the symmetric buckling behavior of the serpentine interconnect with strain from 0% to 80% (scale bar: 1 mm). Reproduced from [156] with permission of The Royal Society of Chemistry; (**d**) Diamond-shaped stretchable interconnects based on liquid metal. Reprinted from [144], with the permission of AIP Publishing; (**e**) Honeycomb-patterned polyimide substrate at stretched state. Reprinted with permission from [153]. (**f**) Zigzag Cu conductive tracks on PDMS [100]. Reproduced with permission from © IOP Publishing. All rights reserved; (**g**) Three representative fractal patterns. Reprinted by permission from Macmillan Publishers Ltd.: *Nat. Commun.* [102].

3.5. Buckling

The buckling phenomenon was first found in the metal thin film on PDMS. Due to the large thermal expansion of PDMS, the subsequent cooling processes caused compressive stress in the metal film, which was calculated using Equation (2).

$$\sigma_0 = \frac{E_\mathrm{m}(\alpha_p - \alpha_m)(T_D - T)}{(1 - \nu_m)} \tag{2}$$

where σ_0 was the compressive stress in the metal film. E, α and ν were referred to as Young's modulus, thermal expansion coefficient, and Poisson's ratio, respectively. The subscripts m and p referred to metal film and PDMS, respectively. T and T_D were the real-time temperature and the deposition temperature [1]. The compressive stress developed in the metal film led to buckling. It is worth mentioning that an interfacial adhesion layer of titanium or chromium was critical in making metal film compliant to PDMS. The buckling of metal film on a flat, unconstrained PDMS was however, uncontrollable. Ordered buckling was obtained by molding the PDMS surface with certain patterns or by partial modification of the surface [72].

The idea to utilize the pre-stretched elastomer to get buckled metal film further propelled the application of buckling strategy in stretchable conductors. In this approach, metal film was evaporated on 15% pre-stretched PDMS with an adhesion layer and patterned by standard lift-off process. Upon releasing the PDMS from the fixture, the metal film stayed compliant on the surface of PDMS and buckled. The as-prepared metal film was stretched, with stable resistance up to 25% [92]. Analytical models were proposed to predict the wavelength and amplitude of the buckling [39,70,157,158]. Interested readers can refer to a review on mechanical buckling [159], where the buckling phenomenon was studied in cases of compliant/small strain, compliant/large strain, and delaminating.

With the advents of more and more electrical conductive materials in stretchable conductors, the buckling strategy was explored in stretchable conductors based on CNTs [42,57], graphene [128], AgNWs [113], conductive polymers [133], and a combination of other interesting mechanics [160]. Figure 5b shows the SEM image of the buckling of an elastomer-infiltrated vertically aligned carbon nanotube (VACNT) film. This film was reported as having a very small resistance change (ratio of less than 6%) when stretched to the level of pre-strain (100%).

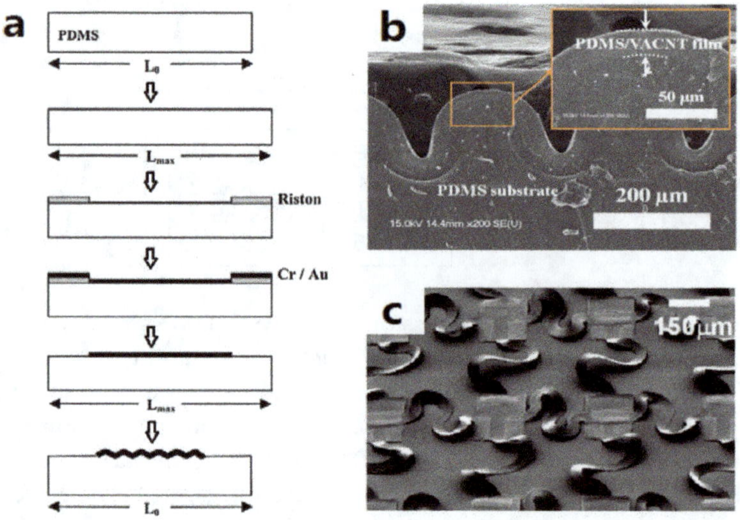

Figure 5. (**a**) The schematic of the fabrication of Au interconnects on PDMS substrate with buckling strategy. © 2004 IEEE. Reprinted, with permission, from [92]; (**b**) Cross-sectional SEM image of a VACNT/PDMS conductive film. The inset shows the magnified SEM image of the composite part. Reprinted with permission from [45]. (**c**) SEM image of an array of stretchable complementary metal-oxide-semiconductor (CMOS) inverters with non-coplanar bridges [160].

Buckling was also reported without the pre-strain of substrate. The buckling of AgNW/PDMS film emerged upon the first stretch/release cycle. Such films could maintain stable conductivity of 5285 S·cm^{-1} at the strain range of 0%–50% [104]. A similar strategy was employed to fabricate laterally buckled CNT ribbons on PDMS [43].

3.6. Out-of-Plane Design

According to some studies of stretchable conductors, the conductive materials were supposed to have good adhesion to the substrate [10,161]. However, it was also pointed out that, for serpentine metallic lines on elastomer, the lines would deform out of plane of the substrate to alleviate the strain [101]. The strong bonding with the substrate would constrain such deformation and thus increase the plastic strain and in-plane shear strain [155].

An innovative approach to fabricate stretchable conductors is to make them free of substrate constraints. Figure 5c shows an array of stretchable complementary metal-oxide-semiconductor (CMOS) inverters with non-coplanar bridges. The islands with devices are well-attached to the substrate, while the interconnecting bridges are lifted off from the substrate to accommodate deformation [160]. Similar structures which contained anchored square islands and S-shaped suspensions were also reported [162].

3.7. Designs in Substrate

Generally, electrical conductors of a stretchable device are prone to failure. That is why most of the mechanics of strechability are focused on the electrical conductors. However, designing the substrate is also a good strategy to broaden the mechanics of stretchable electronics. One way to enhance stretchability of the substrate is to use two substrates with different Young's modulus so that the substrate with a higher modulus can sustain lesser strain [52,53]. For example, in a stretchable device array, stiff PDMS islands were used for device assembly, while soft Ecoflex substrate was designed to sustain the deformation. PDMS relief, consisting of raised islands and recessed trenches, was used as the substrate for rigid GaAs photovoltaics. It was found that stretching the substrate to an overall strain of 20% induced 123% strain in the trenches and only 0.4% strain at the islands [54]. Fabricating the substrate into a wavy shape would largely improve the stretchability of the whole system [24,125,160,163]. Besides that, a novel shape-memory shrinkable polymeric substrate utilized in the fabrication of highly stretchable gold films was reported [164].

4. Patterning Techniques

Patterning is one of the key techniques in the fabrication of both wafer-based electronics and stretchable conductors. The common patterning techniques for stretchable conductors are lithography, screen/stencil printing, direct printing, and transfer printing. The techniques from conventional wafer-based electronics, such as photolithography and lift-off, are subsequently adopted in stretchable electronics [17]. Besides, stencil printing and screen printing are employed to pattern features that do not require high complexity and superfine resolution [19–21]. With the development of additive manufacturing, direct printing emerges as an alternative to the conventional subtractive patterning method [14,22–31]. Transfer printing generally means to transfer functional materials from one substrate to another. With a mold-patterned elastomer stamp, those materials can be selectively transferred from donor substrate to the target substrate [32–34].

4.1. Lithography

Subtractive patterning techniques, such as photolithography and lift-off, are well-developed and dominatingly used in wafer-based electronics. However, in elastomer-based stretchable conductors, the elastomers are mostly incompatible with the chemicals and high energy beams used in these processes. Despite that, the patterning of devices still relies on lithography, where high resolution and complexity is required. In addition to the photo-defining and etching steps, an extra transfer step is needed. The patterning of the conductive materials is initially processed on regular substrates such as silicon wafer or glass and then transferred to elastomeric substrate. The transfer step can be done by directly coating the liquid elastomer precursor, curing, and peeling off (similar to the process shown in Figure 4a) [165]. It is also achievable by picking up and releasing the conducive materials with the help of an intermediate substrate [102,166,167].

4.2. Stencil/Screen Printing

Compared to the lithographic approach, direct deposition of conductive materials through a mask, stencil, or screen onto the substrate is relatively simple and cost-effective. It is suitable for applications that do not require high complexity and superfine resolution. Figure 6a is an SEM image of stencil-printed Ag/PDMS composite with the highest resolution achieved (linewidth of 150 μm with spacing of 100 μm).

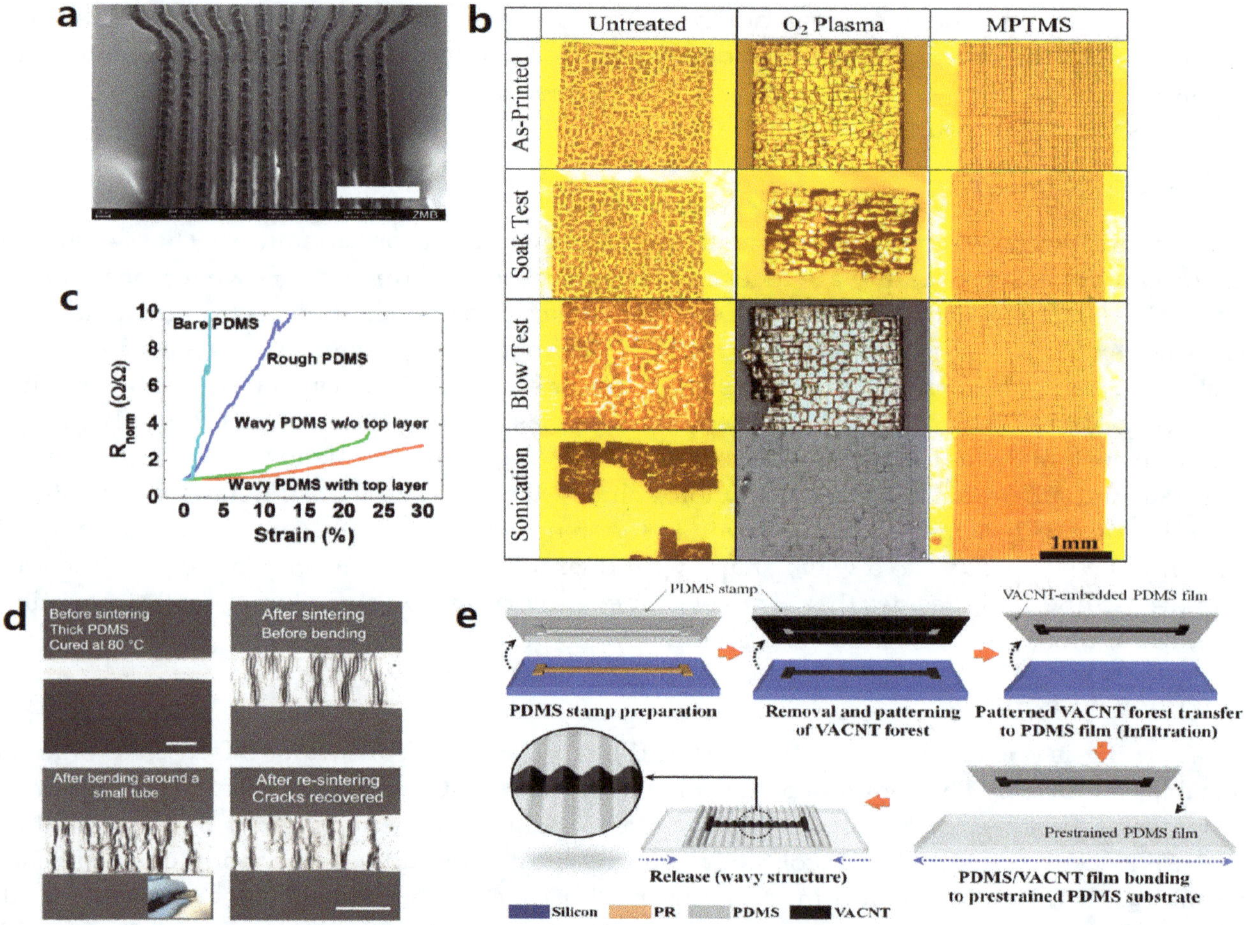

Figure 6. (**a**) SEM image of stencil printed Ag/PDMS tracks with the highest resolution achieved (scale bar: 1 mm). Reprinted by permission from Macmillan Publishers Ltd.: *Sci. Rep.* [19]; (**b**) Photographs from adhesion tests of untreated, O_2-plasma treated, and adhesion-promoter (MPTMS)-treated 2 mm square ink-jet-printed Ag patterns on PDMS under destructive tests. © 2014 IEEE. Reprinted, with permission, from [69]; (**c**) Changes in resistance as a function of strain of the ink-jet-printed silver tracks on PDMS substrate with varied adhesion situations. Reprinted from [24], with the permission of AIP Publishing; (**d**) Morphologies of ink-jet-printed silver track on PDMS before and after sintering, after bending, and after re-sintering (scale bar: 100 μm) [26]. © IOP Publishing. Reproduced with permission. All rights reserved; (**e**) The schematic of fabrication process of PDMS/VACNT-film-based wavy-configured stretchable conductors. Reprinted with permission from [45].

This approach is versatilely utilized with various conductive materials. Gold strip on PDMS was made by evaporating gold on PDMS laminated with patterned Dupont Riston photoresist film [92]. Liquid metals embedded into elastomer were fabricated using masked deposition [168]. A strain sensor array was fabricated by cross-folding screen printing AgNW stripes on PDMS [20]. Ag/PDMS composite was both stencil printed and screen printed to obtain soft PCBs [19]. PEDOT:PSS ink was stencil printed on Ecoflex to fabricate electrochemical sensors [51].

4.3. Direct Printing

Owing to the development of additive manufacturing, conductive inks can be directly printed on various substrates without physical contact. Various printing techniques, including laser-aided direct writing [22], ink-jet printing [23–26], and aerosol-jet printing [14,27–30] have been developed for printed electronics. The maskless processes are well suited for rapid prototyping. Also, they offer a cost-effective way to achieve large-area electronics production with minimal materials waste and without lengthy subtractive processes. Both planar and 3D architectures are achievable by laser-aided direct writing [169]. Ink-jet printing is a well-established technique for graphical printing and has been extensively used for printed electronics [170]. In this technique, the ink drop is driven by thermal, piezoelectric, or electrostatic actuation and delivered through a printing orifice to the substrate at a demanded area [171]. Aerosol-jet printing is a relatively new technique. Instead of printing the ink drop by drop, it prints the aerosol mist generated by ultrasonic or pneumatic atomizer. Aerosol-jet print has the capability of generating smaller feature size [30], as compared to the ink-jet printing and is compatible with 3D non-planar substrates [170]. The broad applications of such direct printing methods have been demonstrated by printing various conductive materials, such as metallic NP inks [24,69], CNTs [28,30,172], conductive polymers [27,173,174], and dielectric materials, such as ion gel [27,174] and polyimide [28], on not only regular flexible plastic substrate but also elastomeric substrates.

One of the issues in the direct printing technique is that of wetting and adhesion between elastomers and conductive inks. PDMS is known to have surface hydrophobicity, which can be modified to hydrophilicity by O_2 plasma [68,69], UV/ozone [24,70,71], and chemical treatment [68]. The surface treatment of PDMS imparts good wetting of PDMS with various inks and promotes the adhesion between PDMS and conductive materials. However, in some cases, such adhesion might not be strong enough and require further deposition of adhesion layer or blending of an adhesion promoter in the initial stages. Super-thin (5 nm) titanium or chromium adhesion layer was largely used to bond gold thin film with PDMS [1,12,92,162,175]. Poly-dopamine adhesion layer was employed to significantly modify the surface property of PDMS and enhance the adhesion between PDMS and spray-deposited AgNWs [112]. Zonyl, as a non-ionic fluorosurfactant, can be mixed in both elastomer precursors and conductive inks, which enhances the interfacial adhesion between elastomers and various conductive materials [7,51,71,176]. Silane adhesion promoter MPTMS (3-mercaptopropyl trimethoxysilane) is widely used in PDMS-based stretchable conductors. It was used as the adhesion promoter between PDMS and molten liquid solder [149], evaporated gold film [165,177], and printed metallic inks as well [25,69]. In Figure 6b, the adhesion tests of the ink-jet- printed silver patterns on PDMS with blank treatment, O_2 plasma treatment, and MPTMS treatment are compared. Only the MPTMS-treated sample survived the three destructive tests. [69]. The influence of the adhesion on the stretchability of ink-jet-printed silver tracks is demonstrated in Figure 6c. The PDMS rough surface obtained from UV/ozone treatment provided better adhesion, thus making the tracks more stretchable. Besides that, a top PDMS layer encapsulated the silver tracks, making them compliant to the wavy PDMS and hence further enhancing the stretchability [24].

Besides the adhesion issue, many other problems, such as the swelling [62], thermal hardening, and large thermal expansion of the elastomers, still exist in the direct printing of conductive inks [26], resulting in poor performance of printed stretchable conductors. Cracking [26,31] and low stretchability are some of the resulting issues found in printed stretchable conductors [14]. As an example, Figure 6d shows the morphologies of ink-jet-printed silver track on PDMS before and after sintering, after bending, and after re-sintering [26], in which the silver track cracked as a result of the thermal expansion problem. All these problems and issues account for the current situation that most of the stretchable conductors are not directly fabricated on elastomers. This should be addressed in order to realize the direct printing of stretchable conductors.

4.4. Transfer Printing

The challenges involved in direct printing of stretchable conductors are discussed above. Transfer printing is an alternative approach, which effectively circumvents such difficulties. In a general sense, transfer means switching the substrate of the conductive patterns. As mentioned in Section 4.1, coating the liquid elastomer precursor onto the donor substrate, curing, and peeling off is a typical transfer process [82,104,165].

Transfer is also achievable via intermediate media. Poly(methyl methacrylate) (PMMA) is one of the common transfer media [46,132]. Conductive materials can be embedded into PMMA like in PDMS and transferred to the target substrate, followed by dissolving PMMA in hot acetone. Some other materials, such as functional tapes, were also reported as the transfer media. Water-soluble tape is an adhesive tape which is dissolvable in water. Conductive materials on donor substrate could be archived by using such tape and releasing them on the target substrate. The tape would totally disappear when dipped in water, leaving only conductive materials on the target substrate [27,41,56]. Thermal release tape is an adhesive tape which loses its adhesion upon heating. Thermal release tape was reported as a temporary holder to transfer NW-based devices [178]. Besides the transfer media, a sacrificial layer is an essential assistant in transfer printing techniques. A sacrificial layer is typically fabricated as a structural support between the donor substrate and the conductive materials. This layer can be removed by solvent or etchant, thus lifting off the conductive materials and making them easily transferable. The commonly used sacrificial materials include copper [101], gold [177,179], PMMA [160], SiO_2 [46,157], and hard-baked photoresist [164,180].

Transfer by PDMS stamp is extensively used in soft lithography, where the conductive materials can be selectively transferred from donor substrate to the target substrate using a mold-patterned PDMS stamp. Transfer printing by such PDMS stamps is either additive or subtractive [32]. The key to the PDMS stamp-based transfer process is controlling the adhesion. The adhesion of the PDMS stamp to the donor or the target substrate can be controlled kinetically, owing to the rate-independent viscoelastic property of PDMS [33]. The adhesion can also be controlled by manipulating the surface chemistry and the interfacial properties of the PDMS stamp. [34].

As an example of transfer printing, Figure 6e illustrates the fabrication of stretchable vertical-aligned carbon nanotube (VACNT) film on PDMS substrate, in which two transfer processes were involved. A subtractive transfer printing technique was used to remove the undesired part of VACNT film. The remaining part was transferred by an intermediate PDMS substrate to a pre-stretched target PDMS substrate.

5. Applicable Devices

Stretchable electronics have been developed into a unique and emerging field of electronics. They contravene the common idea that people would attach to electronics—stiff circuit board and cold metals. They make electronics closer to human. Because of the soft, attachable nature, stretchable electronics show great potential in developing strong intimacy with humans, like conforming to the skin (epidermal electronics) [18,35] and implanting into the body [36,37]. They have an enormous potential in future electronics for biomedical use, such as lightweight physiological monitors and in personalized healthcare [60,61]. Stretchable conductors are the most essential building block of stretchable electronic devices. They could simply serve as interconnects, or as functional electrodes. In this section, the applications of the stretchable conductors are discussed by demonstrating the functional devices based on them.

5.1. Sensors

Stretchable sensors are probably the most widely studied type of devices in the recent development of stretchable electronics. Stretchable sensors with various functions have been fabricated, including strain sensors [181], pressure sensors [20,59,107,182], temperature sensors [18,183], gas sensors [132,184], and UV sensors [53]. These sensors detect the changes in the outside environment and reflect these changes in the electrical properties, such as resistance and capacitance. Sensors are also fabricated based on the performance of devices such as transistors [152] and antennas [21].

Stretchable strain sensors have largely extended the sensing range beyond those of traditional non-stretchable strain sensors from below 1% [136] to several hundred percent [181]. The strain sensors are mostly resistive or capacitive. Stretchable conductors with reversible and repeatable resistance change under strain are suitable for fabricating resistive strain sensors. Carbon-based materials are widely used in this application due to their intrinsic piezoresistivity [46,47]. Network structures fabricated from AgNWs [111] or CNTs [41] are also piezoresistive, with mechanisms of local disconnections and tunneling effects [181]. In capacitive strain sensors, an elastomeric dielectric layer is sandwiched between two stretchable electrodes. At certain strain, the decreased width and thickness of the dielectric layer results in changes of capacitance. Both AgNWs and CNTs were reported as the electrode materials [20,59,80]. Figure 7a shows a resistive stretchable strain sensor based on AgNW/PDMS nanocomposites. More information on the stretchable strain sensors can be found in a recent review [181].

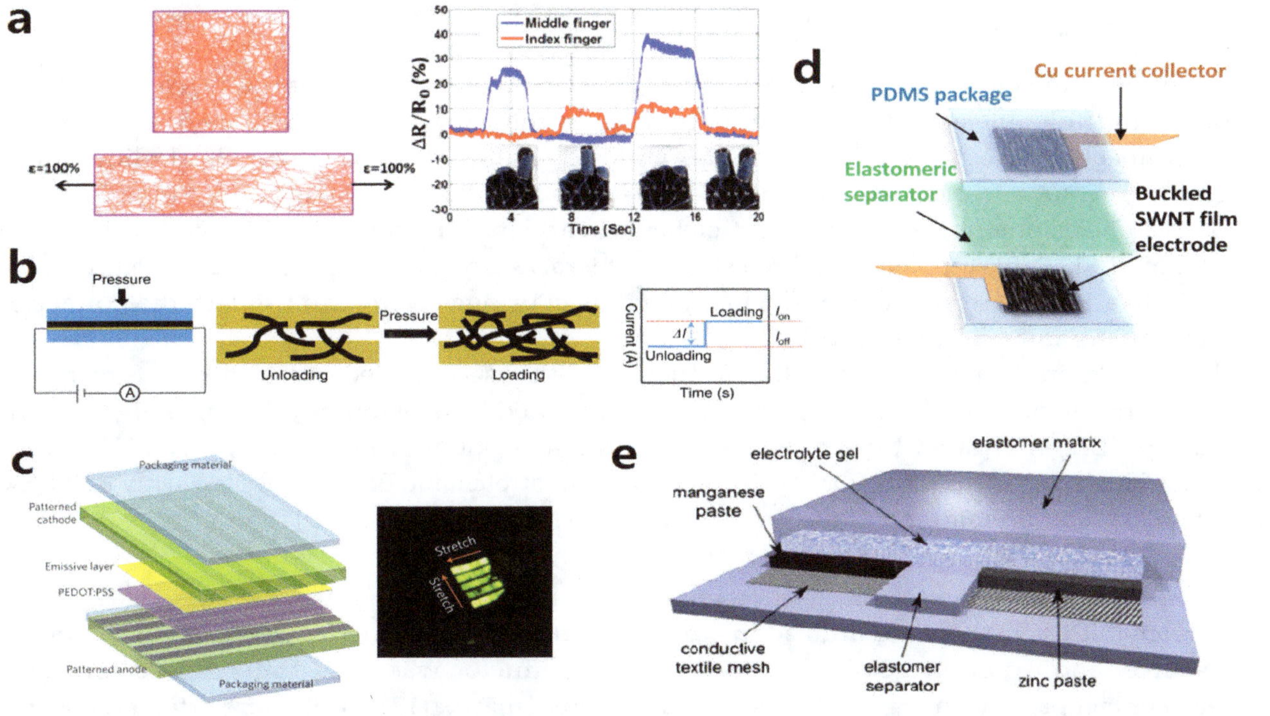

Figure 7. (**a**) Schematic of AgNW network at relaxed and stretched states and the illustration of the motion detect function of an AgNW/PDMS strain sensor. Reprinted with permission from [111]. (**b**) Schematic of the sensing mechanism of an AuNW-based pressure sensor. Reprinted by permission from Macmillan Publishers Ltd.: *Nat. Commun.* [107]; (**c**) Schematic and photograph of an encapsulated fully stretchable elastomeric polymer light-emitting devices (EPLED) display with 25 pixels. Reprinted by permission from Macmillan Publishers Ltd.: *Nat. Photonics* [185]; (**d**) Schematic of the components of a dynamic SWCNT (Single-wall carbon nanotubes)-based stretchable supercapacitor. Reprinted with permission from [57]. (**e**) Schematic of a rechargeable stretchable alkaline manganese cell. Reproduced from [83] with permission of The Royal Society of Chemistry.

Stretchable pressure sensors detect the vertical deformation caused by pressure and send out differential electrical signals. Capacitive pressure sensors have a similar mechanism to capacitive strain sensors. When the capacitor is under pressure, the distance between the two electrodes decreases, causing changes in capacitance. Capacitive sensing arrays based on AgNWs and various elastomers have been demonstrated with the functionality as strain sensor, pressure sensor, and touch sensor [20,59]. A stretchable resistive pressure sensor was fabricated by sandwiching AuNW coated tissue sheet with an interdigitated electrode between two PDMS sheets, as shown in Figure 7b. The applied pressure was determined by measuring the changes in resistance caused by the different loading of AuNWs on the electrode [107].

Noble metals, such as gold and platinum, are resistance-sensitive under varied temperature. Both traditional temperature sensors and stretchable temperature sensors take advantage of this property. However, these noble metals are not intrinsically stretchable. To incorporate them into stretchable temperature sensors, mechanics to enhance their stretchability were employed. Platinum meander electrodes were employed as temperature sensors in an integrated epidermal electronic system [18]. The buckling strategy was used to fabricate stretchable temperature sensors by transferring gold strips onto pre-stretched PDMS [183].

In most of these stretchable sensors, rough structures, such as NW networks and buckled films/strips, are well encapsulated or packaged with elastomers [20,59,183]. However, the influence of such rough structures on the device performance has not yet been thoroughly discussed, and this is expected to be addressed in the near future.

5.2. Light-Emitting Circuits

Light-emitting circuits on elastomeric substrates are designed to meet the demands of flexible, wearable, and foldable displays and light sources. In these circuits, stretchable conductors serve as interconnects between the power supplies and the light-emitting devices. EGaIn liquid metals embedded into soft elastomeric substrate were used as interconnects to light-emitting diode (LED) arrays on stiff elastomeric islands [53]. A stretchable heart-shaped light-emitting circuit was fabricated by installing LED lights onto hand-written stretchable tracks from an adhesive silver-based conductive ink [147]. In addition, serpentine graphene interconnect bridges were also used as interconnects for microscale inorganic LED array [129].

Besides interconnects, stretchable conductors were used as electrodes in thin film polymer light-emitting devices as well. Liang et al. demonstrated an elastomeric polymer light-emitting devices (EPLED) fabricated by sandwiching an emissive polymer layer in AgNW-PU composite electrodes. Such a structure was further developed into a stretchable EPLED display array of 25 pixels. The schematic and photograph of the display array are shown in Figure 7c [185].

5.3. Transistors

Stretchable conductors are utilized in stretchable transistors as interconnects and electrodes for gate, source, and drain. An SWCNT-based elastic conductor was used to form interconnections between contact pads of an organic transistor-based active matrix [121]. Polymer-sorted and unsorted CNTs were used as the semiconductor and conductor, respectively, in a thermoplastic polyurethane (TPU)-based stretchable transistor [186]. Graphene/AgNW hybrid conductors were used as the drain/source electrodes of an oxide semiconductor transistor array [187]. A stretchable and transparent transistor array was fabricated with patterned graphene thin film as the semiconducting channel and source/drain electrodes. The patterned graphene was transferred to the rubber substrate, followed by aerosol printing of ion gel and conductive polymer as the gate dielectric and gate electrode, respectively. Such graphene transistors showed hole and electron mobilities of 1188 ± 136 and 422 ± 52 $cm^2/(V \cdot s)$, respectively, with stable operation at strain up to 5%, even after more than 1000 cycles [27]. Highly stretchable polymer transistors made entirely of stretchable components were fabricated on SBS fiber mat with stacked Au nanosheets as electrodes and poly(3-hexylthiophene) (P3HT) nanofibers as channel material. The transistors provided a hole mobility of 18 $cm^2/(V{\cdot}s)$ at a strain of 70% [188]. Recently, an intrinsically stretchable organic thin film transistor was reported with a healable polymeric semiconductor and CNT/PEDOT:PSS electrodes. Such a transistor was able to sustain a field-effect mobility of 1.12 $cm^2/(V{\cdot}s)$ at a large strain of 100% [189].

5.4. Energy Devices

Almost all of the stretchable electronic devices with amazing functionality were connected to and driven by external power sources. Low cost and compatible internal power supplies for independent stretchable electronic systems are in great demand, as they can facilitate the real-life usages of these stretchable devices. Energy devices are basically divided into energy storage devices and energy conversion devices. Although many research works focused on stretchable energy devices such as supercapacitors [57,58,125,132], batteries [83,84,134,166], solar cells [176], and other environmental energy harvests, challenges still remain to fabricate stretchable energy devices with high performance and reliability. Here, in this part of the review, these devices are briefly introduced with an emphasis on the usage of stretchable conductors. Two recent review papers about stretchable energy storage and conversion devices can be referred to for further information [38,190].

Supercapacitors are commonly used as energy storage devices. Carbon-based nanomaterials are ideal for this application due to their good stability, conductivity, and high surface area. Buckled SWCNT films on PDMS substrate as electrode, together with organic electrolyte and a separator layer were used to fabricate stretchable supercapacitors [57,58]. The schematic of the components of an SWCNT-based supercapacitor is shown in Figure 7d. In addition to this, highly-aligned CNT sheets were fabricated from vertically-grown CNT forest and further used in an all-solid supercapacitor [191]. Micro-supercapacitor array based on P3Ain-wrapped MWCNTs was employed to drive stretchable graphene gas sensors [132]. Graphene microribbons were also utilized as the electrode materials in stretchable supercapacitors [125].

Unlike supercapacitors, batteries store energy in forms of chemical energy. Stretchable batteries of various structures and shapes have been fabricated with several materials till now. Buckled polypyrrole macrofilm cathodes on elastomeric substrate were reported for battery applications [134]. Porous CNT/PDMS nanocomposites fabricated by phase separation were used as the anode in a flexible lithium-ion battery [192]. Intrinsically stretchable and rechargeable cells were fabricated by embedding chemically reactive pastes (zinc anode and manganese dioxide cathode) and electrolyte gels into an elastomer matrix, as shown in Figure 7e [83,84]. Although the abovementioned batteries could easily provide energy by electro-chemical process, they still needed to get recharged by physical contact with external power sources. Xu et al. demonstrated stretchable battery arrays with an integrated wireless recharging system. The wireless recharging system, consisting of wireless coil, Schottky diode, and parallel capacitor, was designed to recharge these batteries without direct physical contact [166].

Energy harvesters exploit environmental energy and generate power for the stretchable devices. Solar cells are one of the well-developed types of energy harvesters. To fabricate stretchable solar cells, various stretching mechanics were exploited to incorporate the stiff photovoltaic materials. Stretchable organic solar cells were fabricated by depositing a PEDOT:PSS electrode, organic photovoltaic materials, and EGaIn as the top contact, sequentially on a pre-stretched PDMS substrate. The 3D wavy buckles appeared in the film upon releasing and imparted the solar cells with reversible stretchability [176,193]. Stretchable dye-sensitized solar cells were fabricated using elastic conductive fiber by spinning MWCNT sheets on rubber fiber. Modified titanium wire was wound onto the fiber as the working electrode, followed by coating with photoactive materials. The devices had open-circuit voltages of 0.71 V and energy conversion efficiency of 7.13% [194]. Besides the solar cell, other energy harvesters that have potential but have not yet been well-studied in stretchable electronics include piezoelectric motion energy harvesters [195,196], RF energy harvesters [60,197,198], the thermal energy harvester [199], and triboelectric energy harvester [200].

6. Summary and Outlook

In this review, we went through the materials, mechanics, and patterning techniques of stretchable conductors and the applicable devices. Stretchable conductors are the basic building block of stretchable electronic devices. Their compliant, deformable virtues have changed the common idea that people would have on rigid silicon-based electronics and opened up a new direction of the next generation of electronics. They have enormous potential in biomedical engineering, which has been demonstrated in many research works [4,18,60,61,94,167,181,201]. Besides that, they have application in broad fields such as wearable communication devices [202], prosthetic electronic skins [203], soft robotics [204], and interactive human–machine interfaces [205]. We foresee a bright future for stretchable electronics as they bring electronics closer to humans.

Despite the huge advancements in stretchable electronics so far, there still exist some issues and challenges that should be solved in order to realize the use of stretchable electronics in everyday life. First, the long-term stable performance of these stretchable electronics, which would be influenced by the stability of conductors and the aging of substrates [206–208], has not yet been demonstrated comprehensively. Second, as many stretchable electronics are aimed at biomedical applications, their biocompatibility should be examined extensively. Although the biocompatibility of some materials has been demonstrated [209], however, thermal management of the stretchable devices, where the heat generated may cause tissue lesioning, are relatively less studied [210,211]. Third, the large mismatches in Young's modulus, elongation at break, and thermal expansion of various materials used in stretchable electronics have posed great challenges in integration with the systems. Several stretching mechanics have been proposed which effectively remedy these integration challenges. However, problems such as cracking and delamination are still reported when the system reaches a certain strain. More fundamental understanding of the causes of failure and interfacial properties between heterogeneous materials is needed in order to achieve better performance of novel stretchable devices. Fourth, some of the stretching mechanisms and structures are hard to implement in large-scale manufacturing, such as pre-strain-induced buckling and out-of-plane design. Efforts are still required to figure out cost-effective, scalable fabrication routes with high reliability, repeatability, and precision for the purpose of commercialization. For stretchable electronics based on nanomaterials, repeatability of the device's performance from batch to batch largely relies on the uniformity of the nanomaterials, which is also challenging in itself. Lastly, despite stretchable electronics with manifold functional devices having been demonstrated, integrated internal power supplies are still in great demand for self-powering stretchable electronic systems.

Based on the current trend, the possible future directions of stretchable electronics are foreseeable. First, the diversity of materials used would continuously expand. The rate of discovery and development of new materials are beyond our expectations. Take graphene as an example, it only took about five years from its first discovery [212] to its first utilization in stretchable conductors [118]. For sure, more and more organic, inorganic, conducting, semiconducting, and dielectric materials would provide opportunities in stretchable electronics. As for the substrate and matrix, besides the widely used PDMS, there are emerging elastomeric materials such as rubber fiber mats. Second, the design of discrete devices and their integration in stretchable systems would grow quickly. Stretchable electronics with sensing functions have gained considerable progresses. However, devices with other functions such as memory [213] and communication [202,214] are still in infancy. There is also an urgent demand for energy devices, and many research groups are working to address this issue. Third, with the increase of complexity and resolution of devices, higher requirements for patterning techniques are expected. Direct printing, as an additive manufacturing method, would satisfy such requirement and offer low cost and high speed in both prototyping and manufacturing. It might be a solution for cost-effective and scalable fabrication of stretchable electronics.

Acknowledgments: This research was supported financially by Interdisciplinary Intercampus Funding Program (IDIC) of University of Missouri System, University of Missouri Research Board (UMRB), Intelligent System Center (ISC) and Material Research Center (MRC) at Missouri University of Science and Technology.

Author Contributions: Xiaowei Yu, Bikram K. Mahajan and Wan Shou conducted the review. Xiaowei Yu wrote the manuscript. Bikram. K. Mahajan and Heng Pan revised and proofread the manuscript. Heng Pan guided on the organization and content.

References

1. Bowden, N.; Brittain, S.; Evans, A.G.; Hutchinson, J.W.; Whitesides, G.M. Spontaneous formation of ordered structures in thin films of metals supported on an elastomeric polymer. *Nature* **1998**, *393*, 146–149.
2. Huang, D.; Liao, F.; Molesa, S.; Redinger, D.; Subramanian, V. Plastic-compatible low resistance printable gold nanoparticle conductors for flexible electronics. *J. Electrochem. Soc.* **2003**, *150*, G412–G417. [CrossRef]
3. Nathan, A.; Ahnood, A.; Cole, M.T.; Lee, S.; Suzuki, Y.; Hiralal, P.; Bonaccorso, F.; Hasan, T.; Garcia-Gancedo, L.; Dyadyusha, A.; et al. Flexible electronics: The next ubiquitous platform. *Proc. IEEE* **2012**, *100*, 1486–1517. [CrossRef]
4. Rogers, J.A.; Someya, T.; Huang, Y. Materials and mechanics for stretchable electronics. *Science* **2010**, *327*, 1603–1607. [CrossRef] [PubMed]
5. Wagner, S.; Bauer, S. Materials for stretchable electronics. *MRS Bull.* **2012**, *37*, 207–213. [CrossRef]
6. Park, M.; Park, J.; Jeong, U. Design of conductive composite elastomers for stretchable electronics. *Nano Today* **2014**, *9*, 244–260. [CrossRef]
7. Cheng, T.; Zhang, Y.; Lai, W.-Y.; Huang, W. Stretchable thin-film electrodes for flexible electronics with high deformability and stretchability. *Adv. Mater.* **2015**, *27*, 3349–3376. [CrossRef] [PubMed]
8. Yao, S.; Zhu, Y. Nanomaterial-enabled stretchable conductors: Strategies, materials and devices. *Adv. Mater.* **2015**, *27*, 1480–1511. [CrossRef] [PubMed]
9. Wang, Y.; Li, Z.; Xiao, J. Stretchable thin film materials: Fabrication, application and mechanics. *J. Electron. Packag.* **2016**, *138*, 020801. [CrossRef]
10. McCoul, D.; Hu, W.; Gao, M.; Mehta, V.; Pei, Q. Recent advances in stretchable and transparent electronic materials. *Adv. Electron. Mater.* **2016**, *2*, 1500407. [CrossRef]
11. Harris, K.D.; Elias, A.L.; Chung, H.-J. Flexible electronics under strain: A review of mechanical characterization and durability enhancement strategies. *J. Mater. Sci.* **2016**, *51*, 2771–2805. [CrossRef]
12. Lacour, S.P.; Wagner, S.; Huang, Z.; Suo, Z. Stretchable gold conductors on elastomeric substrates. *Appl. Phys. Lett.* **2003**, *82*, 2404–2406. [CrossRef]
13. Gray, D.S.; Tien, J.; Chen, C.S. High-conductivity elastomeric electronics. *Adv. Mater.* **2004**, *16*, 393–397. [CrossRef]
14. Rabinowitz, J.; Fritz, G.; Kumar, P.; Lewis, P.; Miller, M.; Dineen, A.; Gray, C. Electrical characterization of traditional and aerosol jet printed conductors under tensile strain. *MRS Adv.* **2016**, *1*, 15–20. [CrossRef]
15. Yang, C.; Gu, H.; Lin, W.; Yuen, M.M.; Wong, C.P.; Xiong, M.; Gao, B. Silver nanowires: From scalable synthesis to recyclable foldable electronics. *Adv. Mater.* **2011**, *23*, 3052–3056. [CrossRef] [PubMed]
16. Park, S.; Vosguerichian, M.; Bao, Z. A review of fabrication and applications of carbon nanotube film-based flexible electronics. *Nanoscale* **2013**, *5*, 1727–1752. [CrossRef] [PubMed]
17. Pease, R.F.; Chou, S.Y. Lithography and other patterning techniques for future electronics. *Proc. IEEE* **2008**, *96*, 248–270. [CrossRef]
18. Kim, D.-H.; Lu, N.; Ma, R.; Kim, Y.-S.; Kim, R.-H.; Wang, S.; Wu, J.; Won, S.M.; Tao, H.; Islam, A.; et al. Epidermal electronics. *Science* **2011**, *333*, 838–843. [CrossRef] [PubMed]
19. Larmagnac, A.; Eggenberger, S.; Janossy, H.; Vörös, J. Stretchable electronics based on Ag-PDMS composites. *Sci. Rep.* **2014**, *4*, 7254. [CrossRef] [PubMed]
20. Yao, S.; Zhu, Y. Wearable multifunctional sensors using printed stretchable conductors made of silver nanowires. *Nanoscale* **2014**, *6*, 2345–2352. [CrossRef] [PubMed]
21. Song, L.; Myers, A.C.; Adams, J.J.; Zhu, Y. Stretchable and reversibly deformable radio frequency antennas based on silver nanowires. *ACS Appl. Mater. Interfaces* **2014**, *6*, 4248–4253. [CrossRef] [PubMed]
22. Lee, M.-T.; Lee, D.; Sherry, A.; Grigoropoulos, C.P. Rapid selective metal patterning on polydimethylsiloxane (PDMS) fabricated by capillarity-assisted laser direct write. *J. Micromech. Microeng.* **2011**, *21*, 095018. [CrossRef]

23. Sun, J.; Jiang, J.; Bao, B.; Wang, S.; He, M.; Zhang, X.; Song, Y. Fabrication of bendable circuits on a polydimethylsiloxane (PDMS) surface by inkjet printing semi-wrapped structures. *Materials* **2016**, *9*, 253. [CrossRef]
24. Chung, S.; Lee, J.; Song, H.; Kim, S.; Jeong, J.; Hong, Y. Inkjet-printed stretchable silver electrode on wave structured elastomeric substrate. *Appl. Phys. Lett.* **2011**, *98*, 153110. [CrossRef]
25. Wu, J.; Wang, R.; Yu, H.; Li, G.; Xu, K.; Tien, N.C.; Roberts, R.C.; Li, D. Inkjet-printed microelectrodes on PDMS as biosensors for functionalized microfluidic systems. *Lab Chip* **2015**, *15*, 690–695. [CrossRef] [PubMed]
26. Kim, Y.; Ren, X.; Kim, J.W.; Noh, H. Direct inkjet printing of micro-scale silver electrodes on polydimethylsiloxane (PDMS) microchip. *J. Micromech. Microeng.* **2014**, *24*, 115010. [CrossRef]
27. Lee, S.-K.; Kim, B.J.; Jang, H.; Yoon, S.C.; Lee, C.; Hong, B.H.; Rogers, J.A.; Cho, J.H.; Ahn, J.-H. Stretchable graphene transistors with printed dielectrics and gate electrodes. *Nano Lett.* **2011**, *11*, 4642–4646. [CrossRef] [PubMed]
28. Wang, K.; Chang, Y.-H.; Zhang, C.; Wang, B. Conductive-on-demand: Tailorable polyimide/carbon nanotube nanocomposite thin film by dual-material aerosol jet printing. *Carbon* **2016**, *98*, 397–403. [CrossRef]
29. Aga, R.; Jordan, C.; Aga, R.S.; Bartsch, C.M.; Heckman, E.M. Metal electrode work function modification using aerosol jet printing. *IEEE Electron Device Lett.* **2014**, *35*, 1124–1126. [CrossRef]
30. Zhao, D.; Liu, T.; Park, J.G.; Zhang, M.; Chen, J.-M.; Wang, B. Conductivity enhancement of aerosol-jet printed electronics by using silver nanoparticles ink with carbon nanotubes. *Microelectron. Eng.* **2012**, *96*, 71–75. [CrossRef]
31. Robinson, A.P.; Minev, I.; Graz, I.M.; Lacour, S.P. Microstructured silicone substrate for printable and stretchable metallic films. *Langmuir* **2011**, *27*, 4279–4284. [CrossRef] [PubMed]
32. Carlson, A.; Bowen, A.M.; Huang, Y.; Nuzzo, R.G.; Rogers, J.A. Transfer printing techniques for materials assembly and micro/nanodevice fabrication. *Adv. Mater.* **2012**, *24*, 5284–5318. [CrossRef] [PubMed]
33. Meitl, M.A.; Zhu, Z.-T.; Kumar, V.; Lee, K.J.; Feng, X.; Huang, Y.Y.; Adesida, I.; Nuzzo, R.G.; Rogers, J.A. Transfer printing by kinetic control of adhesion to an elastomeric stamp. *Nat. Mater.* **2006**, *5*, 33–38. [CrossRef]
34. Loo, Y.-L.; Willett, R.L.; Baldwin, K.W.; Rogers, J.A. Interfacial chemistries for nanoscale transfer printing. *J. Am. Chem. Soc.* **2002**, *124*, 7654–7655. [CrossRef] [PubMed]
35. Wang, S.; Li, M.; Wu, J.; Kim, D.-H.; Lu, N.; Su, Y.; Kang, Z.; Huang, Y.; Rogers, J.A. Mechanics of epidermal electronics. *J. Appl. Mech.* **2012**, *79*, 031022. [CrossRef]
36. Kim, S.J.; Cho, K.W.; Cho, H.R.; Wang, L.; Park, S.Y.; Lee, S.E.; Hyeon, T.; Lu, N.; Choi, S.H.; Kim, D.-H. Stretchable and transparent biointerface using cell-sheet-graphene hybrid for electrophysiology and therapy of skeletal muscle. *Adv. Funct. Mater.* **2016**, *26*, 3207–3217. [CrossRef]
37. Huang, X.; Leduc, C.; Ravussin, Y.; Li, S.; Davis, E.N.; Song, B.; Li, D.; Xu, K.; Accili, D.; Wang, Q.; et al. A differential dielectric affinity glucose sensor. *Lab Chip* **2014**, *14*, 294–301. [CrossRef] [PubMed]
38. Yan, C.; Lee, P.S. Stretchable energy storage and conversion devices. *Small* **2014**, *10*, 3443–3460. [CrossRef] [PubMed]
39. Song, J.; Wang, S. Theory for Stretchable Interconnects. In *Stretchable Electronics*; Someya, T., Ed.; Wiley-VCH Verlag GmbH & Co. KGaA: Weinheim, Germany, 2012; pp. 1–29.
40. Lee, P.; Lee, J.; Lee, H.; Yeo, J.; Hong, S.; Nam, K.H.; Lee, D.; Lee, S.S.; Ko, S.H. Highly stretchable and highly conductive metal electrode by very long metal nanowire percolation network. *Adv. Mater.* **2012**, *24*, 3326–3332. [CrossRef] [PubMed]
41. Yamada, T.; Hayamizu, Y.; Yamamoto, Y.; Yomogida, Y.; Izadi-Najafabadi, A.; Futaba, D.N.; Hata, K. A stretchable carbon nanotube strain sensor for human-motion detection. *Nat. Nanotechnol.* **2011**, *6*, 296–301. [CrossRef] [PubMed]
42. Xu, F.; Wang, X.; Zhu, Y.; Zhu, Y. Wavy ribbons of carbon nanotubes for stretchable conductors. *Adv. Funct. Mater.* **2012**, *22*, 1279–1283. [CrossRef]
43. Zhu, Y.; Xu, F. Buckling of aligned carbon nanotubes as stretchable conductors: A new manufacturing strategy. *Adv. Mater.* **2012**, *24*, 1073–1077. [CrossRef] [PubMed]
44. Cai, L.; Li, J.; Luan, P.; Dong, H.; Zhao, D.; Zhang, Q.; Zhang, X.; Tu, M.; Zeng, Q.; Zhou, W.; Xie, S. Highly transparent and conductive stretchable conductors based on hierarchical reticulate single-walled carbon nanotube architecture. *Adv. Funct. Mater.* **2012**, *22*, 5238–5244. [CrossRef]

45. Shin, U.H.; Jeong, D.-W.; Kim, S.-H.; Lee, H.W.; Kim, J.-M. Elastomer-infiltrated vertically aligned carbon nanotube film-based wavy-configured stretchable conductors. *ACS Appl. Mater. Interfaces* **2014**, *6*, 12909–12914. [CrossRef] [PubMed]
46. Wang, Y.; Yang, R.; Shi, Z.; Zhang, L.; Shi, D.; Wang, E.; Zhang, G. Super-elastic graphene ripples for flexible strain sensors. *ACS Nano* **2011**, *5*, 3645–3650. [CrossRef] [PubMed]
47. Wang, Y.; Wang, L.; Yang, T.; Li, X.; Zang, X.; Zhu, M.; Wang, K.; Wu, D.; Zhu, H. Wearable and highly sensitive graphene strain sensors for human motion monitoring. *Adv. Funct. Mater.* **2014**, *24*, 4666–4670. [CrossRef]
48. Huang, Y.Y.; Terentjev, E.M. Tailoring the electrical properties of carbon nanotube-polymer composites. *Adv. Funct. Mater.* **2010**, *20*, 4062–4068. [CrossRef]
49. Hansen, T.S.; West, K.; Hassager, O.; Larsen, N.B. Highly stretchable and conductive polymer material made from poly(3,4-ethylenedioxythiophene) and polyurethane elastomers. *Adv. Funct. Mater.* **2007**, *17*, 3069–3073. [CrossRef]
50. Niu, X.; Peng, S.; Liu, L.; Wen, W.; Sheng, P. Characterizing and patterning of PDMS-based conducting composites. *Adv. Mater.* **2007**, *19*, 2682–2686. [CrossRef]
51. Bandodkar, A.J.; Nuñez-Flores, R.; Jia, W.; Wang, J. All-printed stretchable electrochemical devices. *Adv. Mater.* **2015**, *27*, 3060–3065. [CrossRef] [PubMed]
52. Kubo, M.; Li, X.; Kim, C.; Hashimoto, M.; Wiley, B.J.; Ham, D.; Whitesides, G.M. Stretchable microfluidic radiofrequency antennas. *Adv. Mater.* **2010**, *22*, 2749–2752. [CrossRef] [PubMed]
53. Yoon, J.; Hong, S.Y.; Lim, Y.; Lee, S.-J.; Zi, G.; Ha, J.S. Design and fabrication of novel stretchable device arrays on a deformable polymer substrate with embedded liquid-metal interconnections. *Adv. Mater.* **2014**, *26*, 6580–6586. [CrossRef] [PubMed]
54. Lee, J.; Wu, J.; Shi, M.; Yoon, J.; Park, S.-I.; Li, M.; Liu, Z.; Huang, Y.; Rogers, J.A. Stretchable GaAs photovoltaics with designs that enable high areal coverage. *Adv. Mater.* **2011**, *23*, 986–991. [CrossRef] [PubMed]
55. Vohra, A.; Filiatrault, H.L.; Amyotte, S.D.; Carmichael, R.S.; Suhan, N.D.; Siegers, C.; Ferrari, L.; Davidson, G.J.E.; Carmichael, T.B. Reinventing butyl rubber for stretchable electronics. *Adv. Funct. Mater.* **2016**, *26*, 5222–5229. [CrossRef]
56. Huang, X.; Liu, Y.; Kong, G.W.; Seo, J.H.; Ma, Y.; Jang, K.-I.; Fan, J.A.; Mao, S.; Chen, Q.; Li, D.; et al. Epidermal radio frequency electronics for wireless power transfer. *Microsyst. Nanoeng.* **2016**, *2*, 16052. [CrossRef]
57. Li, X.; Gu, T.; Wei, B. Dynamic and galvanic stability of stretchable supercapacitors. *Nano Lett.* **2012**, *12*, 6366–6371. [CrossRef] [PubMed]
58. Yu, C.; Masarapu, C.; Rong, J.; Wei, B.; Jiang, H. Stretchable supercapacitors based on buckled single-walled carbon nanotube macrofilms. *Adv. Mater.* **2009**, *21*, 4793–4797. [CrossRef] [PubMed]
59. Hu, W.; Niu, X.; Zhao, R.; Pei, Q. Elastomeric transparent capacitive sensors based on an interpenetrating composite of silver nanowires and polyurethane. *Appl. Phys. Lett.* **2013**, *102*, 083303. [CrossRef]
60. Kim, D.-H.; Ghaffari, R.; Lu, N.; Rogers, J.A. Flexible and stretchable electronics for biointegrated devices. *Annu. Rev. Biomed. Eng.* **2012**, *14*, 113–128. [CrossRef] [PubMed]
61. Zhang, Y. Mechanics and designs of stretchable bioelectronics. In *Stretchable Bioelectronics for Medical Devices and Systems*; Rogers, J.A., Ghaffari, R., Kim, D.-H., Eds.; Springer: Zug, Switzerland, 2016; pp. 53–68.
62. Lee, J.N.; Park, C.; Whitesides, G.M. Solvent compatibility of poly(dimethylsiloxane)-based microfluidic devices. *Anal. Chem.* **2003**, *75*, 6544–6554. [CrossRef] [PubMed]
63. Fortunato, G.; Pecora, A.; Maiolo, L. Polysilicon thin-film transistors on polymer substrates. *Mater. Sci. Semicond. Process.* **2012**, *15*, 627–641. [CrossRef]
64. Dangla, R.; Gallaire, F.; Baroud, C.N. Microchannel deformations due to solvent-induced PDMS swelling. *Lab Chip* **2010**, *10*, 2972–2978. [CrossRef] [PubMed]
65. Chan, E.P.; Crosby, A.J. Fabricating microlens arrays by surface wrinkling. *Adv. Mater.* **2006**, *18*, 3238–3242. [CrossRef]
66. Chan, E.P.; Crosby, A.J. Spontaneous formation of stable aligned wrinkling patterns. *Soft Matter* **2006**, *2*, 324–328. [CrossRef]
67. Mata, A.; Fleischman, A.J.; Roy, S. Characterization of polydimethylsiloxane (PDMS) properties for biomedical micro/nanosystems. *Biomed. Microdevices* **2005**, *7*, 281–293. [CrossRef] [PubMed]

68. Bodas, D.; Khan-Malek, C. Hydrophilization and hydrophobic recovery of PDMS by oxygen plasma and chemical treatment-An SEM investigation. *Sens. Actuators B Chem.* **2007**, *123*, 368–373. [CrossRef]
69. Wu, J.; Roberts, R.C.; Tien, N.C.; Li, D. Inkjet Printed Silver Patterning on PDMS to Fabricate Microelectrodes for Microfluidic Sensing. In Proceedings of the 2014 IEEE Sensors, Valencia, Spain, 2–5 November 2014; pp. 1100–1103.
70. Efimenko, K.; Rackaitis, M.; Manias, E.; Vaziri, A.; Mahadevan, L.; Genzer, J. Nested self-similar wrinkling patterns in skins. *Nat. Mater.* **2005**, *4*, 293–297. [CrossRef] [PubMed]
71. Lipomi, D.J.; Lee, J.A.; Vosgueritchian, M.; Tee, B.C.-K.; Bolander, J.A.; Bao, Z. Electronic properties of transparent conductive films of PEDOT:PSS on stretchable substrates. *Chem. Mater.* **2012**, *24*, 373–382. [CrossRef]
72. Huck, W.T.S.; Bowden, N.; Onck, P.; Pardoen, T.; Hutchinson, J.W.; Whitesides, G.M. Ordering of spontaneously formed buckles on planar surfaces. *Langmuir* **2000**, *16*, 3497–3501. [CrossRef]
73. Xia, Y.; Whitesides, G.M. Soft lithography. *Annu. Rev. Mater. Sci.* **1998**, *28*, 153–184. [CrossRef]
74. Lee, P.; Ham, J.; Lee, J.; Hong, S.; Han, S.; Suh, Y.D.; Lee, S.E.; Yeo, J.; Lee, S.S.; Lee, D.; et al. Highly stretchable or transparent conductor fabrication by a hierarchical multiscale hybrid nanocomposite. *Adv. Funct. Mater.* **2014**, *24*, 5671–5678. [CrossRef]
75. Woo, J.Y.; Kim, K.K.; Lee, J.; Kim, J.T.; Han, C.-S. Highly conductive and stretchable Ag nanowire/carbon nanotube hybrid conductors. *Nanotechnology* **2014**, *25*, 285203. [CrossRef] [PubMed]
76. Shin, M.K.; Oh, J.; Lima, M.; Kozlov, M.E.; Kim, S.J.; Baughman, R.H. Elastomeric conductive composites based on carbon nanotube forests. *Adv. Mater.* **2010**, *22*, 2663–2667. [CrossRef] [PubMed]
77. Engel, J.; Chen, J.; Chen, N.; Pandya, S.; Liu, C. Multi-walled carbon nanotube filled conductive elastomers: Materials and application to micro transducers. In Proceedings of the 19th IEEE International Conference on Micro Electro Mechanical Systems, Istanbul, Turkey, 22–26 January 2006; pp. 246–249.
78. Ma, Z.; Su, B.; Gong, S.; Wang, Y.; Yap, L.W.; Simon, G.P.; Cheng, W. Liquid-wetting-solid strategy to fabricate stretchable sensors for human-motion detection. *ACS Sens.* **2016**, *1*, 303–311. [CrossRef]
79. Park, M.; Im, J.; Shin, M.; Min, Y.; Park, J.; Cho, H.; Park, S.; Shim, M.-B.; Jeon, S.; Chung, D.-Y.; et al. Highly stretchable electric circuits from a composite material of silver nanoparticles and elastomeric fibres. *Nat. Nanotechnol.* **2012**, *7*, 803–809. [CrossRef] [PubMed]
80. Cai, L.; Song, L.; Luan, P.; Zhang, Q.; Zhang, N.; Gao, Q.; Zhao, D.; Zhang, X.; Tu, M.; Yang, F.; et al. Super-stretchable, transparent carbon nanotube-based capacitive strain sensors for human motion detection. *Sci. Rep.* **2013**, *3*, 3048. [CrossRef] [PubMed]
81. Huang, X.; Cheng, H.; Chen, K.; Zhang, Y.; Zhang, Y.; Liu, Y.; Zhu, C.; Ouyang, S.-C.; Kong, G.-W.; Yu, C.; et al. Epidermal impedance sensing sheets for precision hydration assessment and spatial mapping. *IEEE Trans. Biomed. Eng.* **2013**, *60*, 2848–2857. [CrossRef] [PubMed]
82. Hu, W.; Niu, X.; Li, L.; Yun, S.; Yu, Z.; Pei, Q. Intrinsically stretchable transparent electrodes based on silver-nanowire–crosslinked-polyacrylate composites. *Nanotechnology* **2012**, *23*, 344002. [CrossRef] [PubMed]
83. Kettlgruber, G.; Kaltenbrunner, M.; Siket, C.M.; Moser, R.; Graz, I.M.; Schwödiauer, R.; Bauer, S. Intrinsically stretchable and rechargeable batteries for self-powered stretchable electronics. *J. Mater. Chem. A* **2013**, *1*, 5505–5508. [CrossRef]
84. Kaltenbrunner, M.; Kettlgruber, G.; Siket, C.; Schwödiauer, R.; Bauer, S. Arrays of ultracompliant electrochemical dry gel cells for stretchable electronics. *Adv. Mater.* **2010**, *22*, 2065–2067. [CrossRef] [PubMed]
85. Wu, J.; Zang, J.; Rathmell, A.R.; Zhao, X.; Wiley, B.J. Reversible sliding in networks of nanowires. *Nano Lett.* **2013**, *13*, 2381–2386. [CrossRef] [PubMed]
86. Chun, K.-Y.; Oh, Y.; Rho, J.; Ahn, J.-H.; Kim, Y.-J.; Choi, H.R.; Baik, S. Highly conductive, printable and stretchable composite films of carbon nanotubes and silver. *Nat. Nanotechnol.* **2010**, *5*, 853–857. [CrossRef] [PubMed]
87. Ecoflex® Series—Smooth-On. Available online: https://www.smooth-on.com/tb/files/ECOFLEX_SERIES_TB.pdf (accessed on 10 October 2016).
88. Dragon Skin® Series—Smooth-On. Available online: https://www.smooth-on.com/tb/files/DRAGON_SKIN_SERIES_TB.pdf (accessed on 10 October 2016).
89. Solaris®—Smooth-On. Available online: https://www.smooth-on.com/tb/files/Solaris_TB.pdf (accessed on 10 October 2016).

90. VHBTM Tape Specialty Tapes—3M.com. Available online: http://multimedia.3m.com/mws/media/986695O/3m-vhb-tape-specialty-tapes.pdf (accessed on 10 October 2016).
91. Nitrile Butadiene Rubber (NBR)—Romac. Available online: http://www.romac.com/Submittals/RUBBER/NBR-SUB.pdf (accessed on 10 October 2016).
92. Lacour, S.P.; Jones, J.; Suo, Z.; Wagner, S. Design and performance of thin metal film interconnects for skin-like electronic circuits. *IEEE Electron Device Lett.* **2004**, *25*, 179–181. [CrossRef]
93. Lacour, S.P.; Jones, J.; Wagner, S.; Li, T.; Suo, Z. Stretchable interconnects for elastic electronic surfaces. *Proc. IEEE* **2005**, *93*, 1459–1467. [CrossRef]
94. Yeo, W.-H.; Kim, Y.-S.; Lee, J.; Ameen, A.; Shi, L.; Li, M.; Wang, S.; Ma, R.; Jin, S.H.; Kang, Z.; et al. Multifunctional epidermal electronics printed directly onto the skin. *Adv. Mater.* **2013**, *25*, 2773–2778. [CrossRef] [PubMed]
95. Kim, D.-H.; Ahn, J.-H.; Choi, W.M.; Kim, H.-S.; Kim, T.-H.; Song, J.; Huang, Y.Y.; Liu, Z.; Lu, C.; Rogers, J.A. Stretchable and foldable silicon integrated circuits. *Science* **2008**, *320*, 507–511. [CrossRef] [PubMed]
96. He, J.; Nuzzo, R.G.; Rogers, J.A. Inorganic materials and assembly techniques for flexible and stretchable electronics. *Proc. IEEE* **2015**, *103*, 619–632. [CrossRef]
97. Leenen, M.A.M.; Arning, V.; Thiem, H.; Steiger, J.; Anselmann, R. Printable electronics: Flexibility for the future. *Phys. Status Solidi A* **2009**, *206*, 588–597. [CrossRef]
98. Perelaer, J.; Smith, P.J.; Mager, D.; Soltman, D.; Volkman, S.K.; Subramanian, V.; Korvink, J.G.; Schubert, U.S. Printed electronics: The challenges involved in printing devices, interconnects, and contacts based on inorganic materials. *J. Mater. Chem.* **2010**, *20*, 8446–8453. [CrossRef]
99. Lee, K.J.; Jun, B.H.; Kim, T.H.; Joung, J. Direct synthesis and inkjetting of silver nanocrystals toward printed electronics. *Nanotechnology* **2006**, *17*, 2424–2428. [CrossRef]
100. Van der Sluis, O.; Hsu, Y.Y.; Timmermans, P.H.M.; Gonzalez, M.; Hoefnagels, J.P.M. Stretching-induced interconnect delamination in stretchable electronic circuits. *J. Phys. D Appl. Phys.* **2011**, *44*, 34008. [CrossRef]
101. Gonzalez, M.; Axisa, F.; Bulcke, M.V.; Brosteaux, D.; Vandevelde, B.; Vanfleteren, J. Design of metal interconnects for stretchable electronic circuits. *Microelectron. Reliab.* **2008**, *48*, 825–832. [CrossRef]
102. Fan, J.A.; Yeo, W.-H.; Su, Y.; Hattori, Y.; Lee, W.; Jung, S.-Y.; Zhang, Y.; Liu, Z.; Cheng, H.; Falgout, L.; et al. Fractal design concepts for stretchable electronics. *Nat. Commun.* **2014**, *5*, 3266. [CrossRef] [PubMed]
103. Hsu, Y.-Y.; Gonzalez, M.; Bossuyt, F.; Vanfleteren, J.; de Wolf, I. Polyimide-enhanced stretchable interconnects: Design, fabrication, and Characterization. *IEEE Trans. Electron Devices* **2011**, *58*, 2680–2688. [CrossRef]
104. Xu, F.; Zhu, Y. Highly conductive and stretchable silver nanowire conductors. *Adv. Mater.* **2012**, *24*, 5117–5122. [CrossRef] [PubMed]
105. Han, S.; Hong, S.; Ham, J.; Yeo, J.; Lee, J.; Kang, B.; Lee, P.; Kwon, J.; Lee, S.S.; Yang, M.-Y.; et al. Fast plasmonic laser nanowelding for a Cu-nanowire percolation network for flexible transparent conductors and stretchable electronics. *Adv. Mater.* **2014**, *26*, 5808–5814. [CrossRef] [PubMed]
106. Chen, Y.; Ouyang, Z.; Gu, M.; Cheng, W. Mechanically strong, optically transparent, giant metal superlattice nanomembranes from ultrathin gold nanowires. *Adv. Mater.* **2013**, *25*, 80–85. [CrossRef] [PubMed]
107. Gong, S.; Schwalb, W.; Wang, Y.; Chen, Y.; Tang, Y.; Si, J.; Shirinzadeh, B.; Cheng, W. A wearable and highly sensitive pressure sensor with ultrathin gold nanowires. *Nat. Commun.* **2014**, *5*, 3132. [CrossRef] [PubMed]
108. Langley, D.; Giusti, G.; Mayousse, C.; Celle, C.; Bellet, D.; Simonato, J.-P. Flexible transparent conductive materials based on silver nanowire networks: A review. *Nanotechnology* **2013**, *24*, 452001. [CrossRef] [PubMed]
109. Hu, L.; Kim, H.S.; Lee, J.-Y.; Peumans, P.; Cui, Y. Scalable coating and properties of transparent, flexible, silver nanowire electrodes. *ACS Nano* **2010**, *4*, 2955–2963. [CrossRef] [PubMed]
110. De, S.; Higgins, T.M.; Lyons, P.E.; Doherty, E.M.; Nirmalraj, P.N.; Blau, W.J.; Boland, J.J.; Coleman, J.N. Silver nanowire networks as flexible, transparent, conducting films: Extremely high DC to optical conductivity ratios. *ACS Nano* **2009**, *3*, 1767–1774. [CrossRef] [PubMed]
111. Amjadi, M.; Pichitpajongkit, A.; Lee, S.; Ryu, S.; Park, I. Highly stretchable and sensitive strain sensor based on silver nanowire-elastomer nanocomposite. *ACS Nano* **2014**, *8*, 5154–5163. [CrossRef] [PubMed]
112. Akter, T.; Kim, W.S. Reversibly stretchable transparent conductive coatings of spray-deposited silver nanowires. *ACS Appl. Mater. Interfaces* **2012**, *4*, 1855–1859. [CrossRef] [PubMed]
113. Huang, G.-W.; Xiao, H.-M.; Fu, S.-Y. Wearable electronics of silver-nanowire/poly(dimethylsiloxane) nanocomposite for smart clothing. *Sci. Rep.* **2015**, *5*, 13971. [CrossRef] [PubMed]

114. Madaria, A.R.; Kumar, A.; Ishikawa, F.N.; Zhou, C. Uniform, highly conductive, and patterned transparent films of a percolating silver nanowire network on rigid and flexible substrates using a dry transfer technique. *Nano Res.* **2010**, *3*, 564–573. [CrossRef]
115. Liang, J.; Li, L.; Tong, K.; Ren, Z.; Hu, W.; Niu, X.; Chen, Y.; Pei, Q. Silver nanowire percolation network soldered with graphene oxide at room temperature and its application for fully stretchable polymer light-emitting diodes. *ACS Nano* **2014**, *8*, 1590–1600. [CrossRef] [PubMed]
116. Chen, S.-P.; Liao, Y.-C. Highly stretchable and conductive silver nanowire thin films formed by soldering nanomesh junctions. *Phys. Chem. Chem. Phys.* **2014**, *16*, 19856–19860. [CrossRef] [PubMed]
117. Tamai, T. Electrical properties of conductive elastomer as electrical contact material. *IEEE Trans. Compon. Hybrids Manuf. Technol.* **1982**, *5*, 56–61. [CrossRef]
118. Kim, K.S.; Zhao, Y.; Jang, H.; Lee, S.Y.; Kim, J.M.; Kim, K.S.; Ahn, J.-H.; Kim, P.; Choi, J.-Y.; Hong, B.H. Large-scale pattern growth of graphene films for stretchable transparent electrodes. *Nature* **2009**, *457*, 706–710. [CrossRef] [PubMed]
119. Rwei, S.-P.; Ku, F.-H.; Cheng, K.-C. Dispersion of carbon black in a continuous phase: Electrical, rheological, and morphological studies. *Colloid Polym. Sci.* **2002**, *280*, 1110–1115.
120. Kujawski, M.; Pearse, J.D.; Smela, E. Elastomers filled with exfoliated graphite as compliant electrodes. *Carbon* **2010**, *48*, 2409–2417. [CrossRef]
121. Sekitani, T.; Noguchi, Y.; Hata, K.; Fukushima, T.; Aida, T.; Someya, T. A rubberlike stretchable active matrix using elastic conductors. *Science* **2008**, *321*, 1468–1472. [CrossRef] [PubMed]
122. Kim, K.H.; Vural, M.; Islam, M.F. Single-walled carbon nanotube aerogel-based elastic conductors. *Adv. Mater.* **2011**, *23*, 2865–2869. [CrossRef] [PubMed]
123. Zhou, Y.; Bayram, Y.; Du, F.; Dai, L.; Volakis, J.L. Polymer-carbon nanotube sheets for conformal load bearing antennas. *IEEE Trans. Antennas Propag.* **2010**, *58*, 2169–2175. [CrossRef]
124. Zhang, Y.; Sheehan, C.J.; Zhai, J.; Zou, G.; Luo, H.; Xiong, J.; Zhu, Y.T.; Jia, Q.X. Polymer-embedded carbon nanotube ribbons for stretchable conductors. *Adv. Mater.* **2010**, *22*, 3027–3031. [CrossRef] [PubMed]
125. Liu, K.; Sun, Y.; Liu, P.; Lin, X.; Fan, S.; Jiang, K. Cross-stacked superaligned carbon nanotube films for transparent and stretchable conductors. *Adv. Funct. Mater.* **2011**, *21*, 2721–2728. [CrossRef]
126. Shang, Y.; He, X.; Li, Y.; Zhang, L.; Li, Z.; Ji, C.; Shi, E.; Li, P.; Zhu, K.; Peng, Q.; et al. Super-stretchable spring-like carbon nanotube ropes. *Adv. Mater.* **2012**, *24*, 2896–2900. [CrossRef] [PubMed]
127. Li, X.; Cai, W.; An, J.; Kim, S.; Nah, J.; Yang, D.; Piner, R.; Velamakanni, A.; Jung, I.; Tutuc, E.; et al. Large-area synthesis of high quality and uniform graphene films on copper foils. *Science* **2009**, *324*, 1312–1314. [CrossRef] [PubMed]
128. Qi, D.; Liu, Z.; Liu, Y.; Leow, W.R.; Zhu, B.; Yang, H.; Yu, J.; Wang, W.; Wang, H.; Yin, S.; et al. Suspended wavy graphene microribbons for highly stretchable microsupercapacitors. *Adv. Mater.* **2015**, *27*, 5559–5566. [CrossRef] [PubMed]
129. Kim, R.-H.; Bae, M.-H.; Kim, D.G.; Cheng, H.; Kim, B.H.; Kim, D.-H.; Li, M.; Wu, J.; Du, F.; Kim, H.-S.; et al. Stretchable, transparent graphene interconnects for arrays of microscale inorganic light emitting diodes on rubber substrates. *Nano Lett.* **2011**, *11*, 3881–3886. [CrossRef] [PubMed]
130. Sepúlveda, A.T.; Guzman de Villoria, R.; Viana, J.C.; Pontes, A.J.; Wardle, B.L.; Rocha, L.A. Full elastic constitutive relation of non-isotropic aligned-CNT/PDMS flexible nanocomposites. *Nanoscale* **2013**, *5*, 4847–4854. [CrossRef] [PubMed]
131. Zhang, Y.; Ronning, F.; Gofryk, K.; Mara, N.A.; Haberkorn, N.; Zou, G.; Wang, H.; Lee, J.H.; Bauer, E.; McCleskey, T.M.; et al. Aligned carbon nanotubes sandwiched in epitaxial NbC film for enhanced superconductivity. *Nanoscale* **2012**, *4*, 2268–2271. [CrossRef] [PubMed]
132. Yun, J.; Lim, Y.; Jang, G.N.; Kim, D.; Lee, S.-J.; Park, H.; Hong, S.Y.; Lee, G.; Zi, G.; Ha, J.S. Stretchable patterned graphene gas sensor driven by integrated micro-supercapacitor array. *Nano Energy* **2016**, *19*, 401–414. [CrossRef]
133. Watanabe, M.; Shirai, H.; Hirai, T. Wrinkled polypyrrole electrode for electroactive polymer actuators. *J. Appl. Phys.* **2002**, *92*, 4631–4637. [CrossRef]
134. Wang, C.; Zheng, W.; Yue, Z.; Too, C.O.; Wallace, G.G. Buckled, stretchable polypyrrole electrodes for battery applications. *Adv. Mater.* **2011**, *23*, 3580–3584. [CrossRef] [PubMed]
135. Zeng, W.; Shu, L.; Li, Q.; Chen, S.; Wang, F.; Tao, X.-M. Fiber-based wearable electronics: A review of materials, fabrication, devices, and applications. *Adv. Mater.* **2014**, *26*, 5310–5336. [CrossRef] [PubMed]

136. Noh, J.-S. Conductive elastomers for stretchable electronics, sensors and energy harvesters. *Polymers* **2016**, *8*, 123. [CrossRef]
137. Savagatrup, S.; Chan, E.; Renteria-Garcia, S.M.; Printz, A.D.; Zaretski, A.V.; O'Connor, T.F.; Rodriquez, D.; Valle, E.; Lipomi, D.J. Plasticization of PEDOT:PSS by common additives for mechanically robust organic solar cells and wearable sensors. *Adv. Funct. Mater.* **2015**, *25*, 427–436. [CrossRef]
138. Lee, Y.-Y.; Kang, H.-Y.; Gwon, S.H.; Choi, G.M.; Lim, S.-M.; Sun, J.-Y.; Joo, Y.-C. A Strain-Insensitive Stretchable Electronic Conductor: PEDOT:PSS/Acrylamide Organogels. *Adv. Mater.* **2016**, *28*, 1636–1643. [CrossRef] [PubMed]
139. Ding, H.; Zhong, M.; Wu, H.; Park, S.; Mohin, J.W.; Klosterman, L.; Yang, Z.; Yang, H.; Matyjaszewski, K.; Bettinger, C.J. Elastomeric Conducting Polyaniline Formed Through Topological Control of Molecular Templates. *ACS Nano* **2016**, *10*, 5991–5998. [CrossRef] [PubMed]
140. Palleau, E.; Reece, S.; Desai, S.C.; Smith, M.E.; Dickey, M.D. Self-healing stretchable wires for reconfigurable circuit wiring and 3D microfluidics. *Adv. Mater.* **2013**, *25*, 1589–1592. [CrossRef] [PubMed]
141. Dickey, M.D.; Chiechi, R.C.; Larsen, R.J.; Weiss, E.A.; Weitz, D.A.; Whitesides, G.M. Eutectic gallium-indium (EGaIn): A liquid metal alloy for the formation of stable structures in microchannels at room temperature. *Adv. Funct. Mater.* **2008**, *18*, 1097–1104. [CrossRef]
142. Tabatabai, A.; Fassler, A.; Usiak, C.; Majidi, C. Liquid-phase gallium-indium alloy electronics with microcontact printing. *Langmuir* **2013**, *29*, 6194–6200. [CrossRef] [PubMed]
143. Park, C.W.; Moon, Y.G.; Seong, H.; Jung, S.W.; Oh, J.-Y.; Na, B.S.; Park, N.-M.; Lee, S.S.; Im, S.G.; Koo, J.B. Photolithography-based patterning of liquid metal interconnects for monolithically integrated stretchable circuits. *ACS Appl. Mater. Interfaces* **2016**, *8*, 15459–15465. [CrossRef] [PubMed]
144. Kim, H.J.; Son, C.; Ziaie, B. A multiaxial stretchable interconnect using liquid-alloy-filled elastomeric microchannels. *Appl. Phys. Lett.* **2008**, *92*, 011904. [CrossRef]
145. Armand, M.; Endres, F.; MacFarlane, D.R.; Ohno, H.; Scrosati, B. Ionic-liquid materials for the electrochemical challenges of the future. *Nat. Mater.* **2009**, *8*, 621–629. [CrossRef] [PubMed]
146. Cong, H.; Pan, T. Photopatternable conductive PDMS materials for microfabrication. *Adv. Funct. Mater.* **2008**, *18*, 1912–1921. [CrossRef]
147. Hu, M.; Cai, X.; Guo, Q.; Bian, B.; Zhang, T.; Yang, J. Direct pen writing of adhesive particle-free ultrahigh silver salt-loaded composite ink for stretchable circuits. *ACS Nano* **2016**, *10*, 396–404. [CrossRef] [PubMed]
148. Duffy, D.C.; McDonald, J.C.; Schueller, O.J.A.; Whitesides, G.M. Rapid prototyping of microfluidic systems in poly(dimethylsiloxane). *Anal. Chem.* **1998**, *70*, 4974–4984. [CrossRef] [PubMed]
149. Siegel, A.C.; Bruzewicz, D.A.; Weibel, D.B.; Whitesides, G.M. Microsolidics: Fabrication of three-dimensional metallic microstructures in poly(dimethylsiloxane). *Adv. Mater.* **2007**, *19*, 727–733. [CrossRef]
150. Kim, H.-J.; Maleki, T.; Wei, P.; Ziaie, B. A biaxial stretchable interconnect with liquid-alloy-covered joints on elastomeric substrate. *J. Microelectromech. Syst.* **2009**, *18*, 138–146.
151. Vohra, A.; Imin, P.; Imit, M.; Carmichael, R.S.; Meena, J.S.; Adronov, A.; Carmichael, T.B. Transparent, stretchable, and conductive SWNT films using supramolecular functionalization and layer-by-layer self-assembly. *RSC Adv.* **2016**, *6*, 29254–29263. [CrossRef]
152. Someya, T.; Kato, Y.; Sekitani, T.; Iba, S.; Noguchi, Y.; Murase, Y.; Kawaguchi, H.; Sakurai, T. Conformable, flexible, large-area networks of pressure and thermal sensors with organic transistor active matrixes. *Proc. Natl. Acad. Sci. USA* **2005**, *102*, 12321–12325. [CrossRef] [PubMed]
153. Takahashi, T.; Takei, K.; Gillies, A.G.; Fearing, R.S.; Javey, A. Carbon nanotube active-matrix backplanes for conformal electronics and sensors. *Nano Lett.* **2011**, *11*, 5408–5413. [CrossRef] [PubMed]
154. Zhang, Y.; Wang, S.; Li, X.; Fan, J.A.; Xu, S.; Song, Y.M.; Choi, K.-J.; Yeo, W.-H.; Lee, W.; Nazaar, S.N.; et al. Experimental and theoretical studies of serpentine microstructures bonded to prestrained elastomers for stretchable electronics. *Adv. Funct. Mater.* **2014**, *24*, 2028–2037. [CrossRef]
155. Hsu, Y.-Y.; Gonzalez, M.; Bossuyt, F.; Axisa, F.; Vanfleteren, J.; de Wolf, I. The effects of encapsulation on deformation behavior and failure mechanisms of stretchable interconnects. *Thin Solid Films* **2011**, *519*, 2225–2234. [CrossRef]
156. Zhang, Y.; Xu, S.; Fu, H.; Lee, J.; Su, J.; Hwang, K.-C.; Rogers, J.A.; Huang, Y. Buckling in serpentine microstructures and applications in elastomer-supported ultra-stretchable electronics with high areal coverage. *Soft Matter* **2013**, *9*, 8062–8070. [CrossRef] [PubMed]

157. Khang, D.-Y.; Jiang, H.; Huang, Y.; Rogers, J.A. A stretchable form of single-crystal silicon for high-performance electronics on rubber substrates. *Science* **2006**, *311*, 208–212. [CrossRef] [PubMed]
158. Song, J.; Jiang, H.; Choi, W.M.; Khang, D.Y.; Huang, Y.; Rogers, J.A. An analytical study of two-dimensional buckling of thin films on compliant substrates. *J. Appl. Phys.* **2008**, *103*, 014303. [CrossRef]
159. Khang, D.-Y.; Rogers, J.A.; Lee, H.H. Mechanical buckling: Mechanics, metrology, and stretchable electronics. *Adv. Funct. Mater.* **2008**, *19*, 1526–1536. [CrossRef]
160. Kim, D.-H.; Song, J.; Choi, W.M.; Kim, H.-S.; Kim, R.-H.; Liu, Z.; Huang, Y.Y.; Hwang, K.-C.; Zhang, Y.; Rogers, J.A. Materials and noncoplanar mesh designs for integrated circuits with linear elastic responses to extreme mechanical deformations. *Proc. Natl. Acad. Sci. USA* **2008**, *105*, 18675–18680. [CrossRef] [PubMed]
161. Xiang, Y.; Li, T.; Suo, Z.; Vlassak, J.J. High ductility of a metal film adherent on a polymer substrate. *Appl. Phys. Lett.* **2005**, *87*, 161910. [CrossRef]
162. Hung, P.J.; Jeong, K.; Liu, G.L.; Lee, L.P. Microfabricated suspensions for electrical connections on the tunable elastomer membrane. *Appl. Phys. Lett.* **2004**, *85*, 6051–6053. [CrossRef]
163. Hyun, D.C.; Park, M.; Park, C.; Kim, B.; Xia, Y.; Hur, J.H.; Kim, J.M.; Park, J.J.; Jeong, U. Ordered zigzag stripes of polymer gel/metal nanoparticle composites for highly stretchable conductive electrodes. *Adv. Mater.* **2011**, *23*, 2946–2950. [CrossRef] [PubMed]
164. Zhu, Y.; Moran-Mirabal, J. Highly bendable and stretchable electrodes based on micro/nanostructured gold films for flexible sensors and electronics. *Adv. Electron. Mater.* **2016**, *2*, 1500345. [CrossRef]
165. Lee, K.J.; Fosser, K.A.; Nuzzo, R.G. Fabrication of stable metallic patterns embedded in poly (dimethylsiloxane) and model applications in non-planar electronic and lab-on-a-chip device patterning. *Adv. Funct. Mater.* **2005**, *15*, 557–566. [CrossRef]
166. Xu, S.; Zhang, Y.; Cho, J.; Lee, J.; Huang, X.; Jia, L.; Fan, J.A.; Su, Y.; Su, J.; Zhang, H.; et al. Stretchable batteries with self-similar serpentine interconnects and integrated wireless recharging systems. *Nat. Commun.* **2013**, *4*, 1543. [CrossRef] [PubMed]
167. Huang, X.; Liu, Y.; Cheng, H.; Shin, W.-J.; Fan, J.A.; Liu, Z.; Lu, C.-J.; Kong, G.-W.; Chen, K.; Patnaik, D.; et al. Materials and designs for wireless epidermal sensors of hydration and strain. *Adv. Funct. Mater.* **2014**, *24*, 3846–3854. [CrossRef]
168. Kramer, R.K.; Majidi, C.; Wood, R.J. Masked deposition of gallium-indium alloys for liquid-embedded elastomer conductors. *Adv. Funct. Mater.* **2013**, *23*, 5292–5296. [CrossRef]
169. Skylar-Scott, M.A.; Gunasekaran, S.; Lewis, J.A. Laser-assisted direct ink writing of planar and 3D metal architectures. *Proc. Natl. Acad. Sci. USA* **2016**, *113*, 6137–6142. [CrossRef] [PubMed]
170. Seifert, T.; Sowade, E.; Roscher, F.; Wiemer, M.; Gessner, T.; Baumann, R.R. Additive manufacturing technologies compared: Morphology of deposits of silver ink using inkjet and aerosol jet printing. *Ind. Eng. Chem. Res.* **2015**, *54*, 769–779. [CrossRef]
171. Saunders, R.E.; Derby, B. Inkjet printing biomaterials for tissue engineering: Bioprinting. *Int. Mater. Rev.* **2014**, *59*, 430–448. [CrossRef]
172. Vatani, M.; Lu, Y.; Lee, K.-S.; Kim, H.-C.; Choi, J.-W. Direct-write stretchable sensors using single-walled carbon nanotube/polymer matrix. *J. Electron. Packag.* **2013**, *135*, 11009. [CrossRef]
173. Srichan, C.; Saikrajang, T.; Lomas, T.; Jomphoak, A.; Maturos, T.; Phokaratkul, D.; Kerdcharoen, T.; Tuantranont, A. Inkjet printing PEDOT:PSS using desktop inkjet printer. In Proceedings of the 6th International Conference on Electrical Engineering/Electronics, Computer, Telecommunications and Information Technology, Pattaya, Chonburi, Thailand, 6–9 May 2009; Volume 1, pp. 465–468.
174. Jones, C.S.; Lu, X.; Renn, M.; Stroder, M.; Shih, W.-S. Aerosol-jet-printed, high-speed, flexible thin-film transistor made using single-walled carbon nanotube solution. *Microelectron. Eng.* **2010**, *87*, 434–437. [CrossRef]
175. B'fahy, S.; Yunus, S.; Pardoen, T.; Bertrand, P.; Troosters, M. Stretchable helical gold conductor on silicone rubber microwire. *Appl. Phys. Lett.* **2007**, *91*, 141911. [CrossRef]
176. Lipomi, D.J.; Chong, H.; Vosgueritchian, M.; Mei, J.; Bao, Z. Toward mechanically robust and intrinsically stretchable organic solar cells: Evolution of photovoltaic properties with tensile strain. *Sol. Energy Mater. Sol. Cells* **2012**, *107*, 355–365. [CrossRef]
177. Lim, K.S.; Chang, W.-J.; Koo, Y.-M.; Bashir, R. Reliable fabrication method of transferable micron scale metal pattern for poly(dimethylsiloxane) metallization. *Lab Chip* **2006**, *6*, 578–580. [CrossRef] [PubMed]

178. Lee, C.H.; Kim, D.R.; Zheng, X. Fabrication of nanowire electronics on nonconventional substrates by water-assisted transfer printing method. *Nano Lett.* **2011**, *11*, 3435–3439. [CrossRef] [PubMed]
179. Song, L.; Ci, L.; Gao, W.; Ajayan, P.M. Transfer printing of graphene using gold film. *ACS Nano* **2009**, *3*, 1353–1356. [CrossRef] [PubMed]
180. Park, J.; Wang, S.; Li, M.; Ahn, C.; Hyun, J.K.; Kim, D.S.; Kim, D.K.; Rogers, J.A.; Huang, Y.; Jeon, S. Three-dimensional nanonetworks for giant stretchability in dielectrics and conductors. *Nat. Commun.* **2012**, *3*, 916. [CrossRef] [PubMed]
181. Amjadi, M.; Kyung, K.-U.; Park, I.; Sitti, M. Stretchable, skin-mountable, and wearable strain sensors and their potential applications: A review. *Adv. Funct. Mater.* **2016**, *26*, 1678–1698. [CrossRef]
182. Mannsfeld, S.C.B.; Tee, B.C.-K.; Stoltenberg, R.M.; Chen, C.V.H.-H.; Barman, S.; Muir, B.V.O.; Sokolov, A.N.; Reese, C.; Bao, Z. Highly sensitive flexible pressure sensors with microstructured rubber dielectric layers. *Nat. Mater.* **2010**, *9*, 859–864. [CrossRef] [PubMed]
183. Yu, C.; Wang, Z.; Yu, H.; Jiang, H. A stretchable temperature sensor based on elastically buckled thin film devices on elastomeric substrates. *Appl. Phys. Lett.* **2009**, *95*, 141912. [CrossRef]
184. Chung, M.G.; Kim, D.H.; Lee, H.M.; Kim, T.; Choi, J.H.; Seo, D.K.; Yoo, J.-B.; Hong, S.-H.; Kang, T.J.; Kim, Y.H. Highly sensitive NO_2 gas sensor based on ozone treated graphene. *Sens. Actuators B Chem.* **2012**, *166–167*, 172–176. [CrossRef]
185. Liang, J.; Li, L.; Niu, X.; Yu, Z.; Pei, Q. Elastomeric polymer light-emitting devices and displays. *Nat. Photonics* **2013**, *7*, 817–824. [CrossRef]
186. Chortos, A.; Koleilat, G.I.; Pfattner, R.; Kong, D.; Lin, P.; Nur, R.; Lei, T.; Wang, H.; Liu, N.; Lai, Y.-C.; et al. Mechanically Durable and Highly Stretchable Transistors Employing Carbon Nanotube Semiconductor and Electrodes. *Adv. Mater.* **2016**, *28*, 4441–4448. [CrossRef] [PubMed]
187. Lee, M.-S.; Lee, K.; Kim, S.-Y.; Lee, H.; Park, J.; Choi, K.-H.; Kim, H.-K.; Kim, D.-G.; Lee, D.-Y.; Nam, S.; et al. High-performance, transparent, and stretchable electrodes using graphene-metal nanowire hybrid structures. *Nano Lett.* **2013**, *13*, 2814–2821. [CrossRef] [PubMed]
188. Shin, M.; Song, J.H.; Lim, G.-H.; Lim, B.; Park, J.-J.; Jeong, U. Highly stretchable polymer transistors consisting entirely of stretchable device components. *Adv. Mater.* **2014**, *26*, 3706–3711. [CrossRef] [PubMed]
189. Oh, J.Y.; Rondeau-Gagné, S.; Chiu, Y.-C.; Chortos, A.; Lissel, F.; Wang, G.-J.N.; Schroeder, B.C.; Kurosawa, T.; Lopez, J.; Katsumata, T.; et al. Intrinsically stretchable and healable semiconducting polymer for organic transistors. *Nature* **2016**, *539*, 411–415. [CrossRef] [PubMed]
190. Xie, K.; Wei, B. Materials and structures for stretchable energy storage and conversion devices. *Adv. Mater.* **2014**, *26*, 3592–3617. [CrossRef] [PubMed]
191. Chen, T.; Peng, H.; Durstock, M.; Dai, L. High-performance transparent and stretchable all-solid supercapacitors based on highly aligned carbon nanotube sheets. *Sci. Rep.* **2014**, *4*, 3612. [CrossRef] [PubMed]
192. Lee, H.; Yoo, J.-K.; Park, J.-H.; Kim, J.H.; Kang, K.; Jung, Y.S. A stretchable polymer-carbon nanotube composite electrode for flexible lithium-ion batteries: Porosity engineering by controlled phase separation. *Adv. Energy Mater.* **2012**, *2*, 976–982. [CrossRef]
193. Lipomi, D.J.; Tee, B.C.-K.; Vosgueritchian, M.; Bao, Z. Stretchable organic solar cells. *Adv. Mater.* **2011**, *23*, 1771–1775. [CrossRef] [PubMed]
194. Yang, Z.; Deng, J.; Sun, X.; Li, H.; Peng, H. Stretchable, wearable dye-sensitized solar cells. *Adv. Mater.* **2014**, *26*, 2643–2647. [CrossRef] [PubMed]
195. Dagdeviren, C.; Yang, B.D.; Su, Y.; Tran, P.L.; Joe, P.; Anderson, E.; Xia, J.; Doraiswamy, V.; Dehdashti, B.; Feng, X.; et al. Conformal piezoelectric energy harvesting and storage from motions of the heart, lung, and diaphragm. *Proc. Natl. Acad. Sci. USA* **2014**, *111*, 1927–1932. [CrossRef] [PubMed]
196. Qi, Y.; Kim, J.; Nguyen, T.D.; Lisko, B.; Purohit, P.K.; McAlpine, M.C. Enhanced piezoelectricity and stretchability in energy harvesting devices fabricated from buckled PZT ribbons. *Nano Lett.* **2011**, *11*, 1331–1336. [CrossRef] [PubMed]
197. Bito, J.; Hester, J.G.; Tentzeris, M.M. Ambient RF energy harvesting from a two-way talk radio for flexible wearable wireless sensor devices utilizing inkjet printing technologies. *IEEE Trans. Microw. Theory Tech.* **2015**, *63*, 4533–4543. [CrossRef]

198. Kim, S.; Bito, I.; Jeong, S.; Georgiadis, A.; Tentzeris, M.M. A flexible hybrid printed RF energy harvester utilizing catalyst-based copper printing technologies for far-field RF energy harvesting applications. In Proceedings of the 2015 IEEE MTT-S International Microwave Symposium, Phoenix, AZ, USA, 17–22 May 2015; pp. 1–4.
199. Lee, J.-H.; Ryu, H.; Kim, T.-Y.; Kwak, S.-S.; Yoon, H.-J.; Kim, T.-H.; Seung, W.; Kim, S.-W. Thermally induced strain-coupled highly stretchable and sensitive pyroelectric nanogenerators. *Adv. Energy Mater.* **2015**, *5*, 1500704. [CrossRef]
200. Yi, F.; Lin, L.; Niu, S.; Yang, P.K.; Wang, Z.; Chen, J.; Zhou, Y.; Zi, Y.; Wang, J.; Liao, Q.; et al. Stretchable-rubber-based triboelectric nanogenerator and its application as self-powered body motion sensors. *Adv. Funct. Mater.* **2015**, *25*, 3688–3696. [CrossRef]
201. Honda, W.; Harada, S.; Arie, T.; Akita, S.; Takei, K. Wearable, human-interactive, health-monitoring, wireless devices fabricated by macroscale printing techniques. *Adv. Funct. Mater.* **2014**, *24*, 3299–3304. [CrossRef]
202. Hussain, A.M.; Ghaffar, F.A.; Park, S.I.; Rogers, J.A.; Shamim, A.; Hussain, M.M. Metal/polymer based stretchable antenna for constant frequency far-field communication in wearable electronics. *Adv. Funct. Mater.* **2015**, *25*, 6565–6575. [CrossRef]
203. Chortos, A.; Liu, J.; Bao, Z. Pursuing prosthetic electronic skin. *Nat. Mater.* **2016**, *15*, 937–950. [CrossRef] [PubMed]
204. Yeo, J.C.; Yap, H.K.; Xi, W.; Wang, Z.; Yeow, C.-H.; Lim, C.T. Flexible and stretchable strain sensing actuator for wearable soft robotic applications. *Adv. Mater. Technol.* **2016**, *1*, 1600018. [CrossRef]
205. Lim, S.; Son, D.; Kim, J.; Lee, Y.B.; Song, J.-K.; Choi, S.; Lee, D.J.; Kim, J.H.; Lee, M.; Hyeon, T.; Kim, D.-H. Transparent and stretchable interactive human machine interface based on patterned graphene heterostructures. *Adv. Funct. Mater.* **2015**, *25*, 375–383. [CrossRef]
206. Hillborg, H.; Gedde, U.W. Hydrophobicity changes in silicone rubbers. *IEEE Trans. Dielectr. Electr. Insul.* **1999**, *6*, 703–717. [CrossRef]
207. Grassie, N.; Macfarlane, I.G. The thermal degradation of polysiloxanes-I. Poly(dimethylsiloxane). *Eur. Polym. J.* **1978**, *14*, 875–884. [CrossRef]
208. Morra, M.; Occhiello, E.; Marola, R.; Garbassi, F.; Humphrey, P.; Johnson, D. On the aging of oxygen plasma-treated PDMS surfaces. *J. Colloid Interface Sci.* **1990**, *137*, 11–24. [CrossRef]
209. Park, G.; Chung, H.-J.; Kim, K.; Lim, S.A.; Kim, J.; Kim, Y.-S.; Liu, Y.; Yeo, W.-H.; Kim, R.-H.; Kim, S.S.; et al. Immunologic and tissue biocompatibility of flexible/stretchable electronics and optoelectronics. *Adv. Healthc. Mater.* **2014**, *3*, 515–525. [CrossRef] [PubMed]
210. Song, J.; Feng, X.; Huang, Y. Mechanics and thermal management of stretchable inorganic electronics. *Natl. Sci. Rev.* **2016**, *3*, 128–143. [CrossRef] [PubMed]
211. Li, Y.; Gao, Y.; Song, J. Recent advances on thermal analysis of stretchable electronics. *Theor. Appl. Mech. Lett.* **2016**, *6*, 32–37. [CrossRef]
212. Novoselov, K.S.S.; Geim, A.K.K.; Morozov, S.V.; Jiang, D.; Zhang, Y.; Dubonos, S.V.; Grigorieva, I.V.; Firsov, A.A. Electric field effect in atomically thin carbon films. *Science* **2004**, *306*, 666–669. [CrossRef] [PubMed]
213. Lai, Y.-C.; Huang, Y.-C.; Lin, T.-Y.; Wang, Y.-X.; Chang, C.-Y.; Li, Y.; Lin, T.-Y.; Ye, B.-W.; Hsieh, Y.-P.; Su, W.-F.; et al. Stretchable organic memory: Toward learnable and digitized stretchable electronic applications. *NPG Asia Mater.* **2014**, *6*, e87. [CrossRef]
214. Moon, J.-H.; Baek, D.H.; Choi, Y.Y.; Lee, K.H.; Kim, H.C.; Lee, S.-H. Wearable polyimide-PDMS electrodes for intrabody communication. *J. Micromech. Microeng.* **2010**, *20*, 25032. [CrossRef]

16

Liquid-Metal Enabled Droplet Circuits

Yi Ren [1] and Jing Liu [1,2,3,*]

[1] Department of Biomedical Engineering, Tsinghua University, Beijing 100084, China; reny14@mails.tsinghua.edu.cn
[2] Technical Institute of Physics and Chemistry, Chinese Academy of Sciences, Beijing 100190, China
[3] School of Future Technology, University of Chinese Academy of Sciences, Beijing 100049, China
[*] Correspondence: jliubme@tsinghua.edu.cn

Abstract: Conventional electrical circuits are generally rigid in their components and working styles, which are not flexible and stretchable. As an alternative, liquid-metal-based soft electronics offer important opportunities for innovation in modern bioelectronics and electrical engineering. However, their operation in wet environments such as aqueous solution, biological tissue or allied subjects still encounters many technical challenges. Here, we propose a new conceptual electrical circuit, termed as droplet circuit, to fulfill the special needs described above. Such unconventional circuits are immersed in a solution and composed of liquid metal droplets, conductive ions or wires, such as carbon nanotubes. With specifically-designed topological or directional structures/patterns, the liquid-metal droplets composing the circuit can be discrete and disconnected from each other, while achieving the function of electron transport through conductive routes or the quantum tunneling effect. The conductive wires serve as electron transfer stations when the distance between two separate liquid-metal droplets is far beyond that which quantum tunneling effects can support. The unique advantage of the current droplet circuit lies in the fact that it allows parallel electron transport, high flexibility, self-healing, regulation and multi-point connectivity without needing to worry about the circuit break. This would extend the category of classical electrical circuits into newly emerging areas like realizing room temperature quantum computing, making brain-like intelligence or nerve–machine interface electronics, etc. The mechanisms and potential scientific issues of the droplet circuits are interpreted and future prospects in this direction are outlined.

Keywords: droplet circuits; liquid metal; quantum tunneling effect; solution electronics; electron transport; ionic conduction; quantum computing; brain-like intelligence

---

1. Introduction

Since the origin of electricity, it has become a necessity in daily life. Generally speaking, electrical circuits are rigid and continuous in their structures and components. Print circuit boards (PCB) have been commonly used in various situations (Figure 1A). However, classic rigid circuits cannot easily adapt to the human body due to poor flexibility and biocompatibility, limiting their value in the biomedical and health care fields. The increasing advancement of wearable devices and implantable systems has led to significant growth in flexible electronics (Figure 1B) [1]. Polymer nanomaterials, silk fibroin and liquid metal are being gradually adopted in soft electronics. Another main development trend in artificial circuits is molecular electronics. This was first proposed in 1974 by Aviram and Ratner [2], and refers to a field that seeks to fabricate electrical devices and circuits with single molecules and molecular monolayers [3–6]. The fabrication of molecular electrical devices includes single-molecule break junctions and molecular monolayer devices [3]. Tailored by chemical design and synthesis, the function of molecular components can be rather diverse. Deoxyribonucleic acid (DNA) has potential for molecular devices for its unique structure [5]. Figure 1C presents the logic

'AND' and 'OR' gate fabricated by DNA. Until now, molecular components including diodes, switches, memory and transistors have been intensively researched [6]. Those components can be combined to construct molecular-scale electronic computers [7]. Based on the electrical properties of the molecular diode switches, quantum mechanical calculations can also possibly be performed.

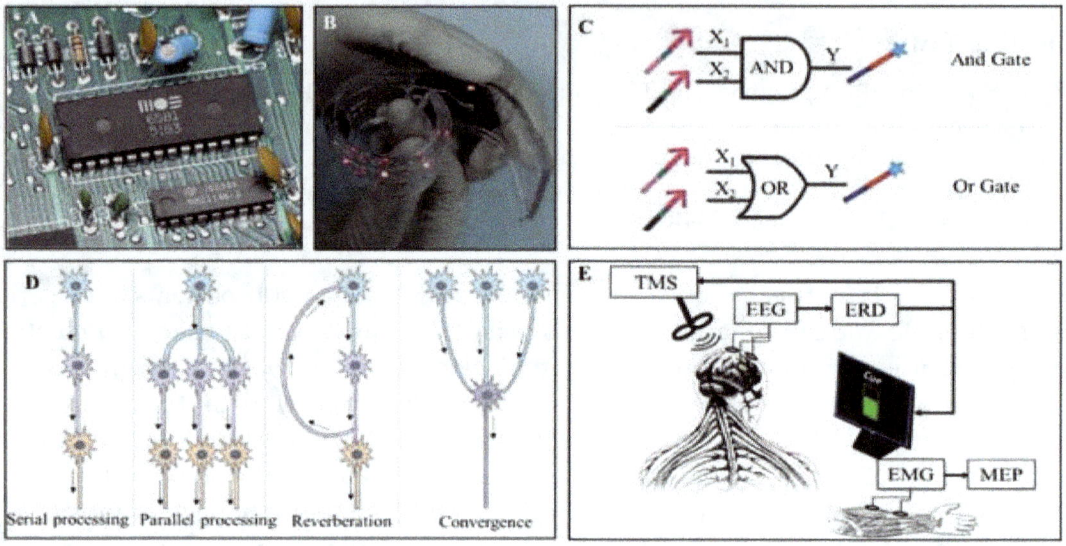

Figure 1. Respective kinds of electrical circuits. (**A**) Traditional rigid print circuit boards (PCB) circuits; (**B**) Soft electronics. Reproduced with permission from [8]; (**C**) Logic AND Gate and OR Gate based on DNA. Reproduced with permission from [6]; (**D**) Several typical types of neural circuits; (**E**) Brain–computer interface. Reproduced with permission from [9]. Note: TMS: transcranial magnetic stimulation; ERD: event-related desynchronization; MEP: motor-evoked potential; EEG: electroencephalogram; EMG: electromyogram.

Besides these artificial circuits, electrical circuits in fact intrinsically exist throughout the human body, as shown in Figure 1D. For instance, Dejean et al. recently studied the neural circuits and cell types that mediate conditioned fear expression and recovery [8]. Moreover, voltage-gated channels are important switches for signal transportation in the central and peripheral neural systems [10]. Keeping the above facts in mind and without losing generality, we can divide electrical circuits into two main categories: biologically-inspired natural circuits and human-made artificial circuits. One more trend, as indicated in Figure 1E, is that a new category is emerging to combine naturally occurring neural circuits and artificial circuits to carry out complex functions. The core of such circuits can generally be called brain–computer interfaces (BCIs) [11–14]. Clearly, BCIs are significant for patients with serious disabilities such as tetraplegia and stroke. They especially mean a lot for future human needs in extending the limits of biological capability. At this stage, BCIs based on rapid serial visual presentation have already been used to detect and recognize objects, providing a viable approach to prompt human–machine systems [14].

Despite the widespread application of currently available circuits, they all unavoidably encounter the possibility of circuit break, which severely affects the normal operation of devices. As a remedy, we here propose an unconventional concept for electrical circuits to tackle the above challenges, which we call droplet circuits. Such an electrical circuit is enabled from liquid-metal droplets (LMDs) and is conductive in discontinuous form. The structure and working style of this kind of circuit are highly analogous and similar to that of the neuro-network system. Therefore it is expected to be very useful in innovating newly emerging areas such as room temperature quantum computing, brain-like intelligence, and brain–machine interfaces, etc.

Liquid metal refers to alloys or metals with a low melting point, which can maintain their liquid phase around room temperature [15]. The electrical properties and biocompatibility of liquid metal have been proven in Yi et al.'s work [15]. In contrast to mercury, which is highly toxic, liquid metal based on gallium is relatively safe for medical applications. Liquid metal has recently been introduced into soft electronics and the biomedical field [15]. Liquid-metal sensors [16,17], memristors [18], diodes [19] and electrodes [20–22] have already been proposed for health monitoring and disease treatment (Figure 2A–C). Recently, some researches has been devoted to the study of liquid-metal droplets (Figure 2D), which show great potential in self-powered devices [23,24] and phagocytosis [25]. Yang et al. introduced millimeter-scale LMDs as thermal switches, unlocking new possible solutions for thermal management [26]. Tang et al. electrically controlled the size and rate of LMD formation [27]. Others fabricated non-stick LMDs by coating polytetrafluoroethylene particles [28] or graphene [29] on NaOH-treated LMDs. Chen et al. found that graphene-coated LMDs can be used as droplet-based floating electrodes [29]. Sivan et al. coated LMDs with n-type and p-type semiconducting nanopowders to study their electronic properties and electrochemical properties [30].

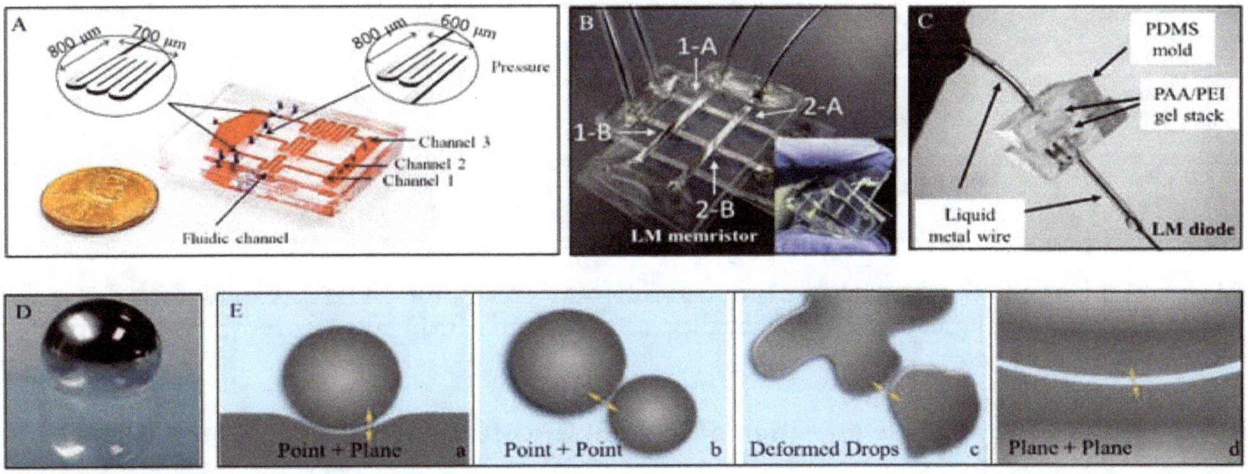

Figure 2. (**A**) Liquid-metal pressure sensor [16]; (**B**) A 2 × 2 crossbar array liquid metal memristor: 1-A, 1-B, 2-A and 2-B refer to the four bars respectively. Reproduced with permission from [18]; (**C**) Liquid-metal diode. Reproduced with permission from [19]; (**D**) Liquid-metal droplet. Reproduced with permission from [28]; (**E**) Four configurations of all-soft liquid metal quantum devices based on tunneling effect. Reproduced with permission from [31].

In contrast to previous studies, where all LMDs were contacted to realize electrical conduction, droplet circuits mean that the circuit is composed of discrete LMDs and can operate well, in the same way as traditional circuits. For this purpose, we would speculate that when the gap between the two LMDs is small enough, the electrical signal can be transported even though they are not connected. This assumption is dependent on the quantum tunneling effect theory, which is a physical phenomenon in which a micro-particle such as an electron can tunnel through a barrier that it classically could not surmount. Thus, droplet circuits can transfer electrical signals even though LMDs are separated in space. Zhao et al. have proposed a transformable soft quantum device based on liquid metal [31]. They found that liquid metal droplets can be adopted to create tunnel junctions and defined four configurations of all-soft quantum devices, as shown in Figure 2E. Based on their research, this article further explores the potential of liquid-metal droplets in constructing droplet circuits and preliminarily demonstrate its probability. Compared to traditional rigid or soft circuits, liquid-metal droplet circuits provide more flexibility without the problem of circuit breaks.

2. Electrical Conduction via the Quantum Tunneling Effect

The quantum tunneling phenomenon cannot be explained by classical mechanics, as it occurs only at the quantum scale. Compared to classical mechanics, matter in quantum mechanics has the properties of waves and particles, involving the Heisenberg uncertainty principle [32–34]. The Heisenberg uncertainty principle states that one can never exactly know the position and speed of a particle at the same time. Thus, events which seem to be impossible in classical mechanics become possible in quantum mechanics. This can be used to explain the quantum tunneling phenomenon.

Quantum tunneling is essential on many occasions, including nuclear fusion in the sun [35], astrochemistry in interstellar clouds, quantum biology, tunnel diodes (Figure 3A) [36], tunnel junctions (Figure 3B) [37], scanning tunneling microscope and quantum computing, etc. A tunnel junction is where two conductors are separated by a thin insulator to create a simple barrier between them. It can be applied to measure voltages and magnetic fields.

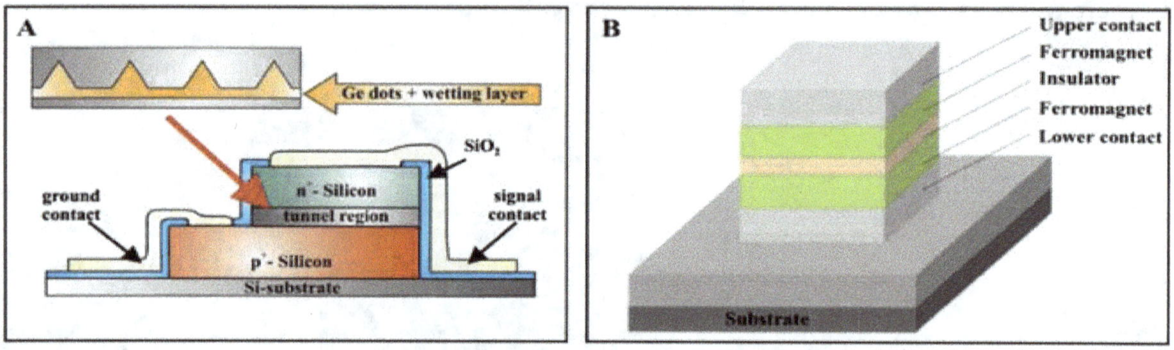

Figure 3. Illustration of the quantum tunneling effect and conventional typical application devices. (**A**) A Ge quantum dots interband tunneling diode. Reproduced with permission from [36]; (**B**) Diagram of a ferromagnet tunnel junction.

3. Fabrication of LMDs and Composition of Droplet Circuits

Until now, the methods for the fabrication of LMDs have been relatively diverse and easily available, including methods like sonication, molding, and flow-focusing [27]. Recently, Yu et al. created a method named suspension 3D printing [38], which can help quickly realize three various dimensional droplet patterns. They successfully patterned LMDs into a self-healing hydrogel (Figure 4A) and studied the relationship between the process parameters, supporting gel concentration, and the deposited micro-droplet geometry. Before that, the present lab also successfully prepared LMDs on a large scale through a fabrication method that does not use channels [39], as shown in Figure 4B. The microscopic image (Figure 4C) presents a single layer of liquid-metal micro-droplets, which were sitting together with a thin interface between each other. From the insert of Figure 4C, we can see that although the distance between the LMDs is extremely small, they are in fact discrete. Furthermore, Tian et al. proposed a microfluidic chip for liquid-metal droplet generation and sorting. Their system could manipulate these neutral liquid-metal droplets in nonconductive fluid [40]. All those studies provide viable methods for the fabrication of LMDs and the construction of droplet circuits, and more efforts should be made to improve the technology.

A rising trend in liquid metal is to cover LMDs with other materials to fabricate multifunctional circuits. For instance, liquid metal can be sealed into polymers to maintain its pattern. Water films and nanoparticles are applied to prevent the formation of an oxide layer, hold the stability and keep them discrete from each other [28,41]. Moreover, by coating LMDs with magnetic particles, they can be induced to move with the external magnetic field. Besides magnetic control, an electrical field can also induce the movement of liquid metal. Therefore, by coating specific materials, we can fabricate liquid-metal droplets with the required properties.

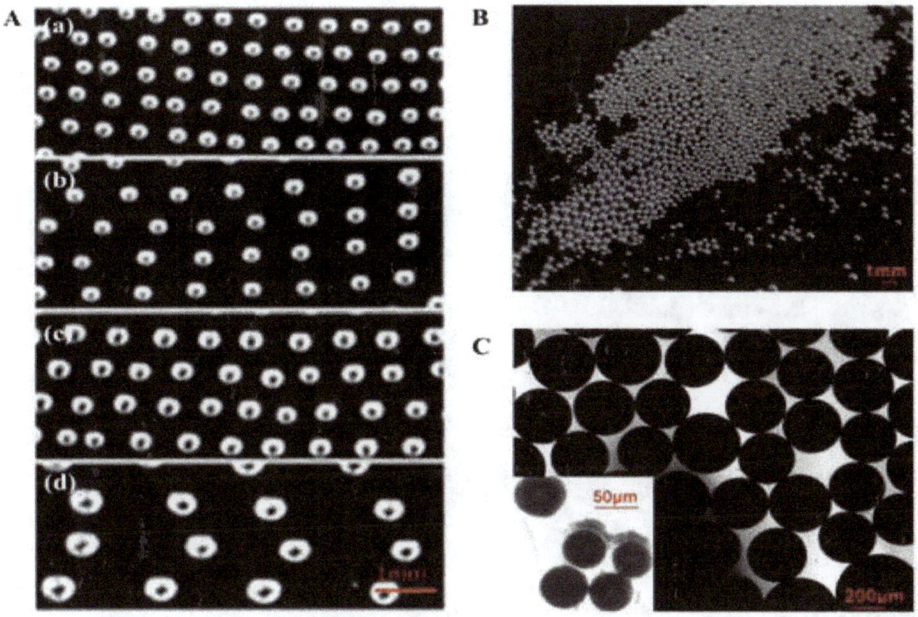

Figure 4. (**A**) Liquid-metal droplets (LMDs) floating in hydrogel with varied diameters; (**a**) 60 μm, (**b**) 90 μm, (**c**) 160 μm, (**d**) 210 μm. Reproduced with permission from [38]; (**B**) LMDs assembly fabricated through injection way. Reproduced with permission from [39]; (**C**) Microscopic image showing a single layer of liquid metal micro-droplets closely sitting together with thin interface between each other. Reproduced with permission from [39].

4. Droplet Circuits in Solution

4.1. Mechanism of Liquid-Metal-Based Droplet Circuits

Droplet circuits are mainly composed of discontinuous LMDs, ions and conductive wires such as carbon nanotubes and operate in electrolyte solution. LMDs and carbon nanotubes are mixed up and cooperate to connect the circuit electrically. Carbon nanotubes have unique electrical, thermal and chemical properties, showing great potential in nanoelectronics [42], especially for transistor applications, owing to their benign carrier mobility and velocity. Therefore, carbon nanotubes can be selected to connect droplet circuits.

Figure 5A shows the electrical conduction of droplet circuits. LMDs in droplet circuits are surrounded by carbon nanotubes and ions and electrons transfer along the pattern of LMDs with the assistance of carbon nanotubes and ions. There are two ways for LMDs to communicate in electrolyte solution, as shown in Figure 5B. When two LMDs are sitting closely enough, a quantum tunneling effect will happen and electrons can possibly transfer from one to another. If the distance of two LMDs becomes a little farther than the tunneling effect can support, carbon nanotubes and ions floating between them serve as transfer stations for electrons to flow. Through the quantum tunneling effect, carbon-nanotube transfer stations and ionic routes, the discrete LMDs are electrically connected and the whole circuit can work well. In general, LMDs are randomly arranged in electrolyte solution. If voltage is applied to the LMDs, they will be organized in order and connect the circuit (Figure 5C).

As the resistance of LMDs is smaller than other regions in electrolyte solution, the current will regularly transfer mainly along LMDs. In addition, one can change the pattern of LMDs to control the conducting direction of the current. That is, given a specific topological or directional design, the liquid-metal droplets composing the circuit can achieve the desired or regulative functions of electron transport through conductive routes or the quantum tunneling effect. Some newly emerging needs, such as room temperature computing or brain-like chips, can possibly be enabled based on such unconventional electrical circuits.

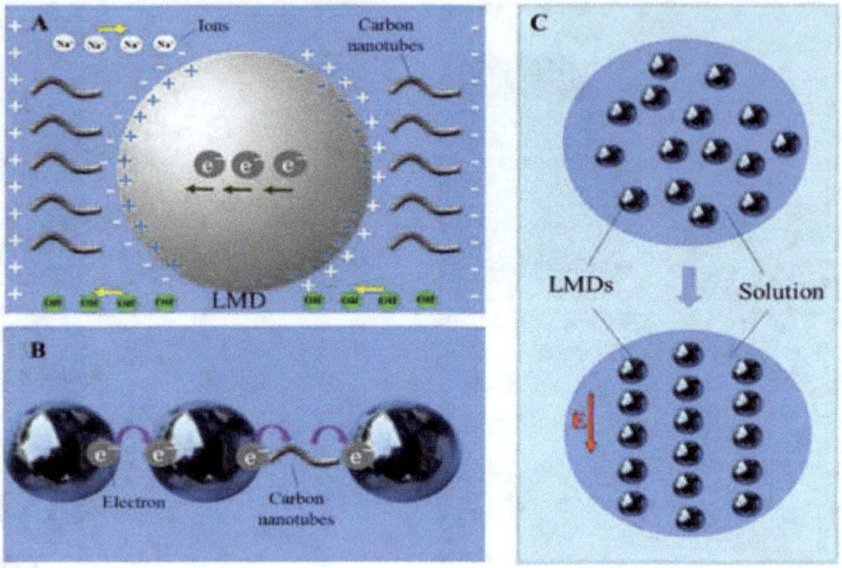

Figure 5. (**A**) The electrical environment of a LMD in NaOH solution: A LMD is surrounded by ions and conductive carbon nanotubes; tunneling effect between two LMDs; (**B**) Combination of two methods of communicating for LMDs in a circuit; (**C**) LMDs arranged with and without voltage.

4.2. Configuration of the Liquid-Metal Droplet Circuit

4.2.1. Low-Dimensional Droplet Circuits

Zero-dimensional (0D) droplet circuits refer to the case where only one LMD works. Figure 6A shows a schematic of a 0D circuit. The LMD is immersed in NaOH solution and replaces part of the wire. When the switch is closed, the lamp can be lit.

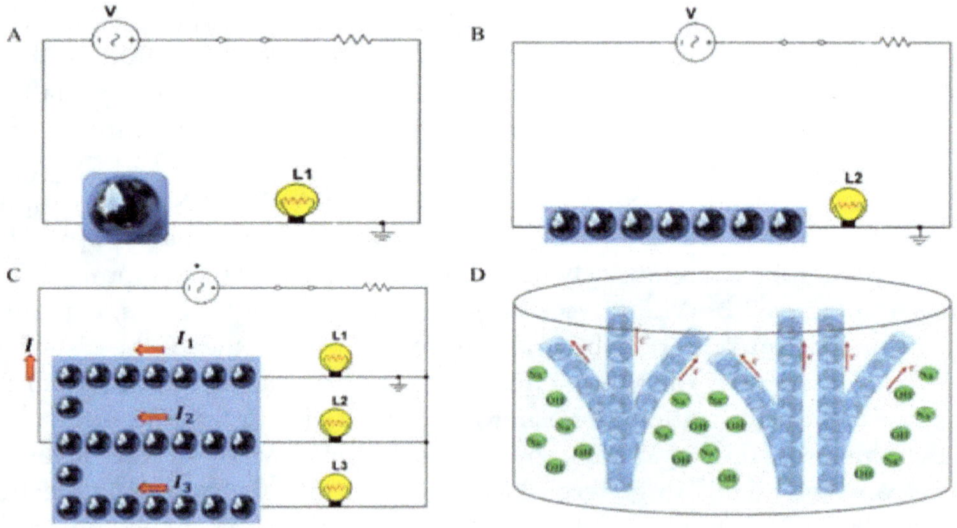

Figure 6. (**A**) Zero-dimensional (0D) droplet circuit; (**B**) 1 D droplet circuit; (**C**) 2 D droplet circuit; (**D**) Topological structure of droplet circuit.

By increasing the number of LMDs and arranging them in a line, a one-dimensional (1D) droplet circuit can be formed. The line of LMDs can be inserted into a traditional circuit as an alternative to the original wire, as shown in Figure 6B. Similarly, one can arrange LMDs in a plane with a specific pattern to realize a two-dimensional (2D) droplet circuit (Figure 6C). In principle, this circuit should also satisfy Kirchhoff's current law:

$$\sum I_{in} = \sum I_{out} \tag{1}$$

Here, I_{in} is the current flowing into a node and I_{out} is the current flowing out from the node.

Kirchhoff's law states that the charge input at a node is equal to the charge output. Therefore, the currents of the circuit in Figure 6C should meet the following equation:

$$I = I_1 + I_2 + I_3 \tag{2}$$

Due to existence of the electrolyte in the solution, charge loss is inevitable. The main loss pathways include the redox chemistry between NaOH and LMDs and the contact between NaOH and carbon nanotubes. However, these are small enough to be neglected since the conductivity of NaOH is much lower than LMDs and carbon nanotubes.

Furthermore, LMDs can be arranged in a special topological structure to fabricate three-dimensional (3D) or even dynamically transformable droplet circuits, as shown in Figure 6D. To form specific structures of 3D droplet circuits, polymer can be applied. LMDs can be sealed in polymer tubes and be arranged as required. Electrical or magnetic fields can be utilized at the same time. These 3D droplet circuits are flexible and able to transport electrons in the desired direction. Topological droplet circuits perfectly imitate the connection between neurons, which is beneficial to artificial neural connections.

The combination of liquid-metal and artificial circuits has been demonstrated before by Zhang et al. [43]. They injected liquid metal in the left and right side of the sciatic nerve near the femur of the interceptive lower part of a bullfrog body. It was found that the two electrodes successfully conducted the electrical stimuli signals to the nerves, which proves the feasibility of applying liquid metal to electrical circuits.

4.2.2. High-Dimensional Droplet Circuits

The further development of droplet circuits may be in more than three dimensions. In this regard, LMDs can be encapsulated into elastic tubes together with carbon nanotubes and electrolyte solution together to replace conventional rigid metal wires, as shown in Figure 7D. Moreover, since the pattern of LMDs can be changed by external magnetic or electrical fields, the circuit will become more fantastic and dynamically change over the time. Applying electrical or magnetic fields to LMD circuits, the structure of the circuits can be changed as required. For instance, LMDs can be induced to move by magnetic field when coating them with ferromagnetic materials (Figure 7A,B) [44]. With the help of aluminum, one can realize reliable motion control of the liquid metal droplets in the electrical field [45]. Figure 7C presents the sequential movement of a liquid-metal droplet propelled by an external electrical field. The velocity of Al/EGaIn and Ni/Al/EGaIn droplets in NaOH solution under different voltages was measured and calculated, as shown in Figure 7E [46]. Al provides a fuel source for LMDs to move, and they observed that Ni is not favorable for the electrical control of LMDs. It is apparent that the LMDs speed up with the increase of the voltage. Moreover, the current of the droplet circuits composed of LMDs in NaOH solution was relatively stable under constant voltages (Figure 7F), demonstrating promising applications of droplet circuits [47]. Recently, Isabela et al. fabricated a magnetocaloric ferrofluid based on Ga liquid metal [48]. They suspend gadolinium nanoparticles in a liquid gallium alloy and found that the material exhibited spontaneous magnetization and a large magnetocaloric effect. This is a significant progress in the magnetic control of liquid metal.

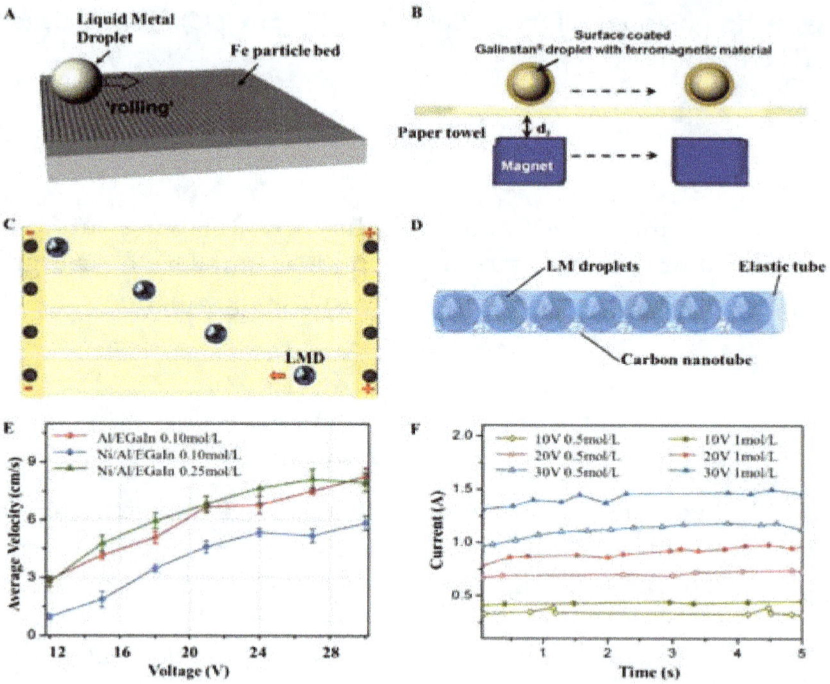

Figure 7. (**A**) Fabrication of Fe-coated LMDs. Reproduced with permission from [44]; (**B**) Surface-coated LMD with ferromagnetic materials induced by magnetic field. Reproduced with permission from [44]; (**C**) Sequential photos of LMD motion in an electrical field; (**D**) Liquid-metal droplet wire; (**E**) Average velocity of LMDs in a straight channel containing NaOH solution under exposure to different voltages, and the concentrations of the NaOH solution [46] (Reproduced with permission); (**F**) The electric current in response to different voltages, and the concentrations of the NaOH solution. Reproduced with permission from [47].

5. Discussion

This article exploits the basic features of droplet circuits and evaluates the probability of functional circuits enabled by liquid-metal droplets. Obviously, droplet circuits display unique advantages that may not be easily offered by conventional strategies. First of all, droplet circuits show a highly parallel electrical transporting capability. In addition, such circuits have excellent flexibility, self-healing capabilities and are stretchable. In contrast to existing liquid-metal soft electronics, LMD circuits consist of spatially transformable discrete liquid-metal droplets. It can tolerate greater elastic deformation and can be shaped into various electrical patterns.

Further, droplet circuits are fault-tolerant and self-error-correcting. Since LMDs can transport electrons while they are disconnected from each other, droplet circuits may not face circuit break errors. Even though the distance between LMDs can be changed by external factors such as strain and twist, carbon nanotubes will keep the circuit electrically connected. It is apparent that droplet circuits appear more robotic than traditional circuits, which is very much like the working style of a biological neurocircuit. The electrical resistance of liquid-metal droplet circuits may be higher due to the discontinuity and cross-talk between individual droplet chains, but since the quantum tunneling effect occurs within an extremely small space, the running time of the whole circuit will be greatly reduced. Particularly, droplet circuits can run in the wet environment of electrolyte solution. Electrical conduction involves not only LMDs and carbon nanotubes, but also ionic conduction. Ions in electrolyte solution play the role of electron transportation along with carbon nanotubes when the quantum tunneling effect of LMDs does not work.

Last but not least, the fabrication of liquid-metal-based droplet circuits is overall not complicated, since LMDs self-assemble in the environment of magnetic or electrical fields. The circuit configurations are in dynamic change at the same time.

From a practical point of view, due to its unique advantages, such as high flexibility and benign electrical conductivity, liquid-metal-based droplet circuits can possibly be applied to make artificial retina or cochlea as an alternative to conventional rigid wires and electrodes. Figure 8A is the presentation of a cochlear implant. The wires and electrodes should be soft and biocompatible enough to ensure minimal damage to biological tissues. Liquid-metal-enabled droplet circuits could perfectly meet these requirements. Nerve repair and neural connection is another promising development for such droplet circuits. Efforts have been made to explore the possibility of liquid-metal neural restoration. Zhang et al. employed liquid metal to reconnect the transected sciatic nerve of a bullfrog (Figure 8B) and found that the measured electroneurographic signals under electric stimulation were similar to those from the intact sciatic nerves [49]. They further put forward three types of nerve conduits to restore damaged peripheral nerves, as shown in Figure 8C. Nerve conduits combined with liquid metal can take the form of microchannels, thin slices or concentric tubes. Unlike the former trial, the current droplet circuit offers more electron transport channels, such as electrically conductive liquid metal, ionic conduction and nanowires, etc. Therefore, it would better serve medical needs. Another potential that is worth of mentioning of droplet circuits lies in its role in constructing computing chips or devices that are different from the classical framework. With discrete LMDs, droplet circuits offer opportunities to perform as quantum processors or to carry out quantum calculating, which may help the design of future quantum computers.

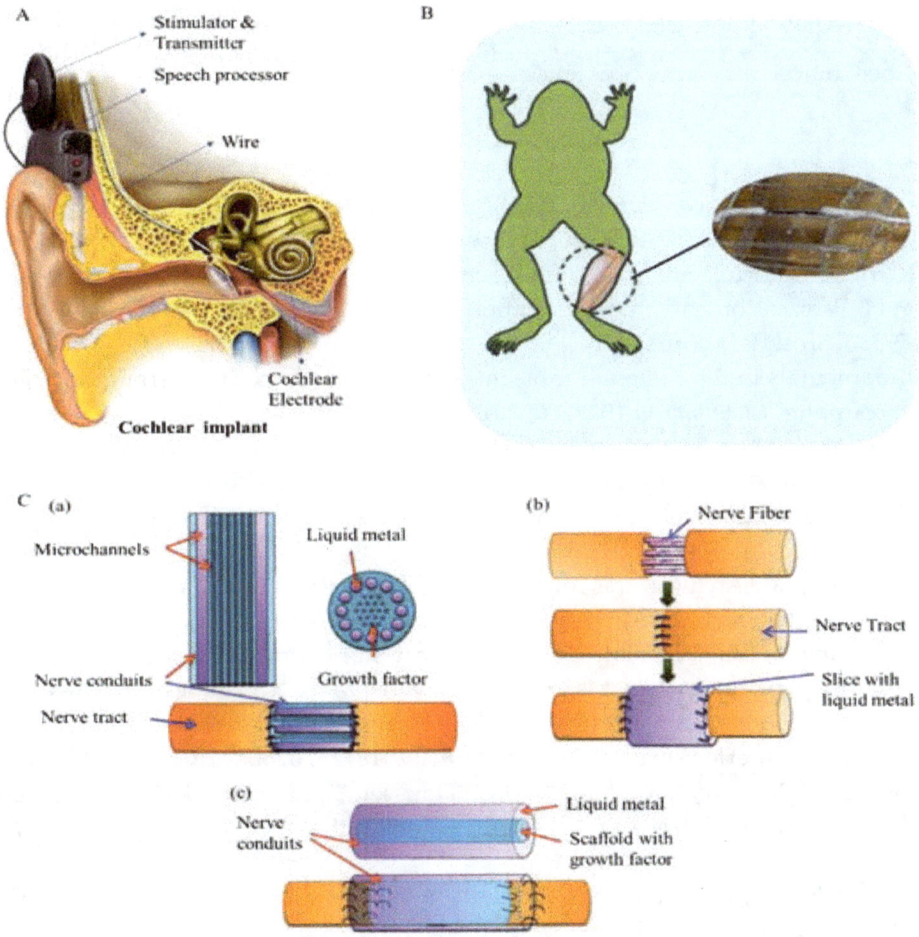

Figure 8. (**A**) The structure of a cochlear implant; (**B**) The transected sciatic nerve of a bullfrog reconnected by liquid metal [49]; (**C**) Three kinds of nerve conduits to repair the injured peripheral nerve: Nerve conduit with microchannels (**a**), a thin slice (**b**) and concentric tubes (**c**) [49]. (Note: Figures reproduced with permission).

6. Conclusions

This work presents a new conceptual droplet circuit that is enabled by liquid metal and demonstrates its potential role in the electronics field. The advantage of droplet circuits lies in that they address well the problem of circuit breaks and allow more flexibility and self-healing features, and show great potential for the development of smart soft electronics, which could imitate well the biological neurocircuits in nature. In addition, droplet circuits show potential for possible applications in future quantum calculation. Such droplet circuits are easily fabricated and self-error-correcting. This shows their promising value for the molding of large-scale application technologies. Further research could be conducted to fabricate stable and nano-scale liquid metal droplets in electrolyte solution and test the electrical properties of those droplets. The main difficulty of liquid-metal droplet circuits is the size of LMDs, which limits the progress of the liquid metal. Micro-scale and nano-scale circuits are currently the main trend. Compared to complementary metal oxide semiconductor (COMS) fabrication technology, the fabrication of liquid metal requires more improvement. Clearly, science, technology and applications in this direction require systematic investigation and integration in the future.

Author Contributions: Y.R. and J.L. analyzed the data and wrote the paper. J.L. conceived and supervised the project.

Acknowledgments: This work is partially supported by the National Natural Science Foundation of China (NSFC) Key Project under Grant No. 91748206, Dean's Research Funding of the Chinese Academy of Sciences and the Frontier Project of the Chinese Academy of Sciences.

References

1. Wang, X.; Liu, J. Recent advancements in liquid metal flexible printed electronics: Properties, technologies, and applications. *Micromachines* **2016**, *7*, 206. [CrossRef]
2. Aviram, A.; Ratner, M.A. Molecular rectifiers. *Chem. Phys. Lett.* **1974**, *29*, 277–283. [CrossRef]
3. Kim, Y.; Song, H. Fabrication and characterization of molecular electronic devices. *J. Nanosci. Nanotechnol.* **2015**, *15*, 921–938. [CrossRef] [PubMed]
4. Fox, M.A. Fundamentals in the design of molecular electronic devices Long-range charge carrier transport and electronic coupling. *Cheminform* **1999**, *30*, 201–207.
5. Delrosso, N.V.; Sarah, H.; Lee, S.; Nathan, D.D. A molecular circuit regenerator to implement iterative strand displacement operations. *Angew. Chem.* **2017**, *129*, 4514–4517. [CrossRef]
6. Son, J.Y.; Song, H. Molecular scale electronic devices using single molecules and molecular monolayers. *Curr. Appl. Phys.* **2013**, *13*, 1157–1171. [CrossRef]
7. Ellenbogen, J.C.; Love, J.C. Architectures for molecular electronic computers, i. logic structures and an adder designed from molecular electronic diodes. *Proc. IEEE* **2000**, *88*, 386–426. [CrossRef]
8. Wang, Q.; Yu, Y.; Yang, J.; Liu, J. Flexible electronics: Fast fabrication of flexible functional circuits based on liquid metal dual-trans printing. *Adv. Mater.* **2015**, *27*, 7109–7116. [CrossRef] [PubMed]
9. Daly, I.; Blanchard, C.; Holmes, N.P. Cortical excitability correlates with the event-related desynchronization during brain-computer interface control. *J. Neural Eng.* **2018**, *15*, 026022. [CrossRef] [PubMed]
10. Dejean, C.; Courtin, J.; Rozeske, R.R.; Bonnet, M.C.; Dousset, V.; Michelet, T.; Herry, C. Neuronal circuits for fear expression and recovery: Recent advances and potential therapeutic strategies. *Biol. Psychiatry* **2015**, *78*, 298–306. [CrossRef] [PubMed]
11. Purves, D.; Augustine, G.J.; Fitzpatrick, D.; Hall, W.C.; Lamantia, A.; Mcnamara, J.O.; Williams, S.M. *Neuroscience*, 3rd ed.; Sinauer Associates, Inc.: Sunderland, MA, USA, 2004.
12. Zhang, Y.; Wang, Y.; Zhou, G.; Jin, J.; Wang, B.; Wang, X.; Cichocki, A. Multi-kernel extreme learning machine for EEG classification in brain-computer interfaces. *Expert Syst. Appl.* **2018**, *96*, 302–310. [CrossRef]
13. Brandman, D.M.; Hosman, T.; Saab, J.; Burkhart, M.C.; Shanahan, B.E.; Ciancibello, J.G.; Sarma, A.A.; Milstein, D.J.; Vargas-Irwin, C.E.; Franco, B.; et al. Rapid calibration of an intracortical brain-computer interface for people with tetraplegia. *J. Neural Eng.* **2018**, *15*, 026007. [CrossRef] [PubMed]

14. Lees, S.; Dayan, N.; Cecotti, H.; Mccullagh, P.; Maguire, L.; Lotte, F.; Coyle, D. A review of rapid serial visual presentation-based brain-computer interfaces. *J. Neural Eng.* **2017**, *15*, 021001. [CrossRef] [PubMed]
15. Yi, L.; Liu, J. Liquid metal biomaterials: A newly emerging area to tackle modern biomedical challenges. *Int. Mater. Rev.* **2017**, *62*, 1–26. [CrossRef]
16. Jung, T.; Yang, S. Highly stable liquid metal-based pressure sensor integrated with a microfluidic channel. *Sensors* **2015**, *15*, 11823–11835. [CrossRef] [PubMed]
17. Yi, L.; Li, J.; Guo, C.; Li, L.; Liu, J. Liquid metal ink enabled rapid prototyping of electrochemical sensor for wireless glucose detection on the platform of mobile phone. *ASME J. Med. Devices* **2015**, *9*, 44507. [CrossRef]
18. Koo, H.J.; So, J.H.; Dickey, M.D.; Velev, O.D. Towards all-soft matter circuits: Prototypes of quasi-liquid devices with memristor characteristics. *Adv. Mater.* **2011**, *23*, 3559–3564. [CrossRef] [PubMed]
19. So, J.H.; Koo, H.J.; Dickey, M.D.; Velev, O.D. Ionic current rectification in soft-matter diodes with liquid-metal electrodes. *Adv. Funct. Mater.* **2012**, *22*, 625–631. [CrossRef]
20. Sun, X.; Yuan, B.; Rao, W.; Liu, J. Amorphous liquid metal electrodes enabled conformable electrochemical therapy of tumors. *Biomaterials* **2017**, *146*, 156–167. [CrossRef] [PubMed]
21. Li, J.; Guo, C.; Wang, Z.; Gao, K.; Shi, X.; Liu, J. Electrical stimulation towards melanoma therapy via liquid metal printed electronics on skin. *Clin. Transl. Med.* **2016**, *5*, 1–7. [CrossRef] [PubMed]
22. Guo, R.; Liu, J. Implantable liquid metal-based flexible neural microelectrode array and its application in recovering animal locomotion functions. *J. Micromech. Microeng.* **2017**, *27*, 104002. [CrossRef]
23. Yuan, B.; Wang, L.; Yang, X.; Ding, Y.; Tan, S.; Yi, L.; He, Z.; Liu, J. Liquid metal machine triggered violin-like wire oscillator. *Adv. Sci.* **2016**, *3*, 1600212. [CrossRef] [PubMed]
24. Zhao, X.; Tang, J.; Liu, J. Surfing liquid metal droplet on the same metal bath via electrolyte interface. *Appl. Phys. Lett.* **2017**, *111*, 101603. [CrossRef]
25. Tang, J.; Zhao, X.; Li, J.; Zhou, Y.; Liu, J. Liquid metal phagocytosis: Intermetallic wetting induced particle internalization. *Adv. Sci.* **2017**, *4*, 1700024. [CrossRef] [PubMed]
26. Yang, T.; Kwon, B.; Weisensee, P.B.; Kang, J.G.; Li, X.; Braun, P.; Miljkovic, N.; King, W.P. Millimeter-scale liquid metal droplet thermal switch. *Appl. Phys. Lett.* **2018**, *112*, 063505. [CrossRef]
27. Tang, S.Y.; Joshipura, I.D.; Lin, Y.; Kalantarzadeh, K.; Mitchell, A.; Khoshmanesh, K.; Dickey, M.D. Liquid-metal microdroplets formed dynamically with electrical control of size and rate. *Adv. Mater.* **2015**, *28*, 604–609. [CrossRef] [PubMed]
28. Chen, Y.; Liu, Z.; Zhu, D.; Handschuh-Wang, S.; Liang, S.; Yang, J.; Kong, T.; Zhou, X.; Liu, Y.; Zhou, X.C. Liquid metal droplets with high elasticity, mobility and mechanical robustness. *Mater. Horiz.* **2017**, *4*, 591–597. [CrossRef]
29. Chen, Y.; Zhou, T.; Li, Y.; Zhu, L.; Handschuh-Wang, S.; Zhu, D.; Zhou, X.; Liu, Z.; Gan, T.; Zhou, X. Robust fabrication of nonstick, noncorrosive, conductive graphene-coated liquid metal droplets for droplet-based, floating electrodes. *Adv. Funct. Mater.* **2018**, *28*, 1706277. [CrossRef]
30. Sivan, V.; Tang, S.Y.; O'Mullane, A.P.; Petersen, P.; Eshtiaghi, N.; Kalantar-Zadeh, K.; Mitchell, A. Liquid metal marbles. *Adv. Funct. Mater.* **2013**, *23*, 144–152. [CrossRef]
31. Zhao, X.; Tang, J.; Yu, Y.; Liu, J. Transformable soft quantum device based on liquid metals with sandwiched liquid junctions. *arXiv* **2017**.
32. Sen, D. The uncertainty relations in quantum mechanics. *Curr. Sci.* **2014**, *107*, 203–218.
33. Menzel, D.H.; Layzer, D. The physical principles of the quantum theory. *Philos. Sci.* **1930**, *16*, 303–324. [CrossRef]
34. Eisberg, R.; Resnick, R.; Brown, J. *Quantum Physics of Atoms, Molecules, Solids, Nuclei, and Particles*; Wiley: Hoboken, NJ, USA, 1974; p. 864.
35. Serway, R.; Serway, R. *College Physics 2*; Cengage Learning: Boston, MA, USA, 2005.
36. Oehme, M.; Karmous, A.; Sarlija, M.; Werner, J.; Kasper, E.; Schulze, J. Ge quantum dot tunneling diode with room temperature negative differential resistance. *Appl. Phys. Lett.* **2010**, *97*, 291. [CrossRef]
37. Ma, Z.; Zhou, P.; Zhang, T.; Liang, K.; Chu, P.K. Resonance-enhanced electroresistance-magnetoresistance effects in multiferroic tunnel junctions. *Mater. Res. Express* **2015**, *2*, 046303. [CrossRef]
38. Yu, Y.; Liu, F.; Zhang, R.; Liu, J. Suspension 3D printing of liquid metal into self-healing hydrogel. *Adv. Mater. Technol.* **2017**, *2*, 1700173. [CrossRef]
39. Yu, Y.; Wang, Q.; Yi, L.; Liu, J. Channelless fabrication for large-scale preparation of room temperature liquid metal droplets. *Adv. Eng. Mater.* **2013**, *16*, 255–262. [CrossRef]

40. Tian, L.; Gao, M.; Gui, L. A microfluidic chip for liquid metal droplet generation and sorting. *Micromachines* **2017**, *8*, 39. [CrossRef]
41. Ding, Y.; Liu, J. Water film coated composite liquid metal marble and its fluidic impact dynamics phenomenon. *Front. Energy* **2016**, *10*, 29–36. [CrossRef]
42. Che, Y.; Chen, H.; Gui, H.; Liu, J.; Liu, B.; Zhou, C. Review of carbon nanotube nanoelectronics and macroelectronics. *Semicond. Sci. Technol.* **2014**, *29*, 073001. [CrossRef]
43. Jin, C.; Zhang, J.; Li, X.; Yang, X.; Li, J.; Liu, J. Injectable 3-D fabrication of medical electronics at the target biological tissues. *Sci. Rep.* **2013**, *3*, 3442. [CrossRef] [PubMed]
44. Kim, D.; Lee, J.B. Magnetic-field-induced liquid metal droplet manipulation. *J. Korean Phys. Soc.* **2015**, *66*, 282–286. [CrossRef]
45. Tan, S.C.; Yuan, B.; Liu, J. Electrical method to control the running direction and speed of self-powered tiny liquid metal motors. *Proc. R. Soc. Lond. A Math. Phys. Eng. Sci.* **2015**, *471*, 32–38. [CrossRef]
46. Zhang, J.; Guo, R.; Liu, J. Self-propelled liquid metal motors steered by a magnetic or electrical field for drug delivery. *J. Mater. Chem. B* **2016**, *4*, 5349–5357. [CrossRef]
47. Zhang, J.; Sheng, L.; Liu, J. Synthetically chemical-electrical mechanism for controlling large scale reversible deformation of liquid metal objects. *Sci. Rep.* **2014**, *4*, 7116. [CrossRef] [PubMed]
48. De Castro, I.A.; Chrimes, A.F.; Zavabeti, A.; Berean, K.J.; Carey, B.J.; Zhuang, J.; Du, Y.; Dou, S.X.; Suzuki, K.; Shanks, R.A.; et al. A gallium-based magnetocaloric liquid metal ferrofluid. *Nano Lett.* **2017**, *17*, 7831–7838. [CrossRef] [PubMed]
49. Zhang, J.; Sheng, L.; Jin, C.; Liu, J. Liquid metal as connecting or functional recovery channel for the transected sciatic nerve. *arXiv* **2014**.

17

Rapid and Effective Electrical Conductivity Improvement of the Ag NW-Based Conductor by using the Laser-Induced Nano-Welding Process

Phillip Lee [1,†], Jinhyeong Kwon [2,†], Jinhwan Lee [2], Habeom Lee [2], Young D. Suh [2], Sukjoon Hong [3,*] and Junyeob Yeo [4,*]

1 Photo-Electronic Hybrids Research Center, National Agenda Research Division, Korea Institute of Science and Technology (KIST), 5 Hwarang-ro 14-gil, Seongbuk-gu, Seoul 02792, Korea; phillip@kist.re.kr
2 Applied Nano and Thermal Science (ANTS) Lab, Department of Mechanical Engineering, Seoul National University, 1 Gwanak-ro, Gwanak-gu, Seoul 00826, Korea; jhs0909k@snu.ac.kr (J.K.); mir.ljh@gmail.com (J.L.); habeom.lee@snu.ac.kr (H.L.); youngduksuh@gmail.com (Y.D.S.)
3 Department of Mechanical Engineering, Hanyang University, 55 Hanyangdaehak-ro, Sangnok-gu, Ansan, Gyeonggi-do 15588, Korea
4 Novel Applied Nano Optics (NANO) Lab, Department of Physics, Kyungpook National University, 80 Daehak-ro, Bukgu, Daegu 41566, Korea
* Correspondence: sukjoonhong@hanyang.ac.kr (S.H.); junyeob@knu.ac.kr (J.Y.)

† These authors contributed equally to this work.

Academic Editor: Zhigang Wu

Abstract: To date, the silver nanowire-based conductor has been widely used for flexible/stretchable electronics due to its several advantages. The optical nanowire annealing process has also received interest as an alternative annealing process to the Ag nanowire (NW)-based conductor. In this study, we present an analytical investigation on the phenomena of the Ag NWs' junction and welding properties under laser exposure. The two different laser-induced welding processes (nanosecond (ns) pulse laser-induced nano-welding (LINW) and continuous wave (cw) scanning LINW) are applied to the Ag NW percolation networks. The Ag NWs are selectively melted and merged at the junction of Ag NWs under very short laser exposure; these results are confirmed by scanning electron microscope (SEM), focused-ion beam (FIB), electrical measurement, and finite difference time domain (FDTD) simulation.

Keywords: laser; laser-induced nano-welding; pulse laser; silver nanowires; silver nanowire percolation networks; transparent flexible conductor

1. Introduction

For almost two decades, the flexible transparent conductor and stretchable conductor have received huge interest from many researchers and industry professionals. Since flexible electronics and stretchable sensors are core devices in wearable computer systems, as a next generation electronics platform, manufacturing/fabrication process technologies as well as flexible electronics compatible materials will be more important in the future.

Meanwhile, among the materials for the transparent conductor, fluorine-doped tin oxide (FTO) thin film is the most popular and famous material in research and industry fields. However, FTO thin film is not an appropriate material for the flexible transparent conductor since it is usually delicate and brittle. Hence, instead of FTO thin film, alternative conducting nanomaterials such as carbon nanotube (CNT) [1,2], graphene [3–5], metal nanoparticle (NP) mesh [6–9], metal nanowires (NWs) [10–19], and metal nano-thin film [20,21] are used in the research field for the flexible transparent conductor and stretchable conductor.

Among the nanomaterials for flexible electronics, silver (Ag) NWs have been widely used as a flexible transparent conductor [10–13,16,22–24] and stretchable conductor [15,17,25] due to various advantages such as high electrical conductivity, high transparency, high ductility and simple fabrication process methods. However, a post thermal annealing process is usually required to increase electrical conductivity since Ag NWs are usually too short (up to ~100 μm) to cover the wide area and are generally covered by capping polymer such as polyvinylpyrrolidone (PVP) which hinders electrical conductivity. The conventional thermal annealing process, such as a hot plate and convection oven, is simple and easily applicable to the Ag NW percolation networks for electrical conductivity improvement, while the thermal annealing process has disadvantages such as oxidation problems and long processing time. In particular, the thermal annealing process is not suitable to the flexible polymer substrate due to the low melting temperature of the polymer. Thus, it is important to conduct a post thermal annealing process with low temperature (below 200 °C) to prevent the damages of the flexible polymer substrate.

Recently, the optical NWs annealing process [14,15,18,19,26–28] was introduced to anneal metal NW percolation networks for the improvement of electrical conductivity. Compared to the conventional thermal annealing process, since optical energy in the optical NWs annealing process is irradiated to the sample over a very short time by a ultraviolet (UV) lamp [14], flash light [26] or scanning laser [18], the optical NWs annealing process is very rapid and suitable for the flexible polymer substrate without any macroscopic damages or deformation of the substrate [29,30] under ambient conditions. In addition, oxidation problems on metal NWs during processing are suppressed due to fast processing [18].

Most previous research [15,18,19,26,27,31] focuses on the fabrication of flexible/stretchable electronics and their applications. However, in this study, we attempt to examine mainly the phenomena of Ag NWs' junction and welding properties when a laser is irradiated to the Ag NW percolation networks. In particular, we focus on how the local laser exposure time (from ns to μs) affects the Ag NWs' percolation networks in laser processing. Thus, we compare two different laser-induced nano-welding (LINW) processes as a post annealing process of Ag NWs: the continuous wave (cw) laser scanning system and the nanosecond (ns) pulse laser system. Although the processing mechanism of two LINW processes is basically the same in terms of using laser energy, there are several similar and different results, which are mentioned in the text.

2. Methods

2.1. Experimental Procedure

Figure 1a shows the preparation of an Ag NW-based conductor sample. In order to prepare a film of Ag NW percolation networks on the substrate, Ag NWs are deposited according to the following procedures. Firstly, Ag NWs are synthesized in a solution via the polyol synthesis method [32]. Afterwards, synthesized Ag NWs are filtered out onto the Teflon filter and transferred onto the glass substrate successively. The diameter of the transferred Ag NW percolation networks is ~36 mm due to the size of the Teflon filter. Since transferred Ag NW percolation networks on the glass substrate are weakly bound, the post thermal annealing process, such as a hot plate and furnace, is required to increase the electrical conductivity of the Ag NW percolation networks. Additionally, the LINW process is conducted to the Ag NW percolation networks as an alternative post annealing process for comparison.

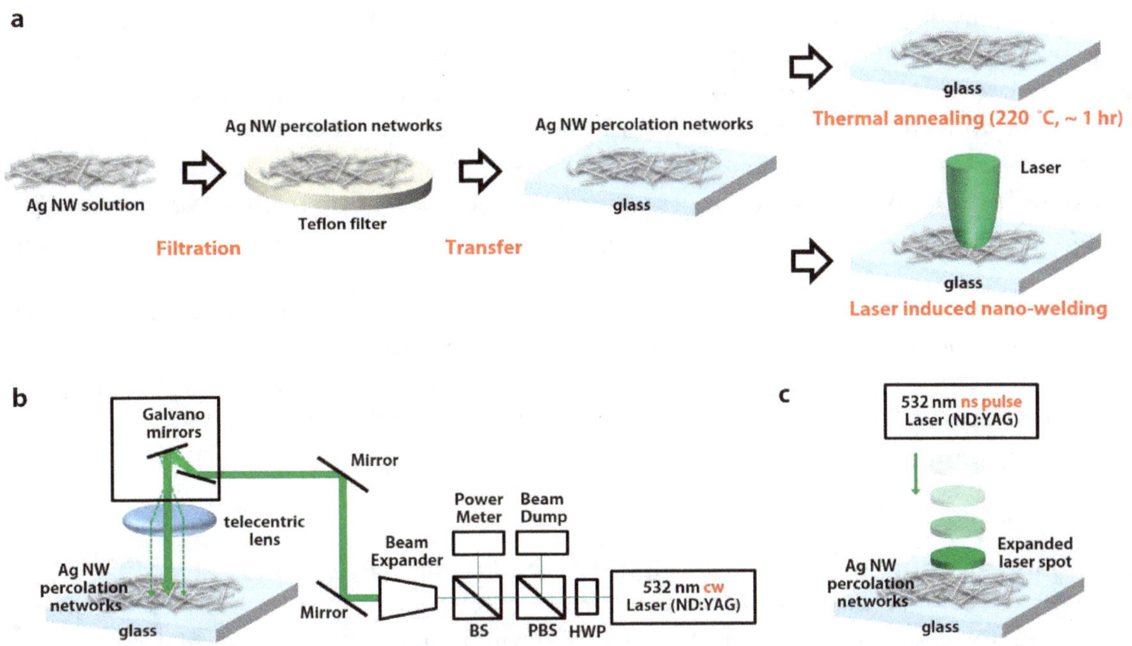

Figure 1. Schematic diagram of the thermal annealing process and laser-induced nano-welding process for Ag nanowires (NWs) percolation networks. (**a**) Sample preparation flow chart. Firstly, Ag NWs are synthesized in a solution by polyol synthesis. Afterwards, they are filtered on the Teflon filter and transferred onto the glass substrate to form Ag NW percolation networks on glass substrate. Finally, the thermal annealing process or laser-induced welding process is applied to Ag NW percolation networks for the improvement of electrical conductivity. The laser-induced nano-welding process through (**b**) the continuous wave (cw) laser scanning system and (**c**) the ns pulse laser system.

2.2. Optical Setup

As we mentioned, two different LINW processes (the cw laser scanning system and the ns pulse laser system) are compared in this study. Figure 1b shows the cw laser scanning system which is combined with Galvano scanning mirrors and a telecentric lens [30,33]. In the cw laser scanning system, a 532 nm cw laser (Millennia 5W, Spectra Physics, Santa Clara, CA, USA) is used on the sample. As shown in Figure 1b, the power of the emitted laser beam is easily controlled through the half wave plate (HWP) and polarized beam splitter (PBS). The beam expander is placed afterwards to enlarge the laser beam for a flat wavefront of laser beam. The angle of the laser beam is deviated by a laser scanner (HurryScan II, Scanlab, Puchheim, Germany) which consists of two electrically driven Galvano mirrors. Afterwards, the laser beam is uniformly focused (10 μm) on the 2D focal plane, without distortion aberration, by a long focal distance f-theta lens (f = 103 mm). The prepared sample that consists of Ag NW percolation networks is placed on the focal plane of the f-theta lens. Figure 1c shows the ns pulse laser system which consists of a 532 nm ns pulse laser (Tempest 300, NewWave, Redwood City, CA, USA) and beam expander. The pulse duration and the repetition rate of the ns laser are 5 ns and 10 Hz, respectively. Since the energy of the applied ns pulse laser is extremely high, only one single shot with proper energy density (mJ/cm^2) is enough for the enhancement of electrical conductivity in Ag NW percolation networks. However, excessive laser energy can ablate Ag NW percolation networks, thus the energy density in the ns pulse laser and the power density (W/cm^2) in the cw laser are carefully controlled by adjustment of the beam waist area through the beam expander and power adjustment, respectively.

2.3. Laser Processing

Firstly, the prepared sample is placed at the focal plane of the applied laser. Afterwards, the laser is irradiated on the sample. In the cw scanning LINW process, the diameter of the prepared sample is ~36 mm. Additionally, the spot size of the focused laser, the laser scanning speed, and the pitch of scanning are 10 μm, 100 mm/s, and 10 μm, respectively. Since the total processing time in the cw scanning LINW process is dependent on sample size and laser scanning speed, fast laser scanning speed is desirable to reduce the processing time. Meanwhile, the laser dwell time, τ ($\tau = 2W_0/v$; τ, W_0, and v are dwell time, beam waist, and scanning speed, respectively), is decreased as the scanning speed increases. Thus, laser energy density (power density by dwell time) and total processing time are in a trade-off relationship, and a laser scanning speed of 100 mm/s is chosen as an optimum scanning speed in this study.

In the case of the ns pulse LINW process, the laser beam is expanded by the beam expander to reduce laser energy density (energy per unit area), since the pulse laser energy is sufficiently high (see Figure 1c). The prepared sample size of Ag NW percolation networks is 6 mm by 6 mm in the ns pulse LINW process, and the extended laser beam illuminates and covers the entire prepared sample area (see Figure S1 in the Supplementary Information).

3. Results and Discussion

Figure 2 shows a photographic image (Figure 2a) and scanning electron microscope (SEM) images (Figure 2b,c) of Ag NW percolation networks after the LINW process. In the case of the conventional thermal annealing process, such as a hot plate [15] and convection oven, the thermal annealing process is often conducted at a low temperature (e.g., below 250 °C) on the flexible substrate due to the low melting temperature of the polymer substrate. Therefore, it is difficult to find meaningful differences in the SEM images of Ag NW percolation networks after the thermal annealing process compared to before the thermal annealing process at a low temperature (below 250 °C).

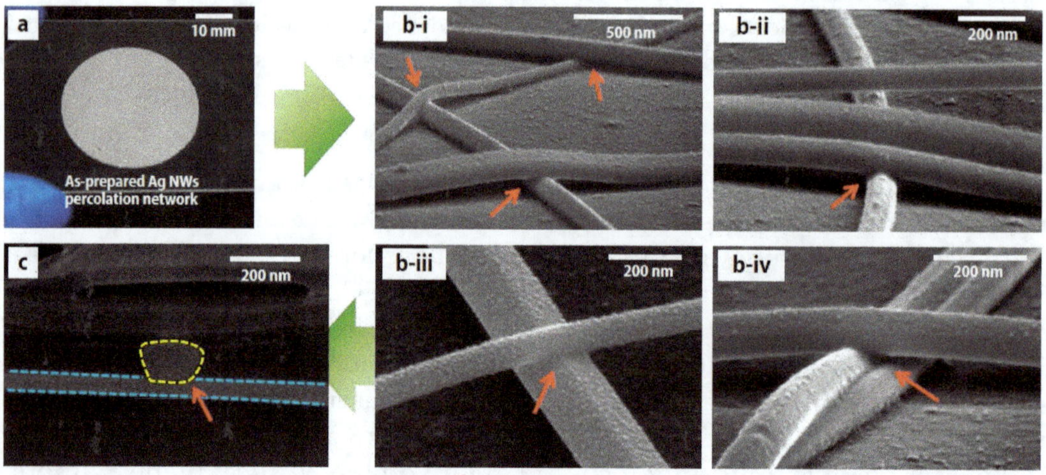

Figure 2. (**a**) Photographic image of the Ag NW percolation networks on the glass substrate (30% transmittance). (**b**) Magnified scanning electron microscope (SEM) images of the Ag NW percolation networks after the laser-induced welding process. (**c**) Cross-sectional SEM image of the junction of the Ag NW percolation networks after the laser-induced welding process. Two Ag NWs are melted and merged at the boundary of two Ag NWs (yellow dots and blue dots). Red arrows represent a junction of Ag NWs in (**b**,**c**).

However, in the case of the LINW process (same for both ns pulse LINW and cw scanning LINW) under ambient conditions (room temperature and atmospheric pressure), melted and merged Ag NWs are found in SEM images after the LINW process, as shown in Figure 2b-i–2b-iv. Since the irradiated laser energy is intensively absorbed to heat up Ag NWs and the laser irradiation time is extremely short (from ns to μs), crossed Ag NWs are only melted and merged at the junction, without damaging the other area of Ag NWs, as shown in Figure 2b.

In order to verify the melting at the junction of Ag NWs, a cross-sectional SEM image at the boundary which is cut by a focused-ion beam (FIB), is examined. It is confirmed that the crossed area of two Ag NWs (yellow and blue dotted lines) are melted and fused at the junction, as shown in Figure 2c. The results for the flexible transparent conductor and stretchable conductor are very noticeable, since melted and merged Ag NW percolation networks will have better electrical/mechanical properties such as electrical conductivity, mechanical elongation, and mechanical strength [13,15,18,19].

Figure 3 shows sheet resistance changes at various times under the thermal annealing process and ns pulse LINW process. The transmittance of prepared Ag NW percolation networks in (a) and (b) are 95% and 96%, respectively. As shown in Figure 3a, the sheet resistance of Ag NW percolation networks is gradually increased and gently dropped below 20 Ω/sq with long processing time (over 1 h) under the thermal annealing process. At first, the sheet resistance is gradually increased due to oxidation formation and the resistance increase of Ag NWs to temperature change, according to their temperature coefficient of resistance [34,35]. Once the temperature of the hot plate reaches 220 °C (~1300 s), the sheet resistance starts to drop due to slight melting at the junction of Ag NW percolation networks. The thermal annealing process ensures stable low sheet resistance in Ag NW percolation networks and easily scales up the sample size. However, long processing time (over 1 h) is generally required to increase the electrical conductivity in the thermal annealing process, since a period of warm-up time for heating is required and only low temperature (below 250 °C) is available for the flexible substrate.

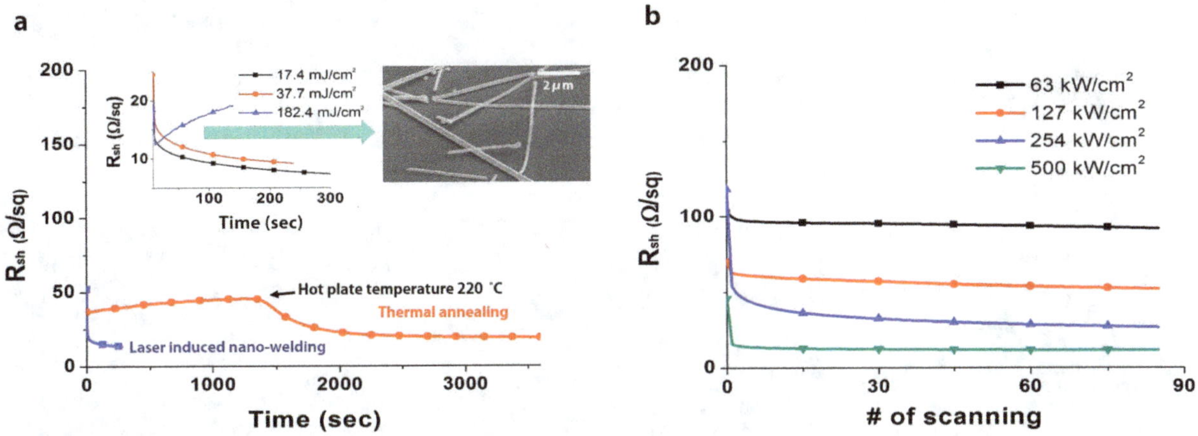

Figure 3. (**a**) Sheet resistance changes at various times in the thermal annealing process and ns pulse laser-induced nano-welding (LINW) process. The inset graph shows a magnified view of the ns pulse LINW process. SEM images show the laser ablation results of Ag NW percolation networks. (**b**) The sheet resistance changes with various numbers of laser scans and various laser power levels in the cw scanning LINW process.

In contrast, the sheet resistance of Ag NW percolation networks drops rapidly in the ns pulse LINW process compared to the thermal annealing process. As shown in the inset graph of Figure 3a, the sheet resistance drops immediately after the start of pulse laser exposure with proper energy density (17.4 mJ/cm^2 and 37.7 mJ/cm^2). Even though a single pulse can be enough to improve the electrical conductivity of Ag NW percolation networks, continued laser pulses (10 Hz repetition rate) are employed in the sample to further improve the conductivity. However, extremely high laser energy density (182.4 mJ/cm^2) can ablate/destroy Ag NW percolation networks (SEM image of Figure 3a), resulting in an increase of the sheet resistance of Ag NWs during the laser exposure—so-called "rebound", as shown in the inset graph (light blue line) of Figure 3a. Thus, in Ag NW percolation networks in the ns pulse LINW process, moderate adjustment of the applied laser energy is required.

Similar to the ns pulse LINW process, in the case of the cw scanning LINW process, the sheet resistance of Ag NW percolation networks drops rapidly below 20 Ω/sq with high power density

(500 kW/cm^2), as shown in Figure 3b. The sheet resistance decreases slightly as the number of scans increases, while the sheet resistance drops considerably with high power density. The spot size ($2W_0$) of the focused laser by the telecentric lens in laser scanning system is ~10 μm. Additionally, the dwell time is 10^{-4} s (10 μm/100 mms^{-1}), thus it is also a very short time compared to the conventional thermal annealing process.

As a result, the LINW process is a more rapid and effective process (due to melting and merging) for improving the electrical conductivity of Ag NW percolation networks than the conventional thermal annealing process. Moreover, it is noticeable that there are no significant differences with respect to the different laser exposure time (from ns to μs) in laser processing. These results are confirmed by SEM images and electrical conductivity measurements. In addition, since the reaction in Ag NWs welding is conducted within an extremely short time—5 ns laser exposure time—this LINW process can be applied to the flexible substrate without any macroscopic damages or deformation of the substrate.

Figure 4 shows SEM images of ablated Ag NWs when the excessive laser energy is applied to the Ag NW percolation networks. Two different LINW processes show fairly different results at excessive laser energy density. Since an extended laser spot is applied to the Ag NW percolation networks in the ns pulse LINW process, entire Ag NWs are heated and melted, resulting in a queue of molten silver micro dots (right SEM images in Figure 4a).

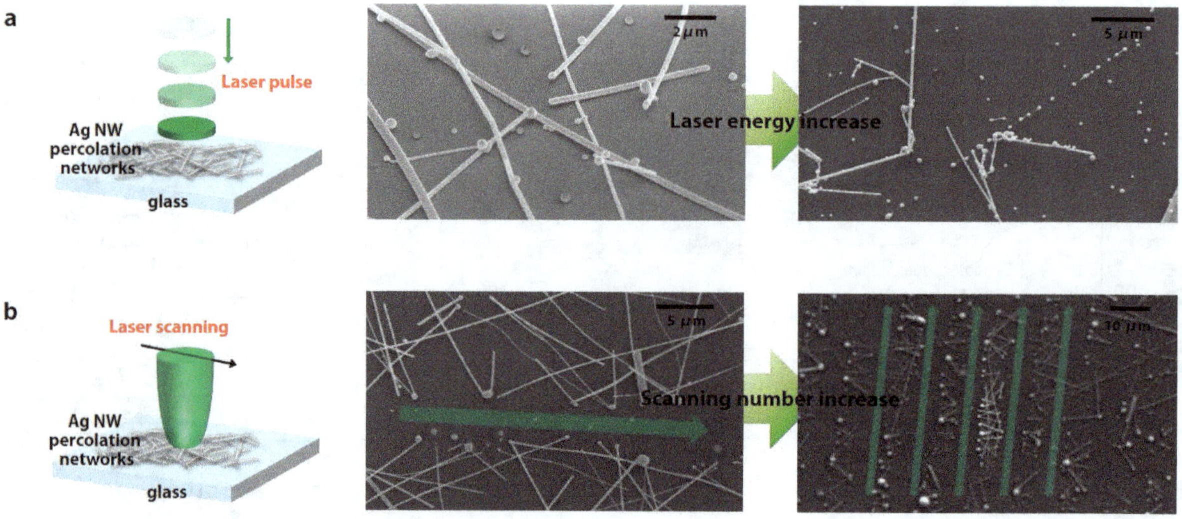

Figure 4. SEM images of Ag NW percolation networks when (**a**) the ns pulse LINW process and (**b**) the cw scanning LINW process are conducted under extremely high power/energy condition.

On the other hand, Ag NWs are selectively melted along the laser scanning direction in the cw scanning LINW process, as shown in Figure 4b (green arrows in SEM images). Since Ag NWs are ablated along the laser scanning direction, the remaining Ag NWs are locally connected to each other, thus this laser ablation technique is applied to fabricate patterned Ag NW mesh for flexible capacitive touch sensors [31].

As shown in previous SEM images, Ag NWs are easily heated and melted in the LINW process when the proper laser energy is irradiated to the Ag NW percolation networks. It is well known that the electromagnetic field enhancements on the surface of Ag NW generate localized thermal heating due to surface plasmon polaritons (SPP) on the surface of Ag NWs [18,36–39]. This behavior can be seen in finite difference time domain (FDTD) simulation (Lumerical), as shown in Figure 5. In this simulation, transverse magnetic (TM) and transverse electric (TE) modes are considered for the two crossed and stacked Ag NWs. In order to simplify the simulation, the diameter size and shape of Ag NWs are fixed at 100 nm and circle, respectively. The complex permittivity of Ag is adopted from Palik, and a simulated pulse covers the wavelength range from 300 to 1000 nm. Total-field/scattering-field (TF-SF), together with perfectly matched layer (PML) formulation, has been employed.

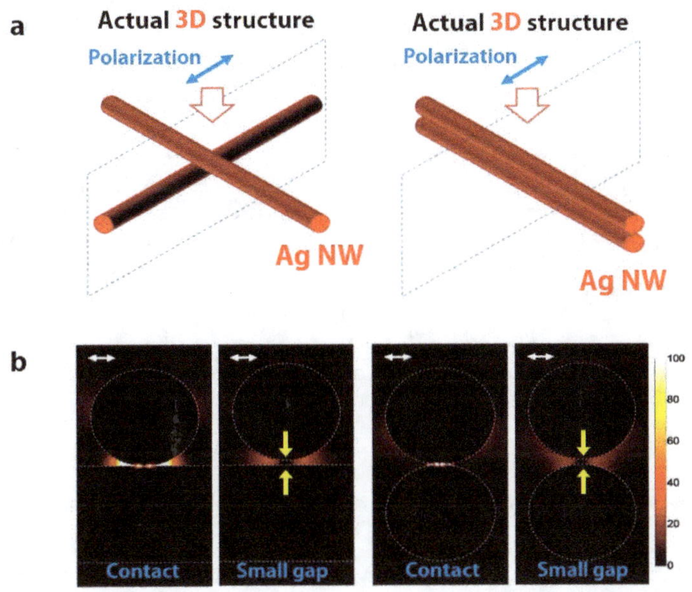

Figure 5. Finite difference time domain (FDTD) simulation at the junction of crossed Ag NWs. (**a**) Simulation layout of two crossed (left) and stacked (right) Ag NWs. (**b**) Electromagnetic field distribution at the junction with various conditions: crossed/contact, crossed/small gap, stacked/contact, stacked/small gap of two Ag NWs. The white arrow is the polarized direction of irradiated light. The yellow arrow indicates a small gap between two Ag NWs.

As shown in Figure 5b, the electromagnetic field enhancements are extremely maximized at the junction of two crossed and stacked Ag NWs. In addition, electromagnetic field enhancement are still maximized near the junction of two Ag NWs, even though two Ag NWs are separated from each other by a small gap. The optical absorption, or the volumetric heat source density generated inside the metal, is calculated by [36]

$$q\left(\vec{r}\right) = (\omega/2)\mathrm{Im}(\varepsilon)\left|E\left(\vec{r}\right)\right|^2 \quad (1)$$

As can be seen from the above equation, the optical absorption is directly proportional to the electrical field intensity, and the simulation result in Figure S2 shows that the optical absorption is concentrated at the regions where the field enhancement adjacent to the surface of the nanowire is the largest. This is the reason why Ag NWs are well melted and merged at the junction of Ag NWs, as shown in Figure 2. Additionally, these results are the reason why 5 nanoseconds is enough to improve the electrical conductivity of Ag NW percolation networks.

In summary, two different laser-induced nano-welding processes (ns pulse LINW and cw scanning LINW), as alternative post annealing processes, are investigated to enhance the electrical conductivity of an Ag NW-based conductor in this study. Thus, various phenomena of Ag NWs are examined when the laser irradiates to the Ag NW percolation networks. Through the various characterizations (SEM, FIB, and electrical measurements) and FDTD simulation, it is confirmed that there are no significant differences with respect to the different laser exposure time (from ns to μs). Additionally, the Ag NWs can be selectively melted and merged at the junction of Ag NWs within less than 5 nanoseconds laser exposure.

These results indicate that the LINW process is expected to apply to the flexible polymer substrate without any macroscopic damages or deformation of the substrate due to the rapid processing time and the effect of localized electromagnetic field enhancements. In addition, since melted and merged Ag NW percolation networks will have better electrical/mechanical properties, we expect that the LINW process will be applied to the fabrication of various flexible/stretchable electronics for better performance.

Acknowledgments: This work is supported by the National Research Foundation of Korea (NRF) Grant funded through Basic Science Research Program (NRF-2016R1C1B1014729), Ministry of Trade Industry & Energy (MOTIE) through development program "Development of high drapability of textile type dye-sensitized solar cell materials and outdoor applications (project No. 10052064)", the research fund of Hanyang University (HY-2016-00000002097), and the ICT & Future and the R&D Convergence Program through the R&D program of MSIP/COMPA (Grant No. 2015K000216).

Author Contributions: P.L., J.K., S.H., and J.Y. designed and conducted experiments. J.L. and H.L. synthesized the silver nanowires. J.K. and Y.D.S. prepared sample and conducted various characterization. S.H. conducted FDTD simulation. P.L., J.K., S.H., and J.Y. co-wrote the manuscript.

References

1. Kaempgen, M.; Duesberg, G.S.; Roth, S. Transparent carbon nanotube coatings. *Appl. Surf. Sci.* **2005**, *252*, 425–429. [CrossRef]
2. Delozier, D.M.; Watson, K.A.; Smith, J.G.; Clancy, T.C.; Connell, J.W. Investigation of aromatic/aliphatic polyimides as dispersants for single wall carbon nanotubes. *Macromolecules* **2006**, *39*, 1731–1739. [CrossRef]
3. Eda, G.; Fanchini, G.; Chhowalla, M. Large-area ultrathin films of reduced graphene oxide as a transparent and flexible electronic material. *Nat. Nanotechnol.* **2008**, *3*, 270–274. [CrossRef] [PubMed]
4. Kim, K.S.; Zhao, Y.; Jang, H.; Lee, S.Y.; Kim, J.M.; Kim, K.S.; Ahn, J.-H.; Kim, P.; Choi, J.-Y.; Hong, B.H. Large-scale pattern growth of graphene films for stretchable transparent electrodes. *Nature* **2009**, *457*, 706–710. [CrossRef] [PubMed]
5. Zheng, Q.; Zhang, B.; Lin, X.; Shen, X.; Yousefi, N.; Huang, Z.-D.; Li, Z.; Kim, J.-K. Highly transparent and conducting ultralarge graphene oxide/single-walled carbon nanotube hybrid films produced by langmuir-blodgett assembly. *J. Mater. Chem.* **2012**, *22*, 25072–25082. [CrossRef]
6. Yeo, J.; Hong, S.; Lee, D.; Hotz, N.; Lee, M.T.; Grigoropoulos, C.P.; Ko, S.H. Next generation non-vacuum, maskless, low temperature nanoparticle ink laser digital direct metal patterning for a large area flexible electronics. *PLoS ONE* **2012**, *7*, e42315. [CrossRef] [PubMed]
7. Hong, S.; Yeo, J.; Kim, G.; Kim, D.; Lee, H.; Kwon, J.; Lee, H.; Lee, P.; Ko, S.H. Nonvacuum, maskless fabrication of a flexible metal grid transparent conductor by low-temperature selective laser sintering of nanoparticle ink. *ACS Nano* **2013**, *7*, 5024–5031. [CrossRef] [PubMed]
8. Liu, Y.-K.; Lee, M.-T. Laser direct synthesis and patterning of silver nano/microstructures on a polymer substrate. *ACS Appl. Mater. Interfaces* **2014**, *6*, 14576–14582. [CrossRef] [PubMed]
9. Tsai, S.-L.; Liu, Y.-K.; Pan, H.; Liu, C.-H.; Lee, M.-T. The coupled photothermal reaction and transport in a laser additive metal nanolayer simultaneous synthesis and pattering for flexible electronics. *Nanomaterials* **2016**, *6*, 12. [CrossRef] [PubMed]
10. Lee, J.-Y.; Connor, S.T.; Cui, Y.; Peumans, P. Solution-processed metal nanowire mesh transparent electrodes. *Nano Lett.* **2008**, *8*, 689–692. [CrossRef] [PubMed]
11. De, S.; Higgins, T.M.; Lyons, P.E.; Doherty, E.M.; Nirmalraj, P.N.; Blau, W.J.; Boland, J.J.; Coleman, J.N. Silver nanowire networks as flexible, transparent, conducting films: Extremely high DC to optical conductivity ratios. *ACS Nano* **2009**, *3*, 1767–1774. [CrossRef] [PubMed]
12. Hu, L.; Kim, H.S.; Lee, J.Y.; Peumans, P.; Cui, Y. Scalable coating and properties of transparent, flexible, silver nanowire electrodes. *ACS Nano* **2010**, *4*, 2955–2963. [CrossRef] [PubMed]
13. Lee, J.; Lee, P.; Lee, H.; Lee, D.; Lee, S.S.; Ko, S.H. Very long Ag nanowire synthesis and its application in a highly transparent, conductive and flexible metal electrode touch panel. *Nanoscale* **2012**, *4*, 6408–6414. [CrossRef] [PubMed]
14. Garnett, E.C.; Cai, W.; Cha, J.J.; Mahmood, F.; Connor, S.T.; Greyson Christoforo, M.; Cui, Y.; McGehee, M.D.; Brongersma, M.L. Self-limited plasmonic welding of silver nanowire junctions. *Nat. Mater.* **2012**, *11*, 241–249. [CrossRef] [PubMed]
15. Lee, P.; Lee, J.; Lee, H.; Yeo, J.; Hong, S.; Nam, K.H.; Lee, D.; Lee, S.S.; Ko, S.H. Highly stretchable and highly conductive metal electrode by very long metal nanowire percolation network. *Adv. Mater.* **2012**, *24*, 3326–3332. [CrossRef] [PubMed]

16. Lee, J.; Lee, P.; Lee, H.B.; Hong, S.; Lee, I.; Yeo, J.; Lee, S.S.; Kim, T.-S.; Lee, D.; Ko, S.H. Room-temperature nanosoldering of a very long metal nanowire network by conducting-polymer-assisted joining for a flexible touch-panel application. *Adv. Funct. Mater.* **2013**, *23*, 4171–4176. [CrossRef]
17. Lee, P.; Ham, J.; Lee, J.; Hong, S.; Han, S.; Suh, Y.D.; Lee, S.E.; Yeo, J.; Lee, S.S.; Lee, D.; et al. Highly stretchable or transparent conductor fabrication by a hierarchical multiscale hybrid nanocomposite. *Adv. Funct. Mater.* **2014**, *24*, 5671–5678. [CrossRef]
18. Han, S.; Hong, S.; Ham, J.; Yeo, J.; Lee, J.; Kang, B.; Lee, P.; Kwon, J.; Lee, S.S.; Yang, M.Y.; et al. Fast plasmonic laser nanowelding for a Cu-nanowire percolation network for flexible transparent conductors and stretchable electronics. *Adv. Mater.* **2014**, *26*, 5808–5814. [CrossRef] [PubMed]
19. Song, W.G.; Kwon, H.-J.; Park, J.; Yeo, J.; Kim, M.; Park, S.; Yun, S.; Kyung, K.-U.; Grigoropoulos, C.P.; Kim, S.; et al. High-performance flexible multilayer MoS_2 transistors on solution-based polyimide substrates. *Adv. Funct. Mater.* **2016**, *26*, 2426–2434. [CrossRef]
20. Paeng, D.; Yoo, J.-H.; Yeo, J.; Lee, D.; Kim, E.; Ko, S.H.; Grigoropoulos, C.P. Low-cost facile fabrication of flexible transparent copper electrodes by nanosecond laser ablation. *Adv. Mater.* **2015**, *27*, 2762–2767. [CrossRef] [PubMed]
21. Qi, D.; Paeng, D.; Yeo, J.; Kim, E.; Wang, L.; Chen, S.; Grigoropoulos, C.P. Time-resolved analysis of thickness-dependent dewetting and ablation of silver films upon nanosecond laser irradiation. *Appl. Phys. Lett.* **2016**, *108*, 211602. [CrossRef]
22. Lee, H.; Hong, S.; Lee, J.; Suh, Y.D.; Kwon, J.; Moon, H.; Kim, H.; Yeo, J.; Ko, S.H. Highly stretchable and transparent supercapacitor by Ag–Au core–shell nanowire network with high electrochemical stability. *ACS Appl. Mater. Interfaces* **2016**, *8*, 15449–15458. [CrossRef] [PubMed]
23. Suh, Y.D.; Jung, J.; Lee, H.; Yeo, J.; Hong, S.; Lee, P.; Lee, D.; Ko, S.H. Nanowire reinforced nanoparticle nanocomposite for highly flexible transparent electrodes: Borrowing ideas from macrocomposites in steel-wire reinforced concrete. *J. Mater. Chem. C* **2017**, *5*, 791–798. [CrossRef]
24. Moon, H.; Lee, H.; Kwon, J.; Suh, Y.D.; Kim, D.K.; Ha, I.; Yeo, J.; Hong, S.; Ko, S.H. Ag/au/polypyrrole core-shell nanowire network for transparent, stretchable and flexible supercapacitor in wearable energy devices. *Sci. Rep.* **2017**, *7*, 41981. [CrossRef] [PubMed]
25. Hong, S.; Lee, H.; Lee, J.; Kwon, J.; Han, S.; Suh, Y.D.; Cho, H.; Shin, J.; Yeo, J.; Ko, S.H. Highly stretchable and transparent metal nanowire heater for wearable electronics applications. *Adv. Mater.* **2015**, *27*, 4744–4751. [CrossRef] [PubMed]
26. Li, R.Z.; Hu, A.; Zhang, T.; Oakes, K.D. Direct writing on paper of foldable capacitive touch pads with silver nanowire inks. *ACS Appl. Mater. Interfaces* **2014**, *6*, 21721–21729. [CrossRef] [PubMed]
27. Spechler, J.A.; Nagamatsu, K.A.; Sturm, J.C.; Arnold, C.B. Improved efficiency of hybrid organic photovoltaics by pulsed laser sintering of silver nanowire network transparent electrode. *ACS Appl. Mater. Interfaces* **2015**, *7*, 10556–10562. [CrossRef] [PubMed]
28. Hong, S.; Lee, H.; Yeo, J.; Ko, S.H. Digital selective laser methods for nanomaterials: From synthesis to processing. *Nano Today* **2016**, *11*, 547–564. [CrossRef]
29. In, J.B.; Hsia, B.; Yoo, J.-H.; Hyun, S.; Carraro, C.; Maboudian, R.; Grigoropoulos, C.P. Facile fabrication of flexible all solid-state micro-supercapacitor by direct laser writing of porous carbon in polyimide. *Carbon* **2015**, *83*, 144–151. [CrossRef]
30. Yeo, J.; Kim, G.; Hong, S.; Kim, M.S.; Kim, D.; Lee, J.; Lee, H.B.; Kwon, J.; Suh, Y.D.; Kang, H.W.; et al. Flexible supercapacitor fabrication by room temperature rapid laser processing of roll-to-roll printed metal nanoparticle ink for wearable electronics application. *J. Power Sources* **2014**, *246*, 562–568. [CrossRef]
31. Hong, S.; Yeo, J.; Lee, J.; Lee, H.; Lee, P.; Lee, S.S.; Ko, S.H. Selective laser direct patterning of silver nanowire percolation network transparent conductor for capacitive touch panel. *J. Nanosci. Nanotechnol.* **2015**, *15*, 2317–2323. [CrossRef] [PubMed]
32. Lee, J.H.; Lee, P.; Lee, D.; Lee, S.S.; Ko, S.H. Large-scale synthesis and characterization of very long silver nanowires via successive multistep growth. *Cryst. Growth Des.* **2012**, *12*, 5598–5605. [CrossRef]
33. Lee, H.; Hong, S.; Kwon, J.; Suh, Y.D.; Lee, J.; Moon, H.; Yeo, J.; Ko, S.H. All-solid-state flexible supercapacitors by fast laser annealing of printed metal nanoparticle layers. *J. Mater. Chem. A* **2015**, *3*, 8339–8345. [CrossRef]
34. Rathmell, A.R.; Bergin, S.M.; Hua, Y.-L.; Li, Z.-Y.; Wiley, B.J. The growth mechanism of copper nanowires and their properties in flexible, transparent conducting films. *Adv. Mater.* **2010**, *22*, 3558–3563. [CrossRef] [PubMed]

35. Rathmell, A.R.; Wiley, B.J. The synthesis and coating of long, thin copper nanowires to make flexible, transparent conducting films on plastic substrates. *Adv. Mater.* **2011**, *23*, 4798–4803. [CrossRef] [PubMed]
36. Bell, A.P.; Fairfield, J.A.; McCarthy, E.K.; Mills, S.; Boland, J.J.; Baffou, G.; McCloskey, D. Quantitative study of the photothermal properties of metallic nanowire networks. *ACS Nano* **2015**, *9*, 5551–5558. [CrossRef] [PubMed]
37. Kang, T.; Yoon, I.; Jeon, K.-S.; Choi, W.; Lee, Y.; Seo, K.; Yoo, Y.; Park, Q.H.; Ihee, H.; Suh, Y.D.; et al. Creating well-defined hot spots for surface-enhanced raman scattering by single-crystalline noble metal nanowire pairs. *J. Phys. Chem. C* **2009**, *113*, 7492–7496. [CrossRef]
38. Prokes, S.M.; Alexson, D.A.; Glembocki, O.J.; Park, H.D.; Rendell, R.W. Effect of crossing geometry on the plasmonic behavior of dielectric core/metal sheath nanowires. *Appl. Phys. Lett.* **2009**, *94*, 093105. [CrossRef]
39. Lei, D.Y.; Aubry, A.; Maier, S.A.; Pendry, J.B. Broadband nano-focusing of light using kissing nanowires. *New J. Phys.* **2010**, 093030. [CrossRef]

Permissions

All chapters in this book were first published in MDPI; hereby published with permission under the Creative Commons Attribution License or equivalent. Every chapter published in this book has been scrutinized by our experts. Their significance has been extensively debated. The topics covered herein carry significant findings which will fuel the growth of the discipline. They may even be implemented as practical applications or may be referred to as a beginning point for another development.

The contributors of this book come from diverse backgrounds, making this book a truly international effort. This book will bring forth new frontiers with its revolutionizing research information and detailed analysis of the nascent developments around the world.

We would like to thank all the contributing authors for lending their expertise to make the book truly unique. They have played a crucial role in the development of this book. Without their invaluable contributions this book wouldn't have been possible. They have made vital efforts to compile up to date information on the varied aspects of this subject to make this book a valuable addition to the collection of many professionals and students.

This book was conceptualized with the vision of imparting up-to-date information and advanced data in this field. To ensure the same, a matchless editorial board was set up. Every individual on the board went through rigorous rounds of assessment to prove their worth. After which they invested a large part of their time researching and compiling the most relevant data for our readers.

The editorial board has been involved in producing this book since its inception. They have spent rigorous hours researching and exploring the diverse topics which have resulted in the successful publishing of this book. They have passed on their knowledge of decades through this book. To expedite this challenging task, the publisher supported the team at every step. A small team of assistant editors was also appointed to further simplify the editing procedure and attain best results for the readers.

Apart from the editorial board, the designing team has also invested a significant amount of their time in understanding the subject and creating the most relevant covers. They scrutinized every image to scout for the most suitable representation of the subject and create an appropriate cover for the book.

The publishing team has been an ardent support to the editorial, designing and production team. Their endless efforts to recruit the best for this project, has resulted in the accomplishment of this book. They are a veteran in the field of academics and their pool of knowledge is as vast as their experience in printing. Their expertise and guidance has proved useful at every step. Their uncompromising quality standards have made this book an exceptional effort. Their encouragement from time to time has been an inspiration for everyone.

The publisher and the editorial board hope that this book will prove to be a valuable piece of knowledge for researchers, students, practitioners and scholars across the globe.

List of Contributors

Chen-Jui Lan, Song-Ling Tsai and Ming-Tsang Lee
Department of Mechanical Engineering, National Chung Hsing University, Taichung 402, Taiwan

Andrew Stier, Eshan Halekote and Andrew Mark
Department of Electrical and Computer Engineering, University of Texas at Austin, Austin, TX 78712, USA

Shutao Qiao and Shixuan Yang
Center for Mechanics of Solids, Structures and Materials, Department of Aerospace Engineering and Engineering Mechanics, University of Texas at Austin, Austin, TX 78712, USA

Kenneth Diller
Department of Biomedical Engineering, University of Texas at Austin, Austin, TX 78712, USA

Nanshu Lu
Department of Electrical and Computer Engineering, University of Texas at Austin, Austin, TX 78712, USA
Center for Mechanics of Solids, Structures and Materials, Department of Aerospace Engineering and Engineering Mechanics, University of Texas at Austin, Austin, TX 78712, USA
Department of Biomedical Engineering, University of Texas at Austin, Austin, TX 78712, USA
Texas Materials Institute, the University of Texas at Austin, Austin, TX 78712, USA

Jong-Hyun Kim and Jiyoung Chang
Department of Mechanical Engineering; University of Utah, Salt Lake City, UT 84112, USA

Qin Zhou
Department of Mechanical Engineering; University of Nebraska–Lincoln, Lincoln, NE 68588, USA

Minghui Luo, Yanhua Liu, Wenbin Huang, Wen Qiao, Yun Zhou, Yan Ye and Lin-Sen Chen
College of Physics, Optoelectronics and Energy & Collaborative Innovation Center of Suzhou Nano Science and Technology, Soochow University, Suzhou 215006, China
Key Laboratory of Advanced Optical Manufacturing Technologies of Jiangsu Province & Key Laboratory of Modern Optical Technologies of Education Ministry of China, Soochow University, Suzhou 215006, China

Behnam Sadri and Debkalpa Goswami
School of Industrial Engineering, Purdue University, 315 N. Grant Street, West Lafayette, IN 47907, USA

Ramses V. Martinez
School of Industrial Engineering, Purdue University, 315 N. Grant Street, West Lafayette, IN 47907, USA
Weldon School of Biomedical Engineering, Purdue University, 206 S. Martin Jischke Drive, West Lafayette, IN 47907, USA

Xuelin Wang
Department of Biomedical Engineering, School of Medicine, Tsinghua University, Beijing 100084, China

Jing Liu
Department of Biomedical Engineering, School of Medicine, Tsinghua University, Beijing 100084, China
Beijing Key Lab of CryoBiomedical Engineering and Key Lab of Cryogenics, Technical Institute of Physics and Chemistry, Chinese Academy of Sciences, Beijing 100190, China
Technical Institute of Physics and Chemistry, Chinese Academy of Sciences, Beijing 100190, China
School of Future Technology, University of Chinese Academy of Sciences, Beijing 100049, China

Jaeho Shin and Hyunmin Cho
Applied Nano and Thermal Science Lab, Department of Mechanical Engineering, Seoul National University, 1 Gwanak-ro, Gwanak-gu, Seoul 151-742, Korea

Habeom Lee and Jinhyeong Kwon
Applied Nano and Thermal Science Lab, Department of Mechanical Engineering, Seoul National University, 1 Gwanak-ro, Gwanak-gu, Seoul 151-742, Korea
Applied Nano and Thermal Science (ANTS) Lab, Department of Mechanical Engineering, Seoul National University, 1 Gwanak-ro, Gwanak-gu, Seoul 00826, Korea

Woo Seop Shin, Hyeon Rack Kim and Sukjoon Hong
Department of Mechanical Engineering, Hanyang University, 55 Hanyangdaehak-ro, Sangnok-gu, Ansan Gyeonggi-do 15588, Korea

Seungyong Han
Department of Mechanical Engineering, Ajou University, San 5, Woncheon-Dong, Yeongtong-Gu, Suwon 16499, Korea

Junyeob Yeo
Novel Applied Nano Optics (NANO) Lab, Department of Physics, Kyungpook National University, 80 Daehak-ro, Bukgu, Daegu 41566, Korea

Kangmin Leng, Kang Wu and Zhigang Wu
State Key Laboratory of Digital Manufacturing Equipment and Technology, Huazhong University of Science and Technology, Wuhan 430074, China

Chuanfei Guo
Department of Materials Science and Engineering, Southern University of Science and Technology, Shenzhen 518055, China

Peng Zhou, Tianhui Li, Tongle Bu, Xueping Liu, Jing Li, Jiang He, Rui Chen, Kunpeng Li and Fuzhi Huang
State Key Laboratory of Advanced Technology for Materials Synthesis and Processing, Wuhan University of Technology, Wuhan 430070, China

Wangnan Li
Hubei Key Laboratory of Low Dimensional Optoelectronic Materials and Devices, Hubei University of Arts and Science, Xiangyang 441053, China

Juan Zhao
School of Automotive Engineering, Wuhan University of Technology, Wuhan 430070, China

Zhaozheng Yu
Department of Engineering Science and Mechanics, The Pennsylvania State University, University Park, State College, PA 16802, USA

Huanyu Cheng
Department of Engineering Science and Mechanics, The Pennsylvania State University, University Park, State College, PA 16802, USA
Materials Research Institute, The Pennsylvania State University, University Park, State College, PA 16802, USA

Ki-Woo Jun, Jong-Nam Kim, Jin-Young Jung and Il-Kwon Oh
Creative Research Initiative Centre for Functionally Antagonistic Nano-Engineering, Department of Mechanical Engineering, Korea Advanced Institute of Science and Technology (KAIST), 291 Daehak-ro, Yuseong-gu, Daejeon 34141, Korea

Bart Plovie
Department of Electronics and Information Systems, Ghent University, Technologiepark 15, 9052 Zwijnaarde, Belgium

Frederick Bossuyt and Jan Vanfleteren
Department of Electronics and Information Systems, Ghent University, Technologiepark 15, 9052 Zwijnaarde, Belgium
IMEC vzw, Kapeldreef 75, 3001 Heverlee, Belgium

Yun Cui and Yufeng Xing
Institute of Solid Mechanics, Beihang University (BUAA), Beijing 100191, China

Yuhang Li
Institute of Solid Mechanics, Beihang University (BUAA), Beijing 100191, China
Key Laboratory of Soft Machines and Smart Devices of Zhejiang Province, Zhejiang University, Hangzhou 310027, China
State Key Laboratory of Digital Manufacturing Equipment and Technology, Huazhong University of Science and Technology, Wuhan 430074, China

Tianzhi Yang
School of Aerospace Engineering, Shenyang Aerospace University, Shenyang 110136, China

Jizhou Song
Key Laboratory of Soft Machines and Smart Devices of Zhejiang Province, Zhejiang University, Hangzhou 310027, China
Department of Engineering Mechanics and Soft Matter Research Center, Zhejiang University, Hangzhou 310027, China

Guang Qian
Key Laboratory of Micro-Inertial Instrument and Advanced Navigation Technology, Ministry of Education, and School of Instrument Science and Engineering, Southeast University, Nanjing 210096, China

Jie Tang and Xiao-Mei Xue
Key Laboratory of Micro-Inertial Instrument and Advanced Navigation Technology, Ministry of Education, and School of Instrument Science and Engineering, Southeast University, Nanjing 210096, China
Suzhou Key Laboratory of Metal Nano-Optoelectronic Technology, Suzhou Research Institute of Southeast University, Suzhou 215123, China

Yi-Ran Liu and Ning Zhao
Suzhou Key Laboratory of Metal Nano-Optoelectronic Technology, Suzhou Research Institute of Southeast University, Suzhou 215123, China
Joint International Research Laboratory of Information Display and Visualization, School of Electronic Science and Engineering, Southeast University, Nanjing 210096, China

Li-Jiang Zhang and Xing-Chang Fu
Joint International Research Laboratory of Information Display and Visualization, School of Electronic Science and Engineering, Southeast University, Nanjing 210096, China

Tong Zhang
Key Laboratory of Micro-Inertial Instrument and Advanced Navigation Technology, Ministry of Education, and School of Instrument Science and Engineering, Southeast University, Nanjing 210096, China
Suzhou Key Laboratory of Metal Nano-Optoelectronic Technology, Suzhou Research Institute of Southeast University, Suzhou 215123, China
Joint International Research Laboratory of Information Display and Visualization, School of Electronic Science and Engineering, Southeast University, Nanjing 210096, China

Xiaowei Yu, Bikram K. Mahajan, Wan Shou and Heng Pan
Department of Mechanical and Aerospace Engineering, Missouri University of Science and Technology, Rolla, MO 65401, USA

Yi Ren
Department of Biomedical Engineering, Tsinghua University, Beijing 100084, China

Phillip Lee
Photo-Electronic Hybrids Research Center, National Agenda Research Division, Korea Institute of Science and Technology (KIST), 5 Hwarang-ro 14-gil, Seongbuk-gu, Seoul 02792, Korea

Jinhwan Lee and Young D. Suh
Applied Nano and Thermal Science (ANTS) Lab, Department of Mechanical Engineering, Seoul National University, 1 Gwanak-ro, Gwanak-gu, Seoul 00826, Korea

Index

www.ingramcontent.com/pod-product-compliance
Lightning Source LLC
LaVergne TN
LVHW081018210926
848485LV00004B/889